Convection in Fluids

FLUID MECHANICS AND ITS APPLICATIONS
Volume 90

Aims and Scope of the Series

The purpose of this series is to focus on subjects in which fluid mechanics plays a fundamental role.

As well as the more traditional applications of aeronautics, hydraulics, heat and mass transfer etc., books will be published dealing with topics which are currently in a state of rapid development, such as turbulence, suspensions and multiphase fluids, super and hypersonic flows and numerical modeling techniques.

It is a widely held view that it is the interdisciplinary subjects that will receive intense scientific attention, bringing them to the forefront of technological advancement. Fluids have the ability to transport matter and its properties as well as to transmit force, therefore fluid mechanics is a subject that is particularly open to cross fertilization with other sciences and disciplines of engineering. The subject of fluid mechanics will be highly relevant in domains such as chemical, metallurgical, biological and ecological engineering. This series is particularly open to such new multidisciplinary domains.

The median level of presentation is the first year graduate student. Some texts are monographs defining the current state of a field; others are accessible to final year undergraduates; but essentially the emphasis is on readability and clarity.

For other titles published in this series, go to
www.springer.com/series/5980

R.Kh. Zeytounian

Convection in Fluids

A Rational Analysis and Asymptotic Modelling

Springer

R.Kh. Zeytounian
Université des Science et Technologies de Lille
France

ISBN 978-94-007-3672-6 e-ISBN 978-90-481-2433-6
Springer Dordrecht Heidelberg London New York

Printed on acid-free paper

Springer is part of Springer Science+Business Media (www.springer.com)

There is no better way
for the derivation of significant
model equations than rational analysis
and asymptotic modeling

Contents

Preface and Acknowledgments

The purpose of this monograph is to present a unified (analytical) approach to the study of various convective phenomena in fluids. Such fluids are mainly considered to be thermally perfect gases or expansible liquids. As a consequence, the main driving force/mechanism is the buoyancy force (Archimedean thrust) or temperature-dependent surface tension inhomogeneities (Marangoni effect). But we take into account, also, in the general mathematical formulation – for instance, in the Bénard problem for a liquid layer heated from below – the effect of an upper deformed free surface, above the liquid layer. In addition, in the case of atmospheric thermal convection, the Coriolis force and stratification effects are also taken into account.

My main motivation in writing this book is to give a rational, analytical, analysis of the main physical effects in each case, on the basis of the full unsteady Navier–Stokes and Fourier (NS–F) equations – for a Newtonian compressible/dilatable, viscous and heat-conductor fluid, coupled with the associated initial and boundary (lower) and free surface (upper) conditions.

This, obviously, is a difficult but necessary task, if we wish to construct a rational modelling process, keeping in mind a coherent numerical simulation on a high speed computer.

It is true that the 'physical approach' can produce valuable qualitative analyses and results for various significant and practical convection phenomena. Unfortunately, an *ad hoc* physical approach would not be able to point the way for a consistent derivation of approximate (leading order) model problems which could be used for a quantitative numerical calculation; this is true especially because such an approach would be unable to provide a rational, logical method for the derivation of an associated second-order model problem with various complementary (e.g., to a usual Boussinesq approximation) effects (such as viscous dissipation and free surface deformation) to buoyancy-driven Rayleigh–Bénard thermal convection.

Concerning this physical approach, which is necessary for full comprehension of the nature of convection, we refer the reader to *Physical Hydrodynamics*, by E. Guyon, J.-P. Hulin, L. Petit and C.D. Mitescu, published by Oxford University Press, Oxford, 2001. On the other hand, in the review paper 'Convective Instability: A Physicist's Approach', by Ch. Normand, Y. Pomeau and M.G. Velarde, published in *Reviews of Modern Physics*, vol. 49, no. 3, pp. 581–624, July 1977, a number of apparently disparate problems from fluid mechanics are thoroughly considered under the unifying heading of natural convection.

Actually, various technologically complex convective flow problems are frequently resolved via massive numerical computations on the basis of *ad hoc* approximate models. It should not be surprising that such a numerical approach leads to a simulation which has little practical interest because of its inconsistency with the experimental results! If one is to use this numerical technique, it is necessary – at least from my point of view – that a rational consistent approach is adopted to make sure that:

> "if, in the fluid dynamics starting equations and boundary/initial conditions, a term is neglected, then, it is essential to be convinced that such a term is really much smaller than the terms retained in the derived approximate model's equations and conditions".

It should be noted that such a rational consistent approach, with an asymptotic modelling process, assures the possibility to obtain – via various similarity rules between small or large non-dimensional parameters governing different physical effects – some criteria for testing the range of validity of these derived approximate models.

My profound conviction is that a rational/analytical-asymptotic modelling is a necessary theoretical basis for research into the solution of a difficult nonlinear problem, before a numerical computation. Both the numerics and modelling are useful and strongly complementary. Our present project is in direct line with our consistent scientific attitude:

Putting a clear emphasis on rigorous – but not strongly formal mathematical – development of consistent approximate model problems for different kinds of convective flows.

However, to acknowledge a certain point of view, I know that some readers do not care much for this rigor and simply want to know: 'what are the relevant model equations and boundary conditions for their problems?' Such a reader will find a special chapter, namely Chapter 8, which is a kind of 'general advice' where, for each of the three particular convections we consider – shallow, deep, and Marangoni – I specify the physical conditions, the limitations for the main parameters, which govern, respectively: buoyancy, viscous dissipation and free surface/surface tension effects. In addition, the leading order model equations and associated boundary conditions for these three cases are specified, and the reader can find our recommendations for taking into account the corresponding second-order, non-Boussinesq, free surface deformation and viscous dissipation effects.

The first, main kind of convective transport (convection) I discuss in this monograph is called natural or free convection, meaning that the fluid (liquid or atmospheric air) flow is a response to a force acting within the body of the fluid. The force is most often gravity (buoyancy) but there are circumstances where some other agency, such as surface (temperature-dependent) tension or other forces – for example the Coriolis force – play a significant or even a primary role.

Convection, as a physical phenomenom, is thoroughly discussed in the survey paper by M.G. Velarde and Ch. Normand, 'Convection', published in *Scientific American*, vol. 243, no. 1, pp. 92–108, July 1980. In this survey paper, 'Convection', the spontaneous upwelling of a heated fluid, can be understood only by untangling the intricate relations among temperature, viscosity, surface tension and other characteristics of the considered fluid flow problem. Natural (or free) convection is defined in contradistinction to forced convection, where the fluid motion is induced by the effect of a heterogeneous temperature field or by a relief as in atmospheric, mesoscale motion, for instance, a lee waves regime (adiabatic and non-viscous) downstream of a mountain!

I have made every effort to present a logical organization of the material and it should be stressed that there is no physics involved, but rather an extensive use of dimensional analysis, similarity rules, asymptotics of NS–F equations with boundary conditions and calculus. Until this is undertood, though even now it is possible (in part!), it will be difficult to convince a detached and possibly skeptical reader of their value as an aid to understanding!

A valuable – again, at least from my point of view – feature of my rational (but not rigorously mathematical) approach is the possibility to derive, consistently, not only the leading-order, limiting first-order, approximate model problem, but also the associated second-order model which takes into account complementary effects.

This gives, curiously, the possibility in many cases to clarify (as this is described in Chapter 8) the conditions required for the validity of the usual derived leading-order model problems. For just this purpose, via non-dimensional analysis and the appearance of reduced parameters/numbers, it is necessary to adequately take into account various similarity rules.

This book has been written for final year undergraduates and graduate students, postgraduate research workers and also for young researchers in fluid mechanics, applied mathematics and theoretical/mathematical physics. However, it is my conviction that anyone who is interested in a systematic and logical account of theoretical aspects of convection in fluids, will find in the present monograph various answers concerning an analytical approach in modelling of the related problems.

The choice of the nine chapters, Chapters 2 to 10, is summarized in Chapter 1 and their ordering is, at least from my point of view, quite natural. The presentation of the material, the relative weight of various arguments and the general style reflects the tastes of the author and his knowledge and ability gained over 50 years of research work in fluid mechanics.

In Chapter 1, devoted to a 'Short Preliminary Comments and Summary of Chapters 2 to 10', the reader can find an 'extended abstract' of the full material included in the other nine chapters. All the papers and books cited in Chapters 1 to 10 are listed at the end of these chapters. In many cases the reader can find (in the sections 'Comments and Complements' before the references in some chapters) various information concerning recent (up to 2008) results linked with convection in fluids.

Fluid mechanics has spawned a myriad of theoretical research projects by numerous fluid dynamicists and applied mathematicians. The richness of the area can be seen in the major questions surrounding Rayleigh-Bénard convection, which itself is an approximate problem resulting from the application of asymptotic/perturbation techniques to the full NS–F equations using Boussinesq approximation for a weakly expansible/dilatable liquid. In the relatively recent survey paper by E. Bodenchatz, W. Pesh and G. Ahlers, published in *Annual Review of Fluid Mechanics*, vol. 32, pp. 709–778, 2000, the reader can find the main results for this RB convection that have been obtained during the past decade, 1990–2000, or so.

I should like to thank to Dr. Christian Ruyer-Quil (from the University of Paris Sud – Orsay) with who I have had during the last years many discussions related to the modelling of thin film problems and also to Dr. B. Scheid (from the Université Libre de Bruxelles, Begium) who gave me the oportunity to visit the 'Microgravity Reseach Center' of Professor J.C. Legros.

My thanks to Professor Manuel G. Velarde, Director of the Unidad de Fluidos in 'Instituto Pluridisciplinar UCM' de Madrid (Spain), for his hospitality in his Unidad de Fluidos and with whom I have had many discussions and collaborations, relative to Marangoni thermocapillary convection, during the years 2000–2004. Together we organized a Summer Course held at CISM (Undine, Italy) in July 2000, devoted to 'Interfacial Phenomena and the Marangoni Effect' and edited in collaboration a CISM Courses and Lectures (No. 428), published by Springer, Wien/New York in 2002.

Finally, my gratitude to Professor René Moreau, as the Series Editor of 'FMIA', who has given me various useful criticism and suggestions, and recommended this book for publication by Springer, Dordrecht.

R.Kh. Zeytounian
Paris, April 2008

Chapter 1
Short Preliminary Comments and Summary of Chapters 2 to 10

1.1 Introduction

During the years 1967–1972 at the ONERA, then 1972–1996 at the University of Lille I, and later, following retirement from this University in the years 1997–2002, at home in Paris-Center, I published more than 20 papers devoted to convection in fluids. As an *Introduction* to this book, I wish to give a short discourse on six of these papers that I consider as particularly valuable results. The interested reader will find all of these quoted papers and books listed at the end of this chapter.

A *first* valuable result was obtained in 1974, namely a rigorous justification, based on an asymptotic approach for low Mach numbers, of the famous Boussinesq approximation and the rational derivation of the associated Boussinesq equations [1]. In chapter 8 of [2], a monograph published in 1990 and devoted to the asymptotic modelling of atmospheric flows, the reader can find a careful derivation and analysis of these Boussinesq equations, valid for atmospheric low velocity motions – the so-called small Mach number/hyposonic case.

A *second* interesting result was published in 1983 in a short note [3], where it seems that, for the first time, there appeared a rigorous formulation of the Rayleigh–Bénard (RB) thermal convection problem using asymptotic techniques. This result opened the door for a consistent derivation of the second-order approximate model equations for Bénard, heated from below, thermal convection (see, for instance, in this book, Sections 3.5 and 3.6, Section 5.3 and 8.1).

In 1989, by means of a careful dimensionless analysis of the exact, full, Bénard problem of thermal instability for a weakly expansible liquid heated from below, as a *third* new result [4], I show also that:

> ... if you have to take into account, in model approximate equations for the Bénard problem, the viscous dissipation term in the temperature equation, then it is necessary to replace the classical shallow convection, (RB) equations, by a new set of equations, called the 'deep convection' (DC-Zeytounian) equations.

These deep convection equations contain a new 'depth parameter' and are derived and analyzed in this book in Chapter 6.

A *fourth* result, which appear as a quantitative criterion for the valuation of the importance of buoyancy in the Bénard problem, is the following alternative [5], obtained in 1997:

> Either the buoyancy is taken into account, and in this case the free-surface deformation effect is negligible and we rediscover the classical Rayleigh–Bénard (RB) shallow convection rigid-free approximate problem or, the free-surface deformation effect is taken into account and, in such a case at the leading-order approximation for a weakly expansible fluid, the buoyancy does not give a significant effect in the Bénard–Marangoni (BM) thermocapillary instability problem.

This alternative is related to the value of the reference Froude number

$$\mathrm{Fr}_d = (\nu_0/d)/(gd)^{1/2},$$

based on the thickness d of the liquid layer, magnitude of the gravity g and constant kinematic viscosity ν_0, and for

$$\text{RB problem:} \quad \mathrm{Fr}_d \ll 1,$$

while for the

$$\text{BM problem:} \quad \mathrm{Fr}_d \approx 1 \Rightarrow d \approx (\nu_0^2/g)^{1/3} \approx 1\ \text{mm}.$$

The small effect of the viscous dissipation, in the RB model problem, gives a complementary criterion for the thickness d (see Chapters 3, 4 and 5).

A *fifth* result is linked with my written lecture notes [6] for the Summer Course held at CISM (Udine, Italy, and coordinated by M.G. Velarde and myself) in July 2000, where I discussed 'Theoretical aspects of interfacial phenomena and Marangoni effect – Modelling and stability'.

Although significant understanding has been achieved, yet surface-tension-gradient-driven BM convection flows, still deserve further studies; in particular, the case of a single Biot number for a conduction motionless state and also for the convection regime (the Biot number is the dimensionless parameter linked with the heat transfer across an upper, liquid-ambient air, free surface) poses many problems, especially in the case when this single Biot number is assumed 'vanishing' in the convection regime!

Concerning the Boussinesq approximation, we refer here to my recent paper [7], 'Foundations of Boussinesq approximation applicable to atmospheric motions', published in 2003 (see Chapter 9 in the present book) as my *sixth*, and last result, to be mentioned here.

However, on the other hand, in addition to the research mentioned above on convection and Boussinesq approximation, during the years 1991–1995 I used a rather new approach to obtain various asymptotically significant models for 'nonlinear long surface waves in shallow water' and 'solitons'. The results of these 'investigations' were written about in two survey papers [8] and [9] in 1994 and 1995 and also in the 1993 monograph [10].

1.2 Summary of Chapters 2 to 10

Chapter 2 is devoted to Navier–Stokes and Fourier (NS-F) systems of equations which are derived from the basic relations for momentum, mass, and energy balance, according to a continuum regime:

$$\rho \frac{\mathrm{d}\mathbf{u}}{\mathrm{d}t} = \rho \mathbf{f} + \nabla \cdot \mathbf{T}, \tag{1.1a}$$

$$\frac{\mathrm{d}\rho}{\mathrm{d}t} = -\rho(\nabla \cdot \mathbf{u}), \tag{1.1b}$$

$$\rho \frac{\mathrm{d}e}{\mathrm{d}t} = -\nabla \cdot \mathbf{q} + \mathbf{T} \cdot (\nabla \mathbf{u}). \tag{1.1c}$$

These three equations (1.1a–c) are the classical conservation laws at any point of continuity in a fluid domain V, where the velocity vector $\mathbf{u}$, the density ρ and the specific internal energy e have piecewise-continuous bounded derivatives.

In equation (1.1a), $\mathbf{f}$ is the body force per unit mass (usually, in convection problems, the gravity force) and $\mathbf{T}$ is the stress tensor (with the components T_{ij}). In equation (1.1c), $\mathbf{q}$ is the heat flux vector with components q_i.

The time derivative, with respect to material motion, is written as

$$\frac{\mathrm{d}}{\mathrm{d}t} := \frac{\partial}{\partial t} + \mathbf{u} \cdot \nabla, \tag{1.1d}$$

and

$$\nabla = \frac{\partial}{\partial x_i}, \quad i = 1, 2, 3,$$

with

$$x_1 = x, \quad x_2 = y, \quad x_3 = z.$$

To obtain the classical, NS–F, Newtonian, system of equations, from (1.1a–c), for $\mathbf{u}$, ρ, p (mechanical pressure) and T (absolute temperature), it is necessary to assume the existence of two equations of state and two constitutive relations for the stress tensor $\mathbf{T}$ and heat flux vector $\mathbf{q}$.

Concerning the equations of state, I consider, mainly, two cases.

First, the case of a thermally perfect gas with two equations of state:

$$e = C_v T, \tag{1.2a}$$

$$p = R\rho T, \tag{1.2b}$$

where C_v is the specific heat at constant volume v $(= 1/\rho)$, and R is the perfect gas constant – the mechanical pressure p being then in a fluid at rest and in the framework of the Newtonian-classical fluid mechanics, such that (see also (1.4a)):

$$T_{ij} = -p\delta_{ij}, \tag{1.2c}$$

with

$$\delta_{ij} = 1, \quad \text{if } i = j \quad \text{and} \quad \delta_{ij} = 0 \quad \text{for } i \neq j, \tag{1.2d}$$

where δ_{ij}, is the well-known Kronecker delta tensor.

Second, the case of an expansible liquid when

$$e = E(v, p), \tag{1.3a}$$

with the following Maxwell relation (see Section 2.1):

$$C_p - C_v = T\left(\frac{\partial p}{\partial T}\right)_v \left(\frac{\partial v}{\partial T}\right)_p, \tag{1.3b}$$

where C_p is the specific heat for a constant pressure p.

In Section 2.2, we give more detailed information concerning 'thermodynamics' for an expansible liquid.

In the framework of Newtonian (classical) fluid mechanics (see, for istance, the very pertinent basic survey paper by Serrin [11]), if we assume that a thermally perfect gas and an expansible liquid can be modelled as a viscous Newtonian fluid, then we can write for the components T_{ij}, of the stress tensor **T**, the following (first) constitutive relation (originally obtained, by de Saint-Venant [12]):

$$T_{ij} = -p\delta_{ij} + 2\mu[d_{ij} - (1/3)\Delta\delta_{ij}], \tag{1.4a}$$

where the d_{ij} are the components of the rate of strain tensor $\mathbf{D}(\mathbf{u})$, such that

$$d_{ij} = (1/2)\left[\frac{\partial u_i}{\partial x_j} + \frac{\partial u_j}{\partial x_i}\right], \tag{1.4b}$$

$$\Delta \equiv d_{kk} = \boldsymbol{\nabla} \cdot \mathbf{u}, \tag{1.4c}$$

and μ is the shear viscosity and depends on thermodynamic pressure P (different, in general, from the mechanical pressure p), and also of the absolute temperature T.

However, here, because the Stokes relation,

$$\lambda \equiv -(2/3)\mu, \tag{1.4d}$$

which gives the second coefficient of viscosity λ, as a function of μ, has been taken into account in the above constitutive relation (1.4a), we have that

$$P \equiv p. \tag{1.4e}$$

Now, if for the heat flux, $\mathbf{q}$, in equation (1.1c), we adopt as (a second) constitutive relation the classical Fourier law:

$$\mathbf{q} = -k\boldsymbol{\nabla}T, \tag{1.5}$$

where k is the thermal conductivity coefficient, then with (1.4a) and (1.5), we have the possibility to write the energy balance equation (1.1c), for the specific internal energy e, in the following form:

$$\rho\frac{\mathrm{d}e}{\mathrm{d}t} = T_{ij}\left(\frac{\partial u_i}{\partial x_j}\right) + \frac{\partial}{\partial x_i}\left(k\frac{\partial T}{\partial x_i}\right), \tag{1.6a}$$

or

$$\rho\frac{\mathrm{d}e}{\mathrm{d}t} = -p\Delta + 2\mu[d_{ij}d_{ij} - (1/3)\Delta^2] + \frac{\partial}{\partial x_i}\left(k\frac{\partial T}{\partial x_i}\right), \tag{1.6b}$$

where

$$(2\mu/\rho)[d_{ij}d_{ij} - (1/3)\Delta^2] \equiv \Phi \tag{1.7}$$

is the rate of (viscous) dissipation of mechanical energy, per unit mass of fluid, due to viscosity.

With the first of equations of state (1.2a) for a thermally perfect gas, since we have $e = C_v T$, it is easy (see Section 2.3) to obtain from (1.6b) an evolution equation for the temperature T, the specific heat, C_v, being usually assumed a constant.

However, for an expansible liquid – in the case of a thermal convection problem – obtaining such a result is more subtle because the density is then a function of pressure and temperature (see Section 2.4). In Section 2.5 the reader can find the free surface jump conditions associated with the NS–F system of equations for an expansible liquid and, in Sections 2.6 and 2.7, a discussion concerning the initial conditions and a short derivation of the Hills and Roberts equations [34] (see also below the summary concerning Chapter 6).

In Chapter 3 we revisit the thermal convection problem considered by Lord Rayleigh in 1916 [13]. Stimulated by the Bénard [14] experiments Lord Rayleigh, in his pioneer 1916 paper, first formulated the theory of convective instability of a layer of fluid: an expansible liquid, with as equation of state:

$$\rho \equiv \rho(T), \tag{1.8a}$$

between two horizontal rigid planes, and derive in an *ad hoc* manner the famous *Rayleigh–Bénard*, (RB), instability model problem.

The starting (approximate) equations in the Rayleigh paper are those obtained by Boussinesq [15] and are valid when

"*the variations of density are taken into account only when they modify the action of gravity force* $\mathbf{g}$ $(= -g\mathbf{k})$"

$\mathbf{k}$ is the unit vector for the vertical axis of z.

The (weakly) expansible liquid layer, for which the fixed thickness is d, is assumed to be bounded by two infinite fixed, rigid horizontal planes, at $z = 0$ and $z = d$, such that

$$T = T_w \quad \text{at} \quad z = 0 \tag{1.8b}$$

and

$$T = T_d \quad \text{at} \quad z = d, \tag{1.8c}$$

such that

$$\Delta T = T_w - T_d > 0. \tag{1.8d}$$

It is well known that the main parameter that drives the thermal convection is the Grashof (Gr) number or Rayleigh (Ra) number,

$$\mathrm{Gr} = \frac{\alpha(T_d)\Delta T g d^3}{\nu_d^2}, \tag{1.9a}$$

$$\mathrm{Ra} = \frac{\alpha(T_d)\Delta T g d^3}{\nu_d \kappa_d}, \tag{1.9b}$$

where Ra ≡ PrGr, with as Prandtl number

$$\mathrm{Pr} = \frac{\nu_d}{\kappa_d}. \tag{1.9c}$$

In (1.9a–c), ν_d and κ_d are, respectively, the constant (at $T = T_d$) kinematic viscosity ν_d [$= \mu(T_d)/\rho(T_d)$] and thermal diffusivity $\kappa_d = [k(T_d)/\rho(T_d)C_v(T_d)]$.

The coefficient of thermal expansion [when $\rho \equiv \rho(T)$] of the liquid is defined as

$$\alpha(T) = -(1/\rho(T))\left[\frac{\mathrm{d}\rho(T)}{\mathrm{d}T}\right]. \tag{1.9d}$$

On the other hand,

$$\varepsilon = \alpha(T_d)\Delta T \approx 5 \times 10^{-3}, \tag{1.10a}$$

which is a small parameter (the expansibility number) for many liquids, and is our main small parameter in derivation of an approximate limit for the model (RB) problem.

In particular, when the square of the Froude number (based on the thickness d)

$$\mathrm{Fr}_d^2 \equiv \frac{(\nu_d/d)^2}{gd}, \tag{1.10b}$$

is small – $\mathrm{Fr}_d^2 \ll 1$ – we obtain for the thickness of the liquid layer, d, the following constraint (a lower bound):

$$d \gg \left(\frac{\nu_d^2}{g}\right)^{1/3} \approx 1 \text{ mm}, \tag{1.11}$$

The main result (according to Zeytounian [5]) in Chapter 3 is that

> the Boussinesq, shallow convection model equations, with the buoyancy as main driving (Achimedean) force, are significant, rational-consistent equations in the framework of the classical RB instability, rigid-rigid problem if, and only if, we assume simultaneously the smallness of both numbers, ε (expansibility) and F_d^2 (square of the Froude number).

In such a case, the limiting process, *à la* Boussinesq,

$$\mathrm{Gr} = \frac{\varepsilon}{\mathrm{Fr}_d^2} \quad \text{fixed, when } \varepsilon \downarrow 0 \text{ and } \mathrm{Fr}_d^2 \downarrow 0, \tag{1.12}$$

is the RB limiting process and, as a consequence, for the validity of the RB model problem (*à la* Rayleigh, derived in Chapter 3) it is necessary to consider a thicker weakly expansible liquid layer than a very thin film layer of the order of the millimetre, as is the case for the Bénard–Marangoni thermocapillary instability problem (considered in Chapter 7).

An important moment in a consistent derivation of shallow convection, RB equations, is strongly linked with an evaluation of the effect of the viscous dissipation, Φ, in energy balance (see, for instance, (1.6b) with (1.7)).

Namely, this evaluation gives an upper bound for the thickness, d, of the weakly expansible liquid layer. More precisely, on the basis of a dimensionless analysis and the derivation of a 'dominant' energy equation for the dimensionless temperature

$$\theta = \frac{(T - T_d)}{\Delta T}, \quad \Delta T = T_w - T_d, \tag{1.13}$$

we obtain that the role of the viscous dissipation is linked with the following 'dissipation number':

$$\mathrm{Di}^* = \frac{\mathrm{Di}}{2\mathrm{Gr}}, \tag{1.14}$$

which was introduced by Turcotte et al. in 1974 [16], where

$$\mathrm{Di} \equiv \varepsilon \mathrm{Bo}. \tag{1.15}$$

In (1.15), the ratio Bo, of two 'thicknesses', d and $C_v(T_d)\Delta T/g$, plays the role of a Boussinesq number:

$$\mathrm{Bo} = \frac{gd}{C_v(T_d)\Delta T}. \tag{1.16}$$

The reader can find a discussion concerning the account of viscous heating effects in a paper by Velarde and Perez Cordon [17]. We observe that, in our 1989 paper [4], the parameter Di is in fact the product of two parameters:

$$\varepsilon \text{ (which is } \ll 1) \text{ by Bo (assumed } \gg 1),$$

and has been denoted by δ (assumed O(1)), which is our 'depth' parameter.

In Section 3.4, the rigid-rigid, *à la* Rayleigh, RB problem is derived and in Sections 3.5 and 3.6, the second-order model equations associated with RB shallow convection equations are obtained in a consistent way. Section 3.7 is devoted to some comments. Concerning the derivation and analysis of the deep thermal convection equations, which take into account the term proportional to dissipation parameter Di*, see Chapter 6 in this monograph.

Chapter 4 has a central place in the present monograph, and the reader can find (in Section 4.2) a full mathematical/analytical rational formulation of the Bénard, heated from below, convection problem and its reduction to a system of non-dimensional 'dominant' equations and conditions, where various reduced parameters are present. In particular, this non-dimensional dominant system takes into account:

(i) the temperature-dependent surface tension,
(ii) the static basic conduction state,
(iii) the deformation of the free surface.
(iv) the heat transfer at the free surface via an usual 'Newton's cooling law'.

This free surface, simulated by the equation

$$z = d + ah(t, x, y) \equiv H(t, x, y),$$

in a Cartesian co-ordinate system (O; x, y, z) in which the gravity vector $\mathbf{g} = -g\mathbf{k}$ acts in the negative z direction and where a is an amplitude, separates the weakly expansible liquid layer from ambient motionless air at constant temperature T_A and constant atmospheric pressure p_A, having a negligible viscosity and density.

We observe that the problem of the upper, free-surface condition for the temperature (in fact, an open problem) is discussed in various parts of this monograph. The temperature-dependent surface tension $\sigma(T)$ is assumed decreasing linearly with temperature. Thus:

$$\sigma(T) = \sigma(T_d) - \gamma_\sigma(T - T_d), \tag{1.17a}$$

where

$$\gamma_\sigma = -\left(\frac{\mathrm{d}\sigma(T)}{\mathrm{d}T}\right)_d \tag{1.17b}$$

is the constant rate of change of surface tension with temperature, which is positive for most liquids.

However we observe that several authors instead of $(T - T_d)$ use $(T - T_A)$, where T_A is the constant ambient air temperature above the deformable free surface of the weakly expansible liquid layer. In such a case, instead of θ given by (1.13), these authors introduce another dimensionless temperature:

$$\Theta = \frac{(T - T_A)}{(T_w - T_A)}. \tag{1.17c}$$

With (1.17a, b) the surface tension effects are expressed by the following two non-dimensional parameters:

$$\text{We} = \frac{\sigma_d d}{\rho_d \nu_d^2}, \tag{1.18a}$$

$$\text{Ma} = \frac{\gamma_\sigma d \Delta T}{\rho_d \nu_d^2}, \tag{1.18b}$$

which are, respectively, the Weber and Marangoni numbers, which play an important role in Bénard–Marangoni (BM) thermocapillary instability problems.

We observe again that, in (1.17a, b) T_d is the constant temperature on the free surface, in the purely, static motionless, basic conduction state, which is obviously (no convection) the level $z = d$ when the (conduction) temperature is simply:

$$T_s(z) = T_w - \beta_s z \tag{1.19a}$$

with

$$\beta_s = -\frac{dT_s(z)}{dz} > 0. \tag{1.19b}$$

Obviously, at $z = d$,

$$T_d = T_w - \beta_s d$$

or

$$\beta_s = \frac{(T_w - T_d)}{d} \equiv \frac{\Delta T}{d}, \tag{1.19c}$$

and the above Marangoni number Ma, according to (1-18b), is expressed via the above β_s,

$$\text{Ma} = \frac{\gamma_\sigma d^2 \beta_s}{\rho_d \nu_d^2}. \tag{1.19d}$$

Concerning Newton's cooling law of heat transfer, written for the basic, motionless conduction temperature $T_s(z)$, we have

$$k(T_d)\frac{dT_s(z)}{dz} + q_s(T_d)[T_s(z) - T_A] = 0, \quad \text{at } z = d; \tag{1.20}$$

when in a basic, motionless conduction state, the thermal conductivity coefficient

$$k = k(T_d) = \text{const.}$$

In (1.20), $q_s(T_d)$ is the unit thermal surface conductance (also a constant). From (1.20) with (1.19a) we obtain:

$$\beta_s = \mathrm{Bi}_s(T_d)\left[\frac{(T_d - T_A)}{d}\right], \tag{1.21a}$$

or

$$\beta_s = \left[\frac{\mathrm{Bi}_s}{(1 + \mathrm{Bi}_s)}\right]\left[\frac{(T_w - T_A)}{d}\right], \tag{1.21b}$$

where

$$\mathrm{Bi}_s(T_d) = \frac{dq_s(T_d)}{k(T_d)}, \tag{1.22}$$

is the conduction Biot number (at $T = T_d = \text{const.}$).

The lower heated plate temperature, $T = T_w \equiv T_s\ (z = 0)$, being a given data in the classical Bénard, heated from below, convection problem, the adverse conduction temperature gradient β_s appears [according to (1.21b)] as a known function of the temperature difference $(T_w - T_A)$, where $T_A < T_w$ is the known constant temperature of the passive (motionless) air far above the free surface, when $\mathrm{Bi}_s(T_d)$ is assumed known, thanks to (1.21a). But for this it is necessary that the constant (conduction) heat transfer $q_s(T_d)$ (the unit thermal surface conductance) was considered as a data! If so,

$$T_s(z = d) = T_d(\equiv T_w - \beta_s d)$$

is the assumed to be determined.

One should realize that β_s is always different from zero in the framework of the Bénard convection problem heated from below!

As a consequence, the above, defined by (1-22), constant conduction Biot number is also always different from zero: $\mathrm{Bi}_s(T_d) \neq 0$*; it characterizes the 'Bénard conduction' effect and makes it possible to determine the purely static basic temperature gradient* β_s.

This seemingly trivial remark is in fact important, because in the mathematical formulation of the full Bénard, heated from below, convection problem, with a deformable free surface, we do not have the possibility to work only with a single conduction, $\mathrm{Bi}_s(T_d) \neq 0$, Biot number. Namely, necessarily a second (but certainly variable) convective Biot number,

$$\mathrm{Bi}_{\mathrm{conv}} = \frac{dq_{\mathrm{conv}}}{k(T_d)}, \tag{1.23a}$$

appears in formulation of the BM problem – unfortunately, in almost all papers devoted to thermocapillary convection (following the paper by Davis published in 1987 [18], we see that this $\mathrm{Bi}_{\mathrm{conv}}$ is 'confused' with $\mathrm{Bi}_s(T_d)$). Indeed, q_{conv} is an unknown and its determination is a difficult and unresolved problem – but here I do not touch this question and I do not for a moment suppose that I shall resolve it – my purpose is to link the formulation of a correct upper, free-surface condition for the dimensionless temperature to the framework of a rigorous modelling of the BM thermocapillary-Marangoni problem.

For convective motion, in principle, again Newton's cooling law can be used, which is usually the case in almost all papers devoted to BM problems (when they follow the Davis papers [18] 'blindly'). In Newton's cooling law, see (1.23b) below, we have assumed (for simplicity, but obviously it is possible also to assume that k is a function of the liquid temperature T) that the thermal conductivity is also a constant, $k_d \equiv k(T_d)$, in convection motion, n being the normal coordinate to a deformable free surface. In such a case, in a convection regime, we write the following jump condition on an upper, free surface for temperature T:

$$-k(T)\frac{\partial T}{\partial n} = q_{\mathrm{conv}}[T - T_A] + Q_0, \quad \text{at } z = H(t, x, y), \tag{1.23b}$$

with $\partial T/\partial n \equiv \nabla T \cdot \mathbf{n}$, as in Davis' (1987) paper [18], where Q_0 is an imposed heat flux to the environment and to be defined! From (1.23b), because on the right-hand side we have as first term $q_{\mathrm{conv}}[T - T_A]$, it seems more judicious (contrary to the Davis approach [18]) to use, as dimensionless temperature, the function Θ defined above by (1.17c), rather than the function θ defined in (1.13)! In such a case, all used physical constants are taken at the constant temperature $T = T_A$.

In the above deformable upper, free-surface boundary condition for the temperature T, (1.23b), written at free surface, $z = H(t, x, y)$, the convective heat transfer (variable?) coefficient q_{conv}, is different from the constant conduction heat transfer, $q_s(T_d)$ which appears in condition (1.20), for the static basic conduction state, and also in the conduction, constant, Biot number (1.22).

As a tentative approach, we can assume that the corresponding variable unknown convection heat transfer coefficient, q_{conv}, in (1.23b), is also temperature, T, dependent! As a consequence, the associated convective Biot number is also a function of the variable liquid temperature T. Namely, as opposed to (1.22), we write, for instance,

$$\mathrm{Bi}_{\mathrm{conv}}(T) = \frac{dq_{\mathrm{conv}}(T)}{k(T_d)}. \tag{1.23c}$$

But another approach may be also:

$$\mathrm{Bi}_{\mathrm{conv}}(H) = \frac{dq_{\mathrm{conv}}(H)}{k(T_d)}, \tag{1.23d}$$

where $H = d + ah(t, x, y)$ is the full (variable) thickness of the convective liquid layer. Indeed, the assumption concerning necessity of the introduction of a variable convective heat transfer coefficient is present in the pioneering paper by Pearson (1958) [19], where a small disturbance analysis is carried out.

If in a conduction (motionless, steady) phase, when the temperature T_d is constant (uniform) along the flat free surface $z = d$, we have obviously, $q = q_s(T_d) = \mathrm{const.}$; unfortunately this is no longer true in a thermocapillary convective regime, because the dimensionless temperature (θ or Θ) at the upper, deformable, free surface, $z = H(t, x, y)$, varies from point to point!

In reality, the heat transfer coefficient and Biot number in a convection regime depend, in general, on the free surface properties of the fluid, the unknown motion of the ambient air near the free surface and also to the spatio-temporal structure of the temperature field – see the discussion in Joseph's 1976 monograph [20, part II], and in Parmentier et al.'s 1996, very pertinent paper [21], where the problem of two Biot numbers is very well discussed.

As a consequence of the 'co-existence' of two Biot numbers, conduction and convection, the formulation of the upper, free-surface boundary condition, derived from the jump condition (1.23b), for θ, is significantly different than the Davis condition derived in [18]. With two Biot numbers the *correct condition* is given by (1.24c), and the Davis condition is given by (1.25) when, as in Davis [18] we confuse $\mathrm{Bi}_s(T_d)$ with $\mathrm{Bi}_{\mathrm{conv}}$! Indeed Davis, in his paper [18], during the derivation from (1.23b) of a dimensionless condition at deformable upper, dimensionless free surface [$t' = t/(d^2/\nu_d)$, $x' = x/d$, $y' = y/d$],

$$H' = \frac{H}{d} \Rightarrow z' = 1 + \eta h'(t', x', y'), \quad \text{with } \eta = a/d, \tag{1.23e}$$

for θ, given by (1.13), to bind oneself to use the relation (1.21a) which gives the possibility to replace the difference of the temperaure $(T_w - T_d)$ by $(T_w - T_A)$, namely,

$$d\beta_s = \mathrm{Bi}_s(T_d)(T_d - T_A) \Rightarrow \frac{(T_d - T_A)}{(T_w - T_d} = \left[\frac{1}{\mathrm{Bi}_s(T_d)}\right]. \tag{1.24a}$$

In such a case, Davis rewrites the above jump condition (1.23b) in the following dimensionless form (see Davis [18, p. 407, formula (3.2)]):

$$\frac{\partial\theta}{\partial n'} + \mathrm{Bi}_{conv}\left\{\left[\frac{(T_d - T_A)}{(T_w - T_d)}\right] + \theta\right\} + \frac{Q_0}{k\beta_s} = 0 \quad \text{at } z' = 1 + \eta h'(t', x', y'). \tag{1.24b}$$

From (1.24b), with (1.24a), we derive the desired correct condition, if we do not confuse Bi_{conv} (from Newton's cooling law, (1.23b), for the convection) with $\mathrm{Bi}_s(T_d)$, which arises from the relation (1.21a), rewritten above as (1.24a).

Namely we obtain the following correct condition:

$$\frac{\partial\theta}{\partial n'} + \left[\frac{\mathrm{Bi}_{\mathrm{conv}}}{\mathrm{Bi}_s(T_d)}\right]\{1 + \mathrm{Bi}_s(T_d)\theta\} + \frac{Q_0}{k\beta_s} = 0, \quad \text{at } z' = 1 + \eta h'(t', x', y'). \tag{1.24c}$$

But this above correct condition, (1.24c), is unfortunately not the condition that Davis derived in [18]! Only after the confusion (by a curious oversight?) of the conduction Biot number, $\mathrm{Bi}_s(T_d)$, with the Biot number for the convection $\mathrm{Bi}_{\mathrm{conv}}$, in (1.24c), and the consideration of a single 'surface Biot number B', did Davis obtain the upper, free-surface condition for the dimensionless temperature θ in the dimensionless form:

$$\frac{\partial\theta}{\partial n'} + 1 + B\theta = 0, \quad \text{at } z' = 1 + \eta h'(t', x', y'), \tag{1.25}$$

when $Q_0 = 0$ – the precise (conduction or convection) meaning of the B, in (1.25), being unclear!

It should be observed also that the appearance of a single, constant (*in fact, only, conduction*) Biot number, simultaneously in a conduction motionless basic state (which makes it possible to evaluate the corresponding value of the purely static basic temperature gradient β_s, according to (1.21b)) and in formulation of the thermocapillary convective Marangoni flow problem – via the upper, at $z' = H'(t', x', y')$, condition (1.25) for θ – leads to a very ambiguous situation.

This is a particularly unfortunate case, when this single (in fact conduction) Biot number is taken equal to zero. From this point of view, the results of Takashima's 1981 paper [22], concerning the linear Marangoni convection – in the case of a zero (conduction?) Biot number – must be accurately reconsidered (at least in a logical derivation process).

This *two Biot problem* deserves, obviously, further attention and I hope that the reader will consider our present discussion as a first step in the explanation of this intriguing question.

Section 4.3 is devoted to a rational analysis and asymptotic modelling of the above Bénard, heated from below, convection problem, taking into account mainly the results of Section 4.2.

In the last section of Chapter 4 (Section 4.4), we give some complements and concluding remarks concerning, first, again, the upper, free-surface condition for the temperature, then, a second discussion is devoted to long-scale evolution of thin liquid films (the models based on the long-wave approximation are also considered in Chapter 7), and a third short discussion concerns the various problems related to liquid films (falling down an inclined or vertical plane or inside a vertical circular or else hanging below a solid ceiling and also over a substrate with topography). Finally, we see now that three significant convection cases deserve interest, namely:

1. *shallow-thermal*, when $\mathrm{Fr}_d^2 \ll 1$,
2. *deep-thermal*, when $\mathrm{Di} \equiv \varepsilon \mathrm{Bo} \approx 1$,
3. *Marangoni-thermocapillary*, when $\mathrm{Fr}_d^2 \approx 1$,

which are considered in Chapters 5, 6 and 7.

Indeed, a fourth special case,

4. *ultra-thin film*, when $\mathrm{Fr}_d^2 \gg 1$,

deserves also a careful investigation – for instance when in a long-wave approximation: $d/\lambda \ll 1 \Rightarrow (d/\lambda)\mathrm{Fr}_d^2 \Rightarrow F^2 = \lambda^2 dg/{\nu_A}^2 = \mathrm{O}(1)$ – but in the present book we do not discuss this fourth case. In Chapter 8 the above three cases are also considered.

In Chapter 5, Section 5.2, we first derive the usual shallow RB convection model equations, where the main driving force is buoyancy – this derivation being performed via the RB limiting process (1.12) as in Chapter 3. In Section 5.3, second-order model equations associated to RB equations are derived. But, in Chapter 5, unlike Chapter 3, a new (curious) problem emerges because of the presence of the term $(\eta/\mathrm{Fr}_d^2)h'$ [where the ratio, $\eta = a/d$ is the upper, free-surface amplitude parameter, see (1.23e)], in the dominant (dimensionless) free surface upper boundary condition for $(p - p_A)$, rewritten with dimensionless pressure π defined by the relation

$$\pi = \left(\frac{1}{\mathrm{Fr}_d^2}\right)\left\{[(p - p_A)/gd\rho_d] + z' - 1\right\}, \quad z' = z/d. \tag{1.26}$$

As a consequence:

The free surface upper boundary condition for the dimensionless pressure π is asymptotically (at the leading order) consistent with the RB limiting process (1.12), only for a small free surface amplitude, $\eta \ll 1$, such that:

$$\frac{\eta}{\mathrm{Fr}_d^2} \equiv \eta^* \approx 1, \tag{1.27a}$$

when

$$\eta \text{ and } \mathrm{Fr}_d^2 \text{ both tend to zero.} \tag{1.27b}$$

In this case, for the RB model limit problem, according to (1.12), the upper, free-surface boundary conditions, at the leading order, are written for a non-deformable free surface $z' = 1$.

Besides, in the framework of a rational formulation of the RB leading-order model problem, a new amazing result is the derivation – at the leading order – from the jump condition for the pressure (see, for instance, the relation (2.42a) in Chapter 2) of an equation for the deformation of the free surface, $h'(t', x', y')$. Namely, for the unknown $h'(t', x', y')$ we obtain the partial differential equations

$$\frac{\partial^2 h'}{\partial x'^2} + \frac{\partial^2 h'}{\partial y'^2} - \left(\frac{\eta^*}{\mathrm{We}^*}\right) h' = -\left(\frac{1}{\mathrm{We}^*}\right) \pi_{\mathrm{sh}}, \quad \text{at } z' = 1, \tag{1.28a}$$

where

$$\mathrm{We}^* = \eta\,\mathrm{We} \approx 1, \tag{1.28b}$$

because usually the Weber number is large, $\mathrm{We} \gg 1$.

In the right-hand side of (1.28a), the term $\pi_{\mathrm{sh}}\,(t', x', y', z' = 1)$, together with $\mathbf{u}_{\mathrm{sh}}$ and θ_{sh}, is known when the solution (subscript 'sh') of the RB shallow convection problem is obtained.

Thus, we verify that, for a rational derivation of the RB, shallow rigid-free thermal convection model problem, it is necessary to assume the existence of three similarity relations:

$$\frac{\varepsilon}{\mathrm{Fr}_d^2} = \mathrm{Gr}, \tag{1.29a}$$

and

$$\frac{\eta}{\mathrm{Fr}_d^2} \equiv \eta^*, \tag{1.29b}$$

$$\frac{\eta}{\mathrm{Cr}} = \mathrm{We}^*, \tag{1.29c}$$

with *four* simultaneous limiting processes:

$$\varepsilon \downarrow 0, \quad \mathrm{Fr}_d^2 \downarrow 0, \quad \eta \downarrow 0, \tag{1.29d}$$

and the *crispation* (or capillary) number

$$\mathrm{Cr} \equiv \frac{1}{\mathrm{We}} \downarrow 0. \tag{1.29e}$$

Owing only to our rational analysis and asymptotic approach is it possible to derive on the one hand, equation (1.28a) for deformation of the free surface, $h'(t', x', y')$, and, on the other hand, a second-order consistent model problem – associated with the leading-order RB model problem – which takes into account the second-order (proportional to $\mathrm{Fr}_d^2 \ll 1$) terms (see Sections 3.5 and 3.6, and Section 5.3).

Indeed, in the RB model problem we have the possibility to partially take into account the Marangoni and Biot effects on the upper, non-deformable, free surface. Recently (in 1996, see [23]) such a model problem has been considered by Dauby and Lebon, but without any justification or discussion.

Finally, we observe that in the case of the RB shallow convection model problem, when the dimensionless parameter Bo, defined by (1.16), is fixed and of the order of unity,

$$d_{\approx} \frac{C_v(T_d)\Delta T}{g}, \tag{1.30a}$$

we can write (or identify) the squared Froude number with a low squared ('liquid') Mach number M_L^2, via the chain rule:

$$\mathrm{Fr}_d^2 \approx \frac{(\nu_d/d)^2}{C_v(T_d)\Delta T} = \left[\frac{(\nu_d/d)}{(C_v(T_d)\Delta T)^{1/2}}\right]^2 \equiv \mathrm{M}_\mathrm{L}^2. \tag{1.30b}$$

Therefore, for our weakly expansible liquid, instead of Fr_d^2, we can use M_L^2, which is the ratio of the reference (intrinsic) velocity $U_L = \nu_d/d$ to the pseudo-sound speed, $C_L = [C_v(T_d)\Delta T]^{1/2}$, for the liquid. The above approach has been used recently in our 2006 book (see [24, chapter 7, section 7.2.3]).

In Section 5.4, an amplitude equation *à la* Newel–Whitehead, is asymptotically derived and Section 5.5 is devoted to instability and route to chaos (to 'temporal' turbulence), in RB thermal shallow convection, via the three main scenarios (Ruelle–Takens, Feigenbaum and Pomeau–Manneville) in the framework of a finite-dimensional dynamical system approach. The last section of that chapter, Section 5.6, is devoted to some comments.

Chapter 6 is devoted to the so-called 'deep thermal convection' problem, first discovered in 1989 [4], and analyzed by Zeytounian, Errafyi, Charki, Franchi and Straughan during the years 1990–1996 (see [25–32]). Indeed, the above discussion concerning the RB problem (in Chapter 5) shows that the RB model problem is valid (operative) only in a (Boussinesq) liquid layer of thickness d such that [see (1.11) and (1.30a)]:

$$\left(\frac{\nu_d^2}{g}\right)^{1/3} \ll d \approx \frac{C_v(T_d)\Delta T}{g} \equiv d_{\text{sh}}, \tag{1.31a}$$

because, according to (1.15), only when Bo ≈ 1 do we have a small dissipation number such that

$$\text{Di} \approx \varepsilon, \tag{1.31b}$$

and the term proportional to εBo (linked with the viscous dissipation) disappears, at the leading order, when we derive RB model equations via the Boussinesq limiting process (1.12).

Therefore, on the contrary, the condition:

$$\text{Bo} \gg 1, \quad \text{such that Di} \equiv \varepsilon\text{Bo fixed, of the order unity,} \tag{1.31c}$$

characterizes the deep convection (DC) problem.

Obviously, (1.31c) is a direct consequence of the relation (1.14) for Di*, because in limiting process (1.12), the Grashof number, Gr, is fixed and of order unity.

In such a case, with (1.31c), the dissipation number Di* is also of the order unity and the viscous dissipation term is operative equally with the buoyancy term in thermal convection equations. As a consequence, in the deep convection problem we have for the thickness d, of the liquid layer, the estimate

$$d \approx \frac{C_v(T_d)}{g\alpha(T_d)} \equiv d_{\text{depth}}, \tag{1.32}$$

and

$$\text{Di} \equiv \frac{g\alpha(T_d)d}{C_v(T_d)} \tag{1.33}$$

is our depth parameter (denoted by δ in our 1989 paper, see [4]).

The formulation of a deep convection (DC) problem is necessary when the thickness d of the liquid layer satisfies the constraint (1.32), Di, given by (1.33), being a significant parameter. Finally, the deep convection model problem is derived, in a rational way, via the following, DC, limiting process:

$$\text{Gr} = \frac{\varepsilon}{\text{Fr}_d^2} \quad \text{fixed and Di} \equiv \varepsilon\text{Bo fixed,} \tag{1.34a}$$

when

$$\varepsilon \downarrow 0, \quad \text{Fr}_d^2 \downarrow 0 \quad \text{and} \quad \text{Bo} \uparrow \infty. \tag{1.34b}$$

In two papers [25,26], by Errafyi and Zeytounian the reader can find a linear theory for deep convection and various routes to chaos in the framework of deep convection unsteady two-dimensional equations.

On the other hand, in two papers [27, 28] by Charki and Zeytounian, the reader can find the derivation of a Lorenz deep system of equations and the Landau-Ginzburg amplitude equation for deep convection. Then, in three papers by Charki [29–31], the reader will find also some rigorous mathematical results – stability, existence and uniqueness of the solution for the initial value problem and for the steady-state problem.

The deep convection problem is derived in Section 6.2, and in Section 6.3 a linear theory is presented. Section 6.4 is devoted to an investigation of three main routes to chaos (mentioned above) and in Section 6.5 some comments are given concerning the rigorous mathematical results of Charki, Richardson and Franchi and Straughan.

Concerning these deep convection equations, we observe that in the paper by Franchi and Straughan [32], a nonlinear energy stability analysis of our 1989 deep convection equations is given. Finally, in the book by Straughan [33], the reader can find a derivation and discussion concerning deep convection in the framework of the theory of Hills and Roberts [34]. Unfortunately the equations derived by these authors in an *ad hoc* manner (with some 'compressible' effects) are not consistent (see also Section 3.6).

Chapter7 is devoted entirely to thermocapillary – Marangoni convection – the so-called Bénard–Marangoni (BM) – thin film problem.

It is now well known (mainly thanks to Pearson [17]) that Bénard convective cells are primarily induced by the temperature-dependent surface tension gradients resulting from the temperature variations along the free surface (the so-called Marangoni/thermocapillary effect) – in the leading order, both the buoyancy and viscous dissipation effects are neglected, but free surface deformations are taken into account, the model equations are those which govern an imcompressible viscous liquid – the temperature field being present via the upper, free-surface conditions where appears the Marangoni, Weber and Biot (convective) numbers.

On the other hand, the classical Rayleigh–Bénard (RB) thermal convection problem (considered in Chapter 5) is produced mainly by the buoyancy – the influence of a deformable free surface being neglected at the leading order for a weakly expansible liquid in a not very thin layer, according to (1.31a).

Naturally, in the general/full nonlinear (NS–F) convection, heated from below Bénard problem for an expansible viscous liquid – considered from the start in Chapter 4 – in the derived dimensionless dominant equations and upper, free-surface conditions, both buoyancy and Marangoni, Weber, Biot effects are operative. But for a weakly expansible liquid, in a thin (of order of the millimetre) layer, when $\mathrm{Fr}_d^2 \approx 1$, the deformable free surface influence is operative and the temperature-dependent surface tension, via the Marangoni number, has a driving effect. The buoyancy force, however, is negligible at the leading order.

As a consequence: it is not consistent (from an asymptotic point of view, at least in the leading-order, limiting, case) to take into account fully the above three effects – thermocapillarity, buoyancy and free surface deformation – simultaneously, for a weakly expansible viscous liquid.

The buoyancy is operative only in the RB thermal convection rigid-free problem. Conversely, the effects linked with the deformable upper, free surface are operative only in the Bénard–Marangoni (BM) thermocapillary thin film problem.

The main cause of this curious (leading-order) aspect of the full Bénard, heated from below, problem for a weakly expansible liquid, is the consequence of the presence of Fr_d^2 in the definition (as a denominator) of the Grashof number,

$$\mathrm{Gr} = \frac{\alpha(T_d)\Delta T}{\mathrm{Fr}_d^2}, \tag{1.35a}$$

where the expansibility number is assumed always to be a small parameter,

$$\varepsilon = \alpha(T_d)\Delta T \ll 1! \tag{1.35b}$$

The only possibility for a full account of the deformation of the free surface, separating the weakly expansible liquid layer from the ambient, passive, motionless air, is directly related to the condition

$$\mathrm{Fr}_d^2 \approx 1 \tag{1.35c}$$

or

$$d \approx \left(\frac{\nu_d^2}{g}\right)^{1/3} = d_{\mathrm{BM}} \approx 1 \text{ mm}. \tag{1.35d}$$

In this case, it is not necessary to assume (in upper, free-surface, dominant conditions derived in Chapter 4) that the free surface amplitude parameter, η, is a small parameter [see (1.27a) and (1.27b)], as is the case for the RB model problem. But, with (1.35c), the buoyancy term, proportional to the Grashof number, is in fact of the order of the small expansibility parameter, ε, and does not appear in the leading-order, limiting case ($\varepsilon \to 0$), in equations governing the BM problem.

The BM model problem, derived at the leading order, from the full dominant Bénard problem (with upper, free-surface, dominant conditions) is formulated in Chapter 4, via the following incompressible limiting process:

$$\varepsilon \to 0, \quad \mathrm{Fr}^2 \approx 1 \text{ fixed}. \tag{1.36}$$

Concerning the influence of the viscous dissipation term, since

$$d_{\mathrm{BM}} = \left(\frac{{\nu_d}^2}{g}\right)^{1/3},$$

according to (1.35d) then, this viscous dissipation term is negligible, according to (1.14), (1.15) and the definition of $\mathrm{Di}^* \approx \mathrm{Bo}/2$ (because $\mathrm{Gr} \approx \varepsilon$), if

$$\mathrm{Bo} \ll 1 \tag{1.37a}$$

or

$$d_{\mathrm{BM}} \ll \frac{\Delta T C_v(T_d)}{g}. \tag{1.37b}$$

In such a case, we obtain also the following lower bound for $\Delta T = T_w - T_d > 0$:

$$\Delta T \gg \frac{(g\nu_d)^{2/3}}{C_v(T_d)}. \tag{1.37c}$$

The BM leading-order equations are, in fact, the usual *Navier viscous incompressible* equations, for the limiting values of the velocity vector $\mathbf{u}_{\mathrm{BM}}$ and perturbation of the pressure π_{BM}, and the (uncoupled with Navier equations) Fourier simple equation for the dimensionless temperature θ_{BM}. The coupling (with $\mathbf{u}_{\mathrm{BM}}$ and π_{BM}), being realized via the upper, free-surface conditions at the deformable free surface (see Section 7.2, where the full BM problem is formulated). In Section 7.3 we return to full formulation of the BM dimensionless thermocapillary convection model (given in Section 7.2) keeping in mind (thanks to a long-wave approximation, $\lambda \gg d_{\mathrm{BM}}$, where λ is a horizontal wavelength) to obtaining a simplified 'BM long-wave reduced model problem'. In Section 7.4, thanks to the results of the preceding section, we derive accurately a 'new' lubrication equation for the thickness of the thin liquid film. In particular, taking into account our, 'two Biot numbers' (for conduction motionless steady-state and convection regime) approach, we show that the consideration of a variable convective Biot number (for instance, a function of the thickness of the liquid film) give the possibility to take into account, in the derived 'new' lubrication equation, the thermocapillary/Marangoni effect, even if the convective Biot number is vanishing! Since most experiments and theories are focussed on thermocapillary instabilities of a freely falling vertical two-dimensional film, the reader can find a formulation of this problem in Section 7.5. This makes it possible to carry out an asymptotic detailed derivation of a generalized, *à la Benney* equation and, then to Kuramuto–Sivashinsky (KS) and KS–KdV (dissipative Korteweg and de Vries) one-dimensional evolution equations. In Section 7.5 we also discuss obtaining the averaged 'integral boundary layer' (IBL) model problems and derive one such, a non-isothermal IBL model system of three equations (see also Section 10.4).

Section 7.6 is devoted to various aspects of the linear and weakly nonlinear stability analysis of thermocapillary convection. In Section 7.7 ('Some Complementary Remarks'), various results derived in Sections 7.4 and 7.5, with $\Theta = (T - T_A)/(T_w - T_A)$, are re-considered and compared with the results obtained when the dimensionless temperature is given by $\theta = (T - T_d)/(T_w - T_d)$. In such a case it is necessary to take into account that the upper, free-surface condition, $\partial\Theta/\partial n' + \mathrm{Bi}_{\mathrm{conv}}\Theta = 0$ at $z' = H'(t', x', y')$, associated with Θ, must be replaced by (for θ)

$$\frac{\partial\theta}{\partial n'} + 1 + \mathrm{Bi}_{\mathrm{conv}}\theta = 0 \quad \text{at } z' = H'(t', x', y'), \tag{1.38}$$

when a judicious choice of Q_0 is made. Namely, if we linearize our upper, free-surface condition (1.24c) for θ, then we easily observe that this linearized condition which emerges from (1.24c) is compatible, at the order η, with a linear condition for θ' (when $\theta = 1 - z + \eta\theta' + \cdots$), only if $Q_0 \equiv k\beta_s[1 - (\mathrm{Bi}_{\mathrm{conv}}/\mathrm{Bi}_s)]$ and in such a case, instead of (1.24c) we obtain the above condition (1.38) for θ with $\mathrm{Bi}_{\mathrm{conv}}$ (instead of the conduction Biot number in Davis [18]).

On the one hand, associated with θ, the dimensionless temperature $\theta_S(z')$, for the steady motionless conduction state, satisfies the upper condition

$$\frac{\mathrm{d}\theta_S}{\mathrm{d}z'} + 1 + \mathrm{Bi}_s(T_d)\theta_S = 0 \quad \text{at } z' = 1,$$

with $\theta_S(z') = 1 - z'$. On the other hand, associated with Θ, the dimensionless temperature $\Theta_S(z')$, for the same steady motionless conduction state, satisfies the upper condition:

$$\frac{\mathrm{d}\Theta_S}{\mathrm{d}z'} + \mathrm{Bi}_s(T_d)\Theta_S = 0 \quad \text{at } z' = 1,$$

with

$$\Theta_S(z') = 1 - \left[\frac{\mathrm{Bi}_s}{1 + \mathrm{Bi}_s}\right] z'.$$

In our 1998 survey paper [35], the reader can find a detailed theory for the Bénard–Marangoni thermocapillary instability problem. We also mention the 12 more recent papers, published in the special double issue of the

Journal of Engineering Mathematics in 2004 [36]. We quote from the preface (pp. 95–97) written by the guest-editors of the *Journal of Engineering Mathematics* (Editor-in-Chief H.K. Kuiken). These 12 papers

> ... demonstrate the state of the art (but, unfortunately, rather in an 'ad-hoc' manner) in describing thin-film flows, and illustrate both the wide variety of mathematical methods that have been employed and the broad range of their applications. Despite the significant advances that have been made in recent years there are still many challenges to be tackled and unsolved problems to be addressed, and we anticipate that liquid films will be a lively and active research area for many years to come.

In Chapter 8 the reader can find a 'summing up' of the three cases related to the Bénard, heated from below, convection problem (discussed in Chapters 5, 6 and 7). In this short chapter, the reader can find, first, an 'interconnection sketch' which illustrates the relations between these three main facets of Bénard convection. First, for the RB model problem (considered in Section 5.2) we give anew the consistent conditions and constraints for derivation of the associated shallow equations and conditions. Then, in Section 8.3, for the deep thermal convection problem (considered in Section 6.3) we give the main results of our rational approach. Third, in Section 8.4, for the Marangoni thin viscous film problem (considered in Section 7.4) the full Bénard–Marangoni model problem is again briefly discussed. This chapter is written especially for the readers who do not care much for rigor, and just want to know, what are the relevant model equations and constraints for their convection problem!

In Chapter 9, atmospheric thermal convection problems are briefly considered. It is necessary to observe that the main mechanism of convective flow in the atmosphere is responsible for the global-wide circulation of the atmosphere, which is a driving motion important for long-range forecasting. It is, also, a disruption of normal convective transport that periodically leaves cities such as Los Angeles and Madrid smogbound under a temperature inversion. On the contrary, the Boussinesq approximation (see [7]), which gives the possibility to consider a Boussinesquian (*à la Boussinesq*) fluid motion, is actually, perhaps, the most widely used simplification in various atmospheric – meso or local scales – thermal convection problems, the (dry atmospheric) air being assumed as a thermally perfect gas.

A very good illustration of the plurality of the Boussinesq approximation is the numerous survey papers in various volumes of the *Annual Review of Fluid Mechanics* (edited in Stanford, USA) where this approximation is the basis for mathematical formulation for various convective problems – for example, convection involving thermal and salt fields [38]. It is interesting to observe that already in 1891 Oberbeck [39] uses a Boussinesq type approximation in meteorological studies of the Hadley thermal regime for the trade-winds arising from the deflecting effect of the Earth's rotation.

In atmosphere problems an important parameter is the *Rossby number* (Ro) or *Kibel number* (Ki); each characterizes the effect of the Coriolis force. If the vector of rotation of the Earth $\mathbf{\Omega}$ is directed from south to north according to the axis of the poles, it can be expressed as follows (see, for instance, our book [40] published in 1991 on Meteorological Fluid Mechanics):

$$\mathbf{\Omega} = \Omega_0 \mathbf{e}, \quad \text{with } \mathbf{e} = \sin\varphi \mathbf{k} + \cos\varphi \mathbf{j}, \tag{1.39}$$

where φ is the algebraic latitude of the observation point P° on the Earth's surface, around which the atmospheric convection motion is analyzed. We observe that $\varphi > 0$ in the northern hemisphere and $\varphi^\circ \approx 45^\circ$ is the usual reference value for φ, the unit vectors being directed to the east, north and zenith, in the opposite direction from the 'force of gravity' $\mathbf{g}$ $(= -g\mathbf{k}$ – more precisely the gravitational acceleration modified by centrifugal force), and are denoted by $\mathbf{i}, \mathbf{j}$ and $\mathbf{k}$. If, now, the reference (atmospheric), time, velocity, horizontal and vertical lengths are: t°, U°, L_0, h°, and $a^\circ \approx 6300$ km is the radius of the Earth, then

$$\mathrm{Ro} = \frac{U^\circ}{f^\circ L^\circ}, \tag{1.40a}$$

$$\mathrm{Ki} = \frac{1}{t^\circ f^\circ}, \tag{1.40b}$$

$$\delta = \frac{L_0}{a_0}, \tag{1.40c}$$

$$\lambda = \frac{h_0}{L_0}. \tag{1.40d}$$

are four main dimensionless parameters in the analysis of the atmospheric convection motion. In (1.40a, b), $f^\circ = 2\Omega^0 \sin\varphi^\circ$ is the Coriolis parameter, δ is the sphericity parameter and λ is the hydrostatic parameter.

A very significant limiting case for study of atmospheric convection (in a thin atmospheric layer) is linked with the following (so-called ' quasi-hydrostatic') limiting process (considered in Section 9.2):

$$\lambda \downarrow 0 \quad \text{and} \quad \mathrm{Re} = \frac{U^\circ L^\circ}{\nu^\circ} \to \infty, \quad \text{with } \lambda^2 \mathrm{Re} \equiv \mathrm{Re}_\perp \text{ fixed.} \tag{1.41}$$

The atmospheric convection problems are mainly related to small Mach number motions

$$\mathrm{M} = \frac{U^\circ}{[\gamma R T_0]^{1/2}} \ll 1, \tag{1.42}$$

because, in thermal boundary conditions on the ground, we have a small rate of temperature $(\Delta T)_0$ relative to the constant reference temperature T_0,

$$\tau = \frac{(\Delta T)_0}{T_0} \gg 1 \quad \text{such that } \tau/\mathrm{M} = \tau^* = \mathrm{O}(1); \tag{1.43}$$

a Boussinesq limit process is also considered when τ and M both tend to zero with the similarity rule (1.43). But, in Chapter 9, I study only some particular (mainly meso or local) convection motions in the atmosphere. Namely, after an Introduction (Section 9.1), we consider the breeze problem via the Boussinesq approximation (in Section 9.2), the infuence of a local temperature field in an atmospheric Ekman layer – via a triple deck asymptotic approach (in Section 9.3) and then, a periodic, double-boundary layer thermal convection over a curvilinear wall (in Section 9.4). In Section 9.5 ('Complements') some other particular atmospheric convection problems are also briefly discussed. We note here the very pertinent book [41] by Turner in 1973, concerning buoyancy effects.

The last chapter is Chapter 10, with nine sections, which gives a miscellany of various convection model problems, as is obvious from the Table of Contents and the short commentary above. After a brief Introduction (Section 10.1) I note in Section 10.2, first, that a very pertinent formulation of the convection problem in the Earth's outer core has been given by Jöhnk and Svendsen [42], and this formulation is briefly discussed. Section 10.3, is devoted to a survey concerning the 'magneto-hydrodynamic, electro, ferro, chemical, solar, oceanic, rotating, and penetrative convections'.

In particular, in the book by Straughan [43], the reader can find various information concerning the 'electro, ferro and magnet-hydrodynamic convections'. Section 10.4 is devoted to the averaged, integral boundary layer (IBL), technique, and the reader can find in two papers by Shkadov [44,45] a pertinent introductory discussion. The papers by Yu et al. [46], Zeytounian [35], Ruyer-Quil and Manneville [47], are devoted to some successful generalizations (for the non-isothermal case) of the basic isothermal averaged Shkadov 1967 model for film flows using long-wave approximation. For the non-isothermal case, first, Zeytounian (see [6, pp. 139–144] and also [35]), has derived a new, more complete, IBL model consisting of three equations in terms of the local film thickness (h), flow rate (q) and mean temperature across the film layer (Θ) – which has been considered in Sections 7.5 and

7.6. This Zeytounian model has been improved by Kalliadasis et al. [48]. In two recent papers [49, 50], the thermocapillary flow is modelled by using a gradient expansion combined with a Galerkin projection with polynomial test functions for both velocity and temperature fields – see, in paper [49] the system of the three equations (6.6a–c) or in paper [50] the system of the three equations (1.1a–c). In Section 10.5, the results of Golovin, Nepommyaschy and Pismen [51] and also Kazhdan et al. [52] is annotated – according to linear theory, there exist two monotonic modes (short-scale mode and long-scale mode) of surface-tension driven, convective instability, which is shown very well in the paper by Golovin, Nepommyaschy and Pismen and also in numerical results of Kazhdan et al. These two types of the Marangoni convection, having different scales, can interact with each other in the course of their nonlinear evolution – near the instability threshold, the nonlinear evolution and interaction between the two modes can be described by a system of two coupled nonlinear equations. Section 10.6, concerns thermosolutal convection (when the density varies both with temperature and concentration/salinity, and the corresponding diffusivities are very different); the reader can find various information in the review paper by Turner [38]. In the paper by Knobloch et al. [53], various facets of the transitions to chaos, in 2D double-diffusive convection are presented; in this paper the reader can also find several pertinent references. In Section 10.7, as a complement of Chapter 9, we consider the so-called 'anelastic approximation for the atmospheric non-adiabatic and viscous thermal convection'. The derivation of these anelastic equations adapted for an atmospheric (deep, non-adiabatic, viscous) convection problem, is inspired from our monograph [2, chap. 10, sec. 2]. In Section 10.8, an interesting convection, initial-boundary value, problem is linked with a thin liquid film over cold/hot rotating disks. This problem has been considered very accurately by Dandapat and Ray in [54]. In Section 10.9, a solitary wave phenomena in convection regime is considered, and, finally, in Section 10.10, some comments and complementary recent results and references concerning convection problems are given and discussed.

References

1. R.Kh. Zeytounian, *Arch. Mech.* (Archiwun Mechaniki Stosowanej) **26**(3), 499–509, 1974.
2. R.Kh. Zeytounian, *Asymptotic Modeling of Atmospheric Flows*. Springer-Verlag, Heidelberg, XII + 396 pp., 1990.

3. R.Kh. Zeytounian, *C.R. Acad. Sc., Paris, Sér. I*, **297**, 271–274, 1983.
4. R.Kh. Zeytounian, *Int. J. Engng. Sci.* **27**(11), 1361–1366, 1989.
5. R.Kh. Zeytounian, *Int. J. Engng. Sci.* **35**(5), 455–466, 1997.
6. R.Kh. Zeytounian, Theoretical aspects of interfacial phenomena and Marangoni effect. In: *Interfacial Phenomena and the Marangoni Effect*, M.G. Velarde and R.Kh. Zeytounian (Eds.), CISM Courses and Lectures, Vol. 428. Springer, Wien/New York, pp. 123–190, 2002.
7. R.Kh. Zeytounian, On the foundations of the Boussinesq approximation applicable to atmospheric motions. Izv. Atmosph. Oceanic Phys. **39**, Suppl. 1, S1–S14, 2003.
8. R.Kh. Zeytounian, A quasi-one-dimensional asymptotic theory for nonlinear water waves. *J. Engng. Math.* **28**, 261–296, 1991.
9. R.Kh. Zeytounian, Nonlinear long waves on water and solitons. *Phys. Uspekhi* (English ed.), **38**(12), 1333–1381, 1995.
10. R.Kh. Zeytounian, *Nonlinear Long Surface Waves in Shallow Water (Model Equations)*. Laboratoire de Mécanique de Lille, Bât. 'Boussinesq', Université des Sciences et Technologies de Lille. Villeneuve d'Asq, France, XXIII + 224 pp., 1993.
11. J. Serrin, Mathematical principles of classical fluid mechanics. In: *Handbuch der Physik*, S. Flügge (Ed.). Springer, Berlin, Vol. VIII/1, pp. 125–263, 1959.
12. A.J.B. Saint-Venant (de), *C.R. Acad. Sci.* **17**, 1240–1243, 1843.
13. Lord Rayleigh, On convection currents in horizontal layer of fluid when the higher temperature is on the under side. *Philos. Mag., Ser. 6* **32**(192), 529–546, 1916.
14. H. Bénard, Les tourbillons cellulaires dans une nappe liquide. *Rev. Générale Sci. Pures Appl.* **11**, 1261–1271 and 1309–1328, 1900. See also: Les tourbillons cellulaires dans une nappe liquide transportant de la chaleur par convection en régime permanent. *Ann. Chimie Phys.* **23**, 62–144, 1901.
15. J. Boussinesq, Théorie analytique de la chaleur, Vol. II. Gauthier-Villars, Paris, 1903.
16. D.L. Turcotte et al., *J. Fluid Mech.* **64**, 369, 1974.
17. R. Perez Cordon and M.G. Velarde, *J. Physique* **36**(7/8), 591–601, 1975.
18. S.H. Davis, Annu. Rev. Fluid Mech. **19**, 403–435, 1987.
19. J.R.A. Pearson, On convection cells induced by surface tension. *J. Fluid Mech.* **4**, 489, 1958.
20. D.D. Joseph, *Stability of Fluid Motions*, Vol. II. Springer, Heidelberg, 1976.
21. P.M. Parmentier, V.C. Regnier and G. Lebond, Nonlinear analysis of coupled gravitational and capillary thermoconvection in thin fluid layers. *Phys. Rev. E* **54**(1), 411–423, 1996.
22. M. Takashima, *J. Phys. Soc. Japan* **50**(8), 2745–2750 and 2751–2756, 1981.
23. P.C. Dauby and G. Lebon, *J. Fluid Mech.* **329**, 25–64, 1996.
24. R.Kh. Zeytounian, *Topics in Hyposonic Flow Theory*. Lecture Notes in Physics, Vol. 672. Springer-Verlag Heidelberg, 2006.
25. M. Errafyi and R.Kh. Zeytounian, *Int. J. Engng. Sci.* **29**(5), 625, 1991.
26. M. Errafyi and R.Kh. Zeytounian, *Int. J. Engng. Sci.* **29**(11), 1363, 1991.
27. Z. Charki and R.Kh. Zeytounian, *Int. J. Engng. Sci.* **32**(10), 1561–1566, 1994.
28. Z. Charki and R.Kh. Zeytounian. *Int. J. Engng. Sci.* **33**(12), 1839–1847, 1995.
29. Z. Charki, Stability for the deep Bénard problem. *J. Math. Sci. Univ. Tokyo* **1**, 435–459, 1994.
30. Z. Charki, *ZAMM* **75**(12), 909–915, 1995.
31. Z. Charki, The initial value problem for the deep Bénard convection equations with data in L_q. *Math. Models Methods Appl. Sci.* **6**(2), 269–277, 1996.
32. F. Franchi and B. Straughan. *Int. J. Engng. Sci.* **30**, 739–745, 1992.

33. B. Straughan, *Mathematical Aspects of Penetrative Convection*. Longman, 1993.
34. R. Hills and P. Roberts, *Stab. Appl. Anal. Continuous Media*, **1**, 205–212, 1991.
35. Kh. Zeytounian, The Bénard–Marangoni thermocapillary-instability problem, *Phys. Uspekhi*, **41**(3), pp. 241-267, March 1998 [English edition].
36. D.G. Crowley, C.J. Lawrence and S. K. Wilson (guest-editors), *The Dynamics of Thin Liquid Film, Journal of Engineering Mathematics* Special Issue, **50**(2–3), 2004.
37. G.A. Shugai and P.A. Yakubenko, Spatio-temporal instability in free ultra-thin films. *Eur. J. Mech. B/Fluids* **17**(3), 371–384, 1998.
38. J.S. Turner, *Annu. Rev. Fluid Mech.* **17**, 11–44, 1985.
39. A. Oberbeck, Ann. Phys. Chem., Neue Folge **7**, 271–292, 1879.
40. R.Kh. Zeytounian, *Meteorological Fluid Mechanics*, Lecture Notes in Physics, Vol. m5. Springer-Verlag, Heidelberg, 1991.
41. J.S. Turner, *Buoyancy Effects in Fluids*. Cambridge, Cambridge University Press, 1973.
42. K. Jöhnk and B. Svendsen, A thermodynamic formulation of the equations of motion and buoyancy frequency for Earth's fluid outer core. *Continuum Mech. Thermodyn.* **8**, 75–101, 1996.
43. B. Straughan, *The Energy Method, Stability, and Nonlinear Convection*. Applied Mathematical Sciences, Vol. 91. Springer-Verlag, New York, 1992.
44. V.Ya. Shkadov, *Izv. Akad. Naouk SSSR, Mech. Zhidkosti i Gaza* **1**, 43–50, 1967.
45. V.Ya. Shkadov, *Izv. Akad. Naouk SSSR, Mech. Zhidkosti i Gaza* **2**, 20–25, 1968.
46. L.-Q. Yu, F.K. Ducker, and A.E. Balakotaiah, *Phys. Fluids* **7**(8), 1886–1902, 1995.
47. C. Ruyer-Quil and P. Manneville, *Eur. Phys. J.* **B6**, 277–292, 1998.
48. S. Kalliadasis, E.A. Demekhin, C. Ruyer-Quil, M.G. Velarde, *J. Fluid Mech.* **492**, 303–338, 2003.
49. C. Ruyer-Quil, B. Scheid, S. Kalliadasis, M.G. Velarde and R.Kh. Zeytounian, *J. Fluid Mech.* **538**, 199–222, 2005.
50. B. Scheid, C. Ruyer-Quil, S. Kalliadasis, M.G. Velarde and R.Kh. Zeytounian, *J. Fluid Mech.* **538**, 223–244, 2005.
51. A.A. Golovin, A.A. Nepommyaschy and L.M. Pismen, *Phys. Fluids* **6**(1), 35–48, 1994.
52. D. Kashdan et al., Nonlinear waves and turbulence in Marangoni convection. *Phys. Fluids* **7**(11), 2679–2685, 1995.
53. E. Knobloch, D.R. Moore, J. Toomre and N.O. Weiss, *J. Fluid Mech.* **166**, 400–448, 1986.
54. B.S. Dandapat and P.C. Ray, *Int. J. Non-Linear Mech.* **28**(5), 489–501, 1993.

Chapter 2
The Navier–Stokes–Fourier System of Equations and Conditions

2.1 Introduction

In the framework of the mechanics of continua, the starting system of equations (local at any point of continuity in a fluid domain) is the one given in Chapter 1 (see equations (1.1a–c)) where the reader also will find some preliminary results on thermodynamics.

Here, the full starting equations are the NS–F equations for compressible and heat conducting fluid. We consider mainly an expansible liquid or a thermally perfect gas. The Bénard, heated from below, convection problem is considered in a weakly expansible liquid layer with a deformable free surface. In the case of atmospheric thermal convection problems, the fluid is a dry air which is assumed to be a thermally perfect gas. Concerning the upper conditions at the deformable free surface, we consider the full jump conditions for pressure with temperature-dependent surface tension and also Newton's cooling law for the temperature.

Our formulation given above makes it possible to take into account both buoyancy and thermocapillary effects, linked with the Rayleigh, Froude, Prandtl, Weber and Marangoni numbers and also the effect linked with deformations of the free surface and heat transfer across this free surface, i.e., the convective Biot number effect. After this short Introduction we give, in Section 2.2, various complementary results from classical thermodynamics which are mainly necessary (especially, in the case of an expansible liquid considered in Section 2.4) to obtain an evolution, *à la* Fourier, equation for the temperature T. In Section 2.3, the full NS–F system of equations for a thermally perfect gas is given and in Section 2.5 the upper, deformable free-surface conditions are derived in detail; again, the problem of the condition for dimensionless temperature is dicussed. In Section 2.6, we give some comments concerning the influence of initial conditions and transient behavior, the last section, Section 2.7 being devoted to a discussion of the Hill and Roberts [20] approach.

From Chapter 1, according to the first, (1.1a), and second, (1.1b), equations of the system (1.1a–c), with (1.4a–c), we have two dynamical (usually named 'Navier–Stokes') equations for the velocity vector **u** and (mechanical) pressure p, namely:

$$\frac{\mathrm{d}\rho}{\mathrm{d}t} + \rho(\nabla \cdot \mathbf{u}) = 0, \tag{2.1a}$$

$$\rho \frac{\mathrm{d}\mathbf{u}}{\mathrm{d}t} + \nabla p = \rho \mathbf{f} + \nabla \cdot [2\mu \mathbf{D}(\mathbf{u})] - (2/3)[\mu(\nabla \cdot \mathbf{u})], \tag{2.1b}$$

where $\mathrm{d}/\mathrm{d}t \equiv \partial/\partial t + \mathbf{u} \cdot \nabla$, and the (Cartesian) components of the rate-of-deformation tensor $\mathbf{D}(\mathbf{u})$ are, according to (1.4b),

$$d_{ij} = \left(\frac{1}{2}\right)\left[\frac{\partial u_i}{\partial x_j} + \frac{\partial u_j}{\partial x_i}\right]. \tag{2.2}$$

Usually for a non-barotropic (baroclinic-trivariate with p, ρ, T, as thermodynamic functions) fluid motion, it is necessary to take into account, with the above two Navier–Stokes equations, (2.1a, b), a general equation of state connecting the three thermodynamic functions under consideration, ρ, p and T; namely:

$$F(\rho, p, T) = 0, \tag{2.3}$$

where T is the (absolute) temperature, a new unknown function, the viscosity coefficient μ, in (2.1b), being often (at least) a function of T also!

As a direct consequence of (2.3), the two Navier–Stokes equations, (2.1a, b), must be complemented by an evolution equation for the temperature T in order to obtain a NS–F closed system of four equations for, **u**, p, ρ and T; this requires some information from thermodynamics.

2.2 Thermodynamics

We assume that the reader of this monograph is familiar with the classical elements of thermodynamics at the level of undergraduate studies. In reality, in the framework of classical/Newtonian fluid mechanics *à la* Serrin (see [1], a pioneering survey paper) when the starting system of equations is the Navier–Stokes and Fourier (NS–F) system, the theory of thermodynamics is very simplified.

Indeed, in 'classical thermodynamics':

The thermodynamics for fluids is mainly related with the formulation of an evolution equation for the (absolute) temperature $T(t, \mathbf{x})$, considered together with pressure $p(t, \mathbf{x})$, density $\rho(t, \mathbf{x})$ and velocity $\mathbf{u}(t, \mathbf{x})$, as an unknown function (of the time t and space coordinate $\mathbf{x}$) of the unsteady compressible, viscous and heat-conducting fluid motion, governed by the Navier–Stokes and Fourier (NS–F) equations.

Classical thermodynamics is concerned with equilibrium states and observation shows that results for equilibrium states are approximately valid for non-equilibrium states (non-uniform) common in practical fluid dynamics.

The state of a given mass of fluid in equilibrium is specified uniquely by two parameters:

$$\text{specific volume,} \;\; v \equiv 1/\rho, \tag{2.4a}$$

and

$$\text{pressure,} \; p = (1/3)T_{ij}, \tag{2.4b}$$

where the T_{ij} are the components of the stress tensor T which appears (in Chapter 1) in the momentum equation (1.1a) and also in energy balance (1.1c), and are given by the constitutive relation (1.4a).

The relation between the temperature T and the two parameters of state, p and v, which we may write also as

$$f(p, v, T) = 0, \tag{2.4c}$$

thereby exhibiting formally the arbitrariness of the choice of these two parameters, is also called an equation of state and is equivalent to the above general equation of state (2.3).

On the one hand, for the specific internal energy $e(t, \mathbf{x})$, which is the solution of the 'mechanics of continua' via energy equation (1.1c), we can write according to (1.6b) the following evolution equation:

$$\rho \frac{\mathrm{d}e}{\mathrm{d}t} = -p(\nabla \cdot \mathbf{u}) + 2\mu\{\mathbf{D}(\mathbf{u}) : \mathbf{D}(\mathbf{u}) - (1/3)(\nabla \cdot \mathbf{u})^2\} + \frac{\partial}{\partial x_i}\left(k \frac{\partial T}{\partial x_i}\right), \tag{2.5}$$

when we use the Fourier law (1.5) for the heat flux vector $\mathbf{q}$ and also (1.4a).

But, on the other hand, if S is the specific entropy, we have also the following classical thermodynamic relation:

$$\mathrm{d}e = T\,\mathrm{d}S - p\,\mathrm{d}v. \tag{2.6}$$

As a consequence, thanks to (2.5), we obtain for the term $T\,\mathrm{d}S/\mathrm{d}t$ the following simpler energy equation:

$$T\,\frac{\mathrm{d}S}{\mathrm{d}t} = \Phi + \left(\frac{1}{\rho}\right)\frac{\partial(k\partial T/\partial x_i)}{\partial x_i}, \tag{2.7a}$$

with [Φ is the rate of viscous dissipation according to (1.7)]

$$\rho\Phi = 2\mu\{\mathbf{D}(\mathbf{u}) : \mathbf{D}(\mathbf{u}) - (1/3)(\nabla\cdot\mathbf{u})^2\}, \tag{2.7b}$$

and instead of (2.3) we write, as general equation of state,

$$\rho = \rho(T, p), \tag{2.8}$$

which characterizes the state of an expansible (or 'dilatable') liquid and is usually used in convection problems.

In fact, the two Navier–Stokes equations, (2.1a, b), with equation (2.7a) for the specific entropy, and state equation (2.8) for the thermodynamic functions, constitute our full Navier–Stokes–Fourier (NS–F) starting 'exact ' system.

However, unfortunately, this system of equations (2.1a, b) and (2.7a), with (2.7b), and (2.8), is not a closed system for $\mathbf{u}$, p, ρ, T and S, since we have four equations for five unknown functions. As a consequence the following necessary step is the introduction of the constant pressure heat capacity, C_p, and the coefficient of thermal expansion, α (mainly, for our expansible liquid). Namely:

$$C_p = T\left(\frac{\partial S}{\partial T}\right)_p \tag{2.9a}$$

and

$$\alpha = -\left(\frac{1}{\rho}\right)\left(\frac{\partial\rho}{\partial T}\right)_p, \tag{2.9b}$$

and we note that four useful identities, known as Maxwell's thermodynamic relations, follow (according to chapter 1 in Batchelor's 1967 book [2]).

For example, to obtain the following two classical Maxwell relations:

$$\left(\frac{\partial S}{\partial p}\right)_T = -\left(\frac{\partial v}{\partial T}\right)_p, \tag{2.10a}$$

and

$$\left(\frac{\partial p}{\partial T}\right)_v = \left(\frac{\partial S}{\partial v}\right)_T, \tag{2.10b}$$

we observe that it is sufficient to form the double derivative, in two different ways, of the functions: $e - TS$ and $e - TS + pv$, respectively, and take into account the thermodynamic relation (2.6), when v and S are regarded as the two independent parameters of state on which all functions of state depend, such that

$$\left(\frac{\partial e}{\partial v}\right)_S = -p, \tag{2.10c}$$

$$\left(\frac{\partial e}{\partial S}\right)_v = T. \tag{2.10d}$$

Obviously we can write (associated with (2.9a)), also for the constant volume, heat capacity:

$$C_v = T\left(\frac{\partial S}{\partial T}\right)_v. \tag{2.11a}$$

Moreover, on regarding S as a function of T and v, we find:

$$\mathrm{d}S = \left(\frac{\partial S}{\partial T}\right)_v \mathrm{d}T + \left(\frac{\partial S}{\partial v}\right)_T \mathrm{d}v$$

or

$$\left(\frac{\partial S}{\partial T}\right)_p = \left(\frac{\partial S}{\partial T}\right)_v + \left(\frac{\partial S}{\partial v}\right)_T \left(\frac{\partial v}{\partial T}\right)_p$$

and it then follows from (2.9a) and (2.11a) and the second Maxwell relation (2.10b) that

$$C_p - C_v = -T\left(\frac{\partial p}{\partial T}\right)_v \left(\frac{\partial v}{\partial T}\right)_p = -T\alpha^2\left(\frac{\partial p}{\partial \rho}\right)_T, \tag{2.11b}$$

because

$$\left(\frac{\partial p}{\partial T}\right)_v = -\left(\frac{\partial p}{\partial T}\right)_v \left(\frac{\partial v}{\partial T}\right)_p.$$

The relation (2.11b) is very interesting for the case of an expansible liquid with a 'full' equation of state (2.8) and shows that the three quantities p, ρ, T are subject to a single remarkable relationship. Now, if

$$\gamma \equiv \frac{C_p}{C_v}, \tag{2.12a}$$

then

$$C_p = \frac{TC_T^2\alpha^2}{(\gamma - 1)} \tag{2.12b}$$

and

$$C_v = \frac{TC_T^2\alpha^2}{\gamma(\gamma - 1)}, \tag{2.12c}$$

where

$$C_T^2 = \gamma\left(\frac{\partial p}{\partial \rho}\right)_T = \left(\frac{\partial p}{\partial \rho}\right)_S \tag{2.12d}$$

is the squared sound speed in the fluid.

But, on the one hand, when p and T are regarded as the two independent parameters of state, on which all functions of state depend, we can write:

$$\mathrm{d}S = \left(\frac{\partial S}{\partial T}\right)_p \mathrm{d}T + \left(\frac{\partial S}{\partial p}\right)_T \mathrm{d}p. \tag{2.13a}$$

As a consequence, with (2.9a) and (2.10a), from the above relation (2.13a), we obtain for $T\,\mathrm{d}S/\mathrm{d}t$ in equation (2.7a), the following relation:

$$T\frac{\mathrm{d}S}{\mathrm{d}t} = C_p\frac{\mathrm{d}T}{\mathrm{d}t} - \left(\frac{\alpha T}{\rho}\right)\frac{\mathrm{d}p}{\mathrm{d}t}. \tag{2.13b}$$

On the other hand, from the equation of state (2.8) $\rho = \rho(T, p)$, we can also write

$$\mathrm{d}\rho = \rho[-\alpha\,\mathrm{d}T + \chi\,\mathrm{d}p], \tag{2.14}$$

where

$$\chi = \left(\frac{1}{\rho}\right)\left[\frac{\partial \rho}{\partial p}\right]_T, \tag{2.15a}$$

and we observe that, as isothermal coefficient of compressibility β, we have

$$\beta = \left(\frac{1}{\rho}\right)\left(\frac{\partial p}{\partial \rho}\right)_T \equiv \frac{1}{\rho^2\chi}. \tag{2.15b}$$

Finally, from (2.11b) with (2.15b), we derive the following remarkable relation:

$$C_p - C_v = -\left(\frac{T}{\rho}\right)\left[\frac{\alpha^2}{\chi}\right]. \tag{2.16}$$

An important conclusion emerges from (2.16):

In order that the difference of two heat capacities be bounded when $\chi \downarrow 0$, it is necessary that the ratio $[\alpha^2/\chi]$ remains bounded!

Under this assumption:

We have the possibility to assume the existence of a similarity rule between the constant values of χ and α^2 (see, for instance, (2.30)).

2.3 NS–F System for a Thermally Perfect Gas

First of all we observe that, from the thermodynamic relation (2.6), we can write

$$T\left(\frac{\partial S}{\partial \rho}\right)_T = \left(\frac{\partial e}{\partial \rho}\right)_T - \frac{p}{\rho^2},$$

and the Maxwell relation (2.10b) allows this to be written as

$$T\left(\frac{\partial p}{\partial T}\right)_p = p - \rho^2\left(\frac{\partial e}{\partial \rho}\right)_T.$$

Now, we have the definition:

A perfect gas is a material for which the internal energy is the sum of the separate energies of the molecules in unit mass and is independent of the distances between the molecules, that is, independent of the density ρ.

Hence for a perfect gas we obtain the following two fundamental relations:

$$e = e(T) \tag{2.17a}$$

and

$$\left(\frac{\partial p}{\partial T}\right)_p = \frac{p}{T}. \tag{2.17b}$$

On the other hand, usually, it is assumed that the molecules are identical, with mass m $(= \rho/N)$, where N is the number density of molecules, and for the pressure we may write

$$p = \frac{N}{\omega} \tag{2.17c}$$

When two different gases are in thermal equilibrium with each other, the corresponding values of ω are equal. Temperature T is a quantity defined as having this same property, and it is therefore natural to seek a connection between the parameter ω and the temperature of the (thermally perfect) gas T.

Namely, if k_B is an absolute constant (known as Boltzmann's constant) then, because it appears that, at constant density, p is proportional to T (Charles's law) we write also

$$\frac{1}{\omega} = k_B T. \tag{2.17d}$$

Finally, for the pressure p we obtain the following equation of state for a thermally perfect gas:

$$p = Nk_B T = \left(\frac{k_B}{\langle m \rangle}\right) \rho T = R\rho T, \tag{2.17e}$$

where $\langle m \rangle$ is the average mass of the molecules of the gas and

$$R = \frac{k_B}{\langle m \rangle},$$

is known as the gas constant.

If such is the case then, for a (thermally) perfect gas, in place of the full equation of state (2.3) we have the following usual two relations:

$$p = R\rho T \tag{2.18a}$$

and also

$$\mathrm{d}e = C_v \,\mathrm{d}T; \tag{2.18b}$$

since $e = e(T)$, C_v can be defined as

$$C_v = \left(\frac{\partial e}{\partial T}\right)_v, \tag{2.18c}$$

which is equivalent to (2.11a).

As a consequence, the various thermodynamic relations derived in Section 2.2 are unnecessary in the case of a thermally perfect gas, and from (2.5), with (2.18b), for temperature T we obtain the following evolution equation for the temperature T:

$$C_v \rho \frac{\mathrm{d}T}{\mathrm{d}t} + p(\nabla \cdot \mathbf{u})$$
$$= 2\mu\{(\mathbf{D}(\mathbf{u}) : \mathbf{D}(\mathbf{u}) - (1/3)(\nabla \cdot \mathbf{u})^2\} + \frac{\partial}{\partial x_i}\left(k \frac{\partial T}{\partial x_i}\right), \tag{2.19}$$

but in the general case the specific heats C_v and C_p both vary with temperature T.

In this book we mainly consider that the dynamic viscosity μ and heat conductivity coefficient k are constant (respectively, μ_d and k_d, as a function of the constant temperature T_d).

In such a case, with (2.19), as an equation for the temperature of a thermally perfect viscous and heat conducting unsteady gas flow, we have the possibility to write a system of four NS–F equations for $\mathbf{u}$, p, ρ and T.

Namely, we have three evolution equations for $\mathbf{u}$, ρ and T:

$$\frac{\mathrm{d}\rho}{\mathrm{d}t} + \rho(\nabla \cdot \mathbf{u}) = 0, \tag{2.20}$$

$$\rho \frac{\mathrm{d}\mathbf{u}}{\mathrm{d}t} + \nabla p = \rho\mathbf{f} + \mu_d[\Delta\mathbf{u} + (1/3)\nabla(\nabla \cdot \mathbf{u})], \tag{2.21}$$

$$C_v \frac{\mathrm{d}T}{\mathrm{d}t} = -\left(\frac{p}{\rho}\right)(\nabla \cdot \mathbf{u}) + \Phi + \left(\frac{k_d}{\rho}\right)\Delta T, \tag{2.22a}$$

with

$$\Phi = 2\left(\frac{\mu_d}{\rho}\right)\left\{\mathbf{D}(\mathbf{u}) : \mathbf{D}(\mathbf{u}) - (1/3)(\nabla \cdot \mathbf{u})^2\right\}, \tag{2.22b}$$

and the usual equation of state for p:

$$p = R\rho T. \tag{2.23}$$

In equation of state (2.23), R is the gas constant ($= 2.870 \times 10^3$ cm^2/sec^2°C for dry air) and we have Carnot's law

$$C_p - C_v = R.$$

On the other hand the coefficient of thermal expansion for a thermally perfect gas is

$$\alpha = \frac{1}{T},$$

and for isothermal coefficient of compressibility (see (2.16) we have

$$\beta = \frac{1}{p}.$$

For the specific entropy, we have the explicit expression

$$S = C_v \log(p\rho^{-\gamma}), \tag{2.24a}$$

which is a consequence of the thermodynamic relation (equivalent to (2.6))

$$T\,\mathrm{d}S = \mathrm{d}h - \left(\frac{1}{\rho}\right)\mathrm{d}p, \tag{2.24b}$$

where (h is the enthalpy)

$$h = e + \left(\frac{p}{\rho}\right), \tag{2.24c}$$

which is derivable from the First Law (conservation of energy) and Second Law (relative to entropy) of thermodynamics.

Naturally, in real conditions, properties of common gases are dependent on T and ρ. As to examples, the reader can find these in the book by Batchelor [2, appendix 1, pp. 594–595], some observed values of the dynamic viscosity μ, kinematic viscosity ν $(= \mu/\rho)$, thermal conductivity k, thermal diffusivity κ $(= k/\rho C_p)$ and Prandtl number Pr $(= \nu/\kappa)$, corresponding to values of temperature T and density ρ.

2.4 NS–F System for an Expansible Liquid

When the temperature of a liquid is increased, with the pressure held constant, the liquid (usually) expands. If the momentum flux alone contributed to the pressure, the consequent fall in density would be such as to keep ρT constant, as in the case of a gas. But the contribution to the pressure from intermolecular forces is more important, and has a less predictable dependence on temperature. Of course for very (ultra) thin films the (long-range) intermolecular interactions (forces) play an important role (taking into account the van der Waals attraction).

In general, measurements show rather smaller values of the coefficient of thermal expansion α (defined as in (2.9b)) for liquids, than the value $1/T$ appropriate to a thermally perfect gas, namely, for water at 15°C,

$$\alpha \approx 1.5 \times 10^{-4}/°\text{C}.$$

But values of α for other common liquids tend to be larger, and range up to about 16×10^{-4}/°C. The value of γ $(= C_p/C_v)$ may be taken as unity for water at temperatures and pressures near the normal values.

Quite small changes of density correspond, at either constant temperature or at constant entropy, to enormous changes in pressure; that is, the coefficient of compressibility for liquids is exceedingly small. For instance, the density of water increases by only 0.5% when the pressure is increased from one to 100 atmospheres at constant (normal) temperature! This great resistance to compression is the important characteristic of liquid, so far as fluid dynamic is concerned, and it enables us to regard them for most purposes as being almost incompressible with high accuracy.

On the contrary, liquids are very sensitive to expansion under the influence of temperature and in the case of an expansible/dilatable liquid the analysis is more subtle, concerning the derivation of an evolution equation for the temperature T.

In reality, in place of the equation of state (2.8), $\rho = \rho(T, p)$, usually it is assumed that the expansible liquid can be described by the following approximate (truncated) law state:

$$\rho = \rho_d[1 - \alpha_d(T - T_d) + \chi_A(p - p_A)], \tag{2.25}$$

where (ρ_d, T_d, p_A) are some constant values for the density, temperature and pressure. In Dutton and Fichtel [3], such an approximate equation of state (2.25) has been adopted by the authors, who attempt to present in a unified theory the cases of liquids and of gases. Indeed, this unification has been realized by Bois in [4], where this unification is presented in a more precise, rational, manner.

In (2.25) the constant coefficients, α_d and χ_A are, respectively:

$$\alpha_d = -\left(\frac{1}{\rho_d}\right)\left[\frac{\partial\rho}{\partial T}\right]_d, \tag{2.26a}$$

$$\chi_A = \left(\frac{1}{\rho_d}\right)\left[\frac{\partial\rho}{\partial p}\right]_A, \tag{2.26b}$$

where $[\partial\rho/\partial T]_d$ and $[\partial\rho/\partial p]_A$ are both constant.

From relation (2.16), with (2.26a, b), we obtain the following remarkable relation between the constant coefficients α_d and χ_A:

$$\frac{\alpha_d^2}{\chi_A} = C_{vd}(\gamma - 1)\left[\frac{\rho_d}{T_d}\right]. \tag{2.27}$$

Now, with the relation (2.13b), from equation (2.7a), we obtain for our expansible liquid the following evolution equation for liquid temperature T:

$$\rho C_p \frac{\mathrm{d}T}{\mathrm{d}t} - \alpha T \frac{\mathrm{d}p}{\mathrm{d}t} - \Phi + k_d \Delta T, \tag{2.28}$$

where the viscous dissipation Φ (per unit of mass) is given by the same expression (2.22b) used for a thermally perfect, viscous and heat conducting, gas.

Here, for the considered liquids, the coefficients χ_A and α_d are usually very small (for water at 15°C, $\alpha_d \approx 10^{-4}$/°C) and for a bounded value of the right-hand side in relation (2.27), when $\chi_A \downarrow 0$, it is necessary that α_d^2/χ_A remain also bounded.

As a consequence we can write a similarity rule between the two small parameters, $\varepsilon = \alpha(T_d)\Delta T$, defined by (1.10a) in Chapter 1, and

$$\ni = g d \rho_d \chi(p_A). \tag{2.29}$$

Namely:

$$\varepsilon^2 = K_0 \ni, \tag{2.30}$$

where the similarity parameter K_0 is fixed – not very much large or small – when both ε and $\ni$ tend to zero.

Usually, in applications, it is sufficient to assume that C_p is only a function of temperature T, such that (see, for instance [4]):

$$C_p = C_{pd}[1 - \alpha_d \Gamma_{pd}(T - T_d)] \tag{2.31}$$

where $\Gamma_{pd} = \text{const}$.

On the other hand, in the left-hand side of equation (2.28) for the liquid's temperature, the coefficient α (coupled with the term $T\,\mathrm{d}p/\mathrm{d}t$) is usually assumed to be a constant and written as α_d, but in an asymptotic modelling approach, when the expansibility parameter ε tends to zero this hypothesis is useless.

Finally, for an expansible liquid we have, as starting full NS–F equations, for $\mathbf{u}$, p, T, the following system of three (2.32a–c) evolution equations:

$$\frac{\mathrm{d}\rho}{\mathrm{d}t} + \rho(\nabla \cdot \mathbf{u}) = 0, \tag{2.32a}$$

$$\rho \frac{\mathrm{d}\mathbf{u}}{\mathrm{d}t} + \nabla p = \rho \mathbf{f} + \mu_d[\Delta \mathbf{u} + (1/3)\nabla(\nabla \cdot \mathbf{u})], \tag{2.32b}$$

$$\rho C_p \frac{\mathrm{d}T}{\mathrm{d}t} - \alpha T \frac{\mathrm{d}p}{\mathrm{d}t} = \Phi + k_d \Delta T, \tag{2.32c}$$

with the approximate equation of state for ρ:

$$\rho = \rho_d \left\{1 - \varepsilon\left[\frac{(T - T_d)}{\Delta T}\right] + \left(\frac{1}{K_0}\right)\varepsilon^2\left[\frac{(p - p_A)}{g\,\mathrm{d}\rho_d}\right]\right\}, \tag{2.32d}$$

where the coefficient C_p, in equation (2.32c), according to (2.31), is given by the relation (the difference of temperature $\Delta T = T_W - T_d$):

$$C_p = C_{pd}\left\{1 - \varepsilon\Gamma_{pd}\left[\frac{(T - T_d)}{\Delta T}\right]\right\}. \tag{2.33}$$

2.5 Upper Free Surface Conditions

In the mathematical formulation of the full Bénard, heated from below, convection problem – considered in detail in the framework of Chapter 4 –

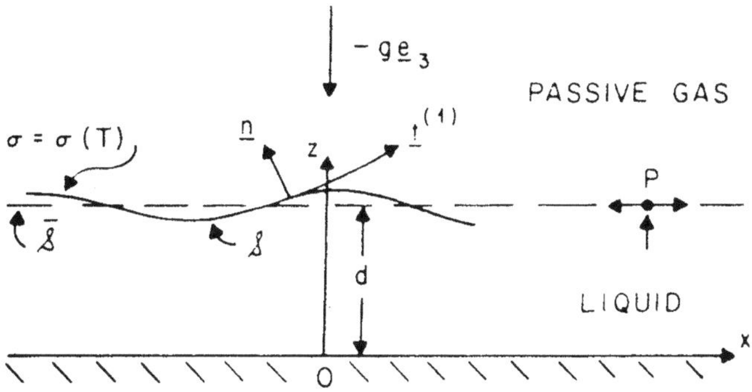

Fig. 2.1 Geometry of the Bénard convection problem, heated from below.

we have in view a physical thermal problem in a layer of expansible liquid which is in contact with a solid heated wall ($z = 0$) of constant temperature, $T = T_w$.

This weakly expansible liquid layer is separated – from a motionless ambient passive atmosphere at constant temperature T_A and constant pressure p_A, having negligible viscosity and density – by an upper (at the level $z = d$) free surface simulated by the following Cartesian equation (see Figure 2.1):

$$z = d + ah(t, x, y) \equiv H(t, x, y). \tag{2.34}$$

Across the free surface given by (2.34) we assume that no mass flows is realized and from the balance of momentum we can write the classical free surface jump condition for the pressure fifference $(p - p_A)$. Namely,

$$\begin{aligned}(p - p_A)\mathbf{n} \;=\; & 2\mu_d[\mathbf{D}(\mathbf{u}) - (1/3)(\nabla \cdot \mathbf{u})] \cdot \mathbf{n} \\ & - 2\sigma(T)K\mathbf{n} - \nabla_{\|}\sigma(T), \quad \text{at } z = H(t, x, y),\end{aligned} \tag{2.35}$$

according to constitutive relation (1.4a) for the stress tensor.

In (2.35), the unit outward normal vector $\mathbf{n}$ is directed from the liquid to the passive ambient air and the surface gradient operator $\nabla_{\|}$ is defined as:

$$\nabla \| = \nabla - (\mathbf{n} \cdot \nabla)\mathbf{n} \tag{2.36a}$$

while the mean curvature is

$$K = -(1/2)[\nabla_{\|} \cdot \mathbf{n}]. \tag{2.36b}$$

We observe that in the free surface jump condition (2.35), the surface viscosities have been neglected. The Marangoni thermocapillary effect is directly connected with the last term in the right-hand side of (2.35) and we can write:

$$\nabla_{\|}\sigma(T) = \left[\frac{\mathrm{d}\sigma(T)}{\mathrm{d}T}\right]\nabla_{\|}T, \tag{2.36c}$$

where according to (1.17a, b) and (1.13)

$$\sigma(T) = \sigma(T_d) - \left[-\frac{\mathrm{d}\sigma(T)}{\mathrm{d}T}\right]_d \Delta T\theta, \tag{2.36d}$$

$\Delta T = T_w - T_d$, when we work with the dimensionless temperature θ defined by the relation (1.13).

As balance of the temperature T, at a free surface we use, as in Chapter 1, the condition (1.23b), which simulates the conservation of heat flux on transport across the upper, deformable free surface in a convection regime. This upper condition (1.23b), for the temperature $T(t, x, y, z)$ of the liquid, at free surface $z = H(t, x, y)$, is in fact, a so-called 'third-mixed type' (or Robin) condition, embracing the classical Dirichlet condition ($T = T_A$ at the free surface, in the case of a perfectly conducting free surface) and Neumann condition (at the free surface $\partial T/\partial n = 0$, for a poor conducting free surface).

We observe that, on the contrary, in the linear (approximate) relation (1.17a)/(2.36d), for the surface tension $\sigma(T)$, we have as the difference of the temperature $(T - T_d)$, where the reference temperature T_d is the interfacial temperature of the basic conduction state, i.e., the constant temperature of the flat film $z = d$. On the one hand, the 'Marangoni effect', linked with the temperature-dependent surface tension, is operative along the free surface, and, on the other hand, the 'Biot effect', linked with the rate of heat loss from the free surface, is operative across this free surface! Indeed, in the case of a convection regime with a deformable upper free surface, it seems judicious to use for both effects (Marangoni and Biot) the difference of the temperature $[T - T_A]$, and work with the dimensionless temperature Θ defined by (1.17c) because, in this convection regime case, the temperature $T = T_d$ at flat film, $z = d$, does not have a real physical sense! Namely, we write in such a case for Θ, in place of (1.23b), that the free surface, $z = H$, is cooled by air currents according to the law:

$$-k(T_d)\frac{\partial\Theta}{\partial n} = q_{\mathrm{conv}}\Theta + \frac{Q_0}{(T_w - T_A)}, \quad \text{at } z = H(t, x, y), \tag{2.37}$$

when $T = T_A + (T_w - T_A)\Theta$.

The justification for such an upper, free-surface condition (2.37), for the temperature of the liquid (with $q_{\mathrm{conv}} \neq q_s(T_d)$), relies on the assumption that heat convection, within the liquid, is so much faster than within air and the heat flux on the free surface, considered from the inside of the fluid, may be approximated by such a difference in temperature according to Newton's cooling law!

It seems that the introduction of a second (different from conduction q_s) convective heat transfer coefficient q_{conv}, in a convection regime, when the condition (1.23b) or (2.37) is used, is reasonable for a more rational and correct formulation of the problem of Marangoni's instability. Of course, we make precise, again, that this rational way does not resolve the untractable heat transfer coupled, air–free surface–liquid, problem, and in particular the determination of the coefficient q_{conv} (for example, as a function of the temperature of the liquid T or of the upper, free-surface deformation $h(t, x, y)$) remains obviously an open problem – but, in return, the mathematical formulation is correct!

The discussion concerning the above upper, free-surface condition (2.37), for the dimensionless temperature Θ, will be complemented in Section 4.4.

Finally, the location of the deformable free surface, (2.34), $z = H(t, x, y)$, is determined via the usual kinematic condition

$$\frac{\mathrm{d}}{\mathrm{d}t}[z - H(t, x, y)] = 0, \quad \text{on } z = H(t, x, y). \tag{2.38}$$

If the truth must be told, in general, the density ρ as a function of T and p, according to equation of state (2.8), must be written as

$$\begin{aligned} \rho = \rho(T, p) = \rho_d \Big\{ & 1 - \alpha(T_d)(T - T_d) + \chi(p_A)(p - p_A) \\ & + (1/2)\left[\alpha^2(T_d) - \left(\frac{\partial \alpha(T)}{\partial T}\right)_{T_d}\right](T - T_d)^2 + \cdots \Big\}, \end{aligned} \tag{2.39}$$

when an expansion in a Taylor's series about some constant thermodynamic reference (fiducial) state (ρ_d, T_d, p_A) is performed. If we consider an ideal liquid, according to the Dutton and Fichtl paper [3], then only three first terms are taken into account in (2.39), as this is the case in Section 2.4 (see (2.25) or (2.32d)).

However, the third term, proportional to pressure difference, $(p - p_A)$, in (2.32d), is in fact a second-order term relative to small parameter ε, when we take into account the similarity rule (2.30).

Often in thermal convection - for instance, in Bénard, heated from below, thermal convection – *à la* Rayleigh's problem – considered in Chapter 3, as approximate equation of state for an expansible liquid, the following simplified, leading-order equation of state is adopted:

$$\rho \approx \rho_d\{1 - \alpha(T_d)(T - T_d)\} \equiv \rho_d(1 - \varepsilon\theta), \tag{2.40}$$

where $\varepsilon = \alpha(T_d)\Delta T$, with $\Delta T = T_w - T_d$, which, at the leading order, is consistent (with an error of order $O(\varepsilon^2)$), if we take into account the relation (2.32d) which is a consequence of the similarity rule (2.30) – the dimensionless temperature, θ, being given by (1.13).

We make precise also that, for an ideal expansible liquid (when (2.25) is assumed), the coefficients $\alpha(T_d) \equiv \alpha_d$ and $\chi(p_A) \equiv \chi_A$ and also (in (2.33)) $C_p(T) \equiv C_{pd}$ are often assumed constant over the range of variation permitted in the fiducial states (ρ_d, T_d, p_A), this is certainly the case at the leading order (when $\varepsilon \downarrow 0$) in an asymptotic modelling approach!

Concerning the approximate, extended, equation of state (2.39), the main problem concerns the influence of the fourth term, proportional to $(T - T_d)^2$, when we want to derive a second-order approximate model. On the other hand, the third term, $\chi_A(p - p_A)$, in (2.39), rewritten for the perturbation of the pressure π, defined by (1.26), has the following approximate form:

$$\chi_A(p - p_A) \approx \varepsilon^2[\mathrm{Fr}_d^2\pi + 1 - z')], \tag{2.41}$$

at least when we assume that, in similarity rule (2.30), K_0 is not very small or not very large.

Now, concerning the upper, free-surface, jump condition, (2.35); this vectorial single condition gives three boundary conditions at the free surface $z = H(t, x, y)$ simulated by equation (2.34).

Let $\mathbf{t}^{(1)}$ and $\mathbf{t}^{(2)}$ be two unit tangent vectors parallel to upper, free surface $z = H(t, x, y)$ given by (2.34) and both orthogonal to unit outward normal vector $\mathbf{n}$ to this free surface, such that

$$\mathbf{t}^{(1)} \cdot \mathbf{n} = 0,$$

and

$$\mathbf{t}^{(2)} \cdot \mathbf{n} = 0.$$

In this case, in place of the single vectorial upper, free-surface, jump condition (2.35), we obtain the following three scalar upper, free-surface boundary conditions:

$$p = p_A + \mu_d[d_{ij}n_in_j - (2/3)(\nabla \cdot \mathbf{u})] + \sigma(T)(\nabla_{\parallel} \cdot \mathbf{n}), \tag{2.42a}$$

$$\mu_d d_{ij}t_i^{(1)}n_j = \left[\frac{\mathrm{d}\sigma(T)}{\mathrm{d}T}\right]\mathbf{t}_i^{(1)}\left(\frac{\partial T}{\partial x_i}\right), \tag{2.42b}$$

$$\mu_d d_{ij} t_i^{(2)} n_j = \left[\frac{\mathrm{d}\sigma(T)}{\mathrm{d}T}\right] \mathbf{t}_i^{(2)} \left(\frac{\partial T}{\partial x_i}\right), \tag{2.42c}$$

written at free surface $z = d + ah(t, x, y) \equiv H(t, x, y)$, where, according to (1.4b),

$$d_{ij} = (1/2)\left[\frac{\partial u_i}{\partial x_j} + \frac{\partial u_j}{\partial x_i}\right].$$

We observe also that

$$\nabla_{\parallel} \cdot \mathbf{n} = -\left(\frac{1}{N^{3/2}}\right)\left\{N_y \frac{\partial^2 H}{\partial x^2} - 2\left(\frac{\partial H}{\partial x}\right)\left(\frac{\partial H}{\partial y}\right)\frac{\partial^2 H}{\partial x \partial y} + N_x \frac{\partial^2 H}{\partial y^2}\right\}, \tag{2.43a}$$

with

$$N = 1 + \left(\frac{\partial H}{\partial x}\right)^2 + \left(\frac{\partial H}{\partial y}\right)^2, \tag{2.43b}$$

$$N_x = 1 + \left(\frac{\partial H}{\partial x}\right)^2, \tag{2.43c}$$

$$N_y = 1 + \left(\frac{\partial H}{\partial y}\right)^2. \tag{2.43d}$$

According to Pavithran and Redeekop [5], the components ($t_i^{(1)}$ and $t_i^{(2)}$) of two tangential vectors, $\mathbf{t}^{(1)}$ and $\mathbf{t}^{(2)}$, in conditions (2.42b, c), and components (n_i) of the outward unit vector, $\mathbf{n}$, to the deformed upper, free surface $z = H(t, x, y)$, are written below in terms of the (x, y, z) Cartesian system of coordinates. Namely we have:

$$\mathbf{t}^{(1)} = \left[\frac{1}{N_x^{1/2}}\right]\left(1; 0; \frac{\partial H}{\partial x}\right); \tag{2.44a}$$

$$\mathbf{t}^{(2)} = \left[\frac{1}{N_x^{1/2} N^{1/2}}\right]\left(-\frac{\partial H}{\partial x}\left(\frac{\partial H}{\partial y}\right); 1 + \left(\frac{\partial H}{\partial x}\right)^2; \frac{\partial H}{\partial y}\right), \tag{2.44b}$$

$$\mathbf{n} = \left[\frac{1}{N^{1/2}}\right]\left(-\frac{\partial H}{\partial x}; -\frac{\partial H}{\partial y}; 1\right). \tag{2.44c}$$

Obviously the convection problem, heated from below, for a liquid layer bounded above by an upper, free surface, $z = H(t, x, y)$, in a convective regime is more difficult mainly because of the complexity of the above upper, free-surface conditions (2.42a–c) with (2.43a–d) and (2.44a–c).

Again, concerning the upper, free-surface condition for the temperature, if now we work with the dimensionless temperature θ, and we choose Newton's cooling law in the form (1.23b), as in Chapter 1, then as condition, instead of (2.37), we have (see (1.24b)):

$$-k(T_d)\frac{\partial\theta}{\partial n} = q_{\text{conv}}\left\{\left[\frac{(T_d - T_A)}{(T_w - T_d)}\right] + \theta\right\} + \frac{Q_0}{(T_w - T_d)},$$
$$\text{at } z = H(t, x, y) \tag{2.45}$$

when $T = T_d + (T_w - T_d)\theta$.

As this has been discussed in detail in Chapter 1 (see (1.21a) and the discussion which follows up Davis' upper condition (1.25) for θ), in the 'useful' 1987 paper by Davis [6], devoted to thermocapillary instability (see, [6, pp. 407–408]), Davis (in the above upper condition (2.45)) takes into account the relation (1.24a)

$$\left[\frac{(T_d - T_A)}{(T_w - T_d)}\right] = \left[\frac{1}{\text{Bi}_s(T_d)}\right],$$

which is a consequence of (1.21a), and introduces in (2.45) a second, conduction Biot number, $\text{Bi}_s(T_d)$. In such a case, the correct result, which follows from (2.45), with (1.24a), is the dimensionless condition (1.24c), namely:

$$\frac{\partial\theta}{\partial n'} + \left[\frac{\text{Bi}_{\text{conv}}}{\text{Bi}_s(T_d)}\right]\{1 + \text{Bi}_s(T_d)\theta\} + \frac{Q_0}{k\beta_s} = 0, \quad \text{at } z' = 1 + \eta h'(t', x', y'), \tag{2.46}$$

where $n' = n/d$ is a non-dimensional normal distance from the free surface and $\text{Bi}_{\text{conv}} = dq_{\text{conv}}/k(T_d)$ is the Biot number in the convection regime! Only after the confusion of the convection (Bi_{conv}) Biot number with the conduction ($\text{Bi}_s(T_d)$) Biot number, did we rediscover the Davis thermal upper surface condition (1.25) for θ – namely (with $Q_0 = 0$ as in the Davis' [6] paper):

$$\frac{\partial\theta}{\partial n'} + 1 + B\theta = 0, \quad \text{at } z' = 1 + \eta h'(t', x', y'), \tag{2.47}$$

when (as in [6]) a single surface Biot number $B = hd/k$ is introduced, where h is the (Davis) unit thermal surface conductance. This condition (2.47) is used in most cases of the theoretical analysis of Bénard–Marangoni thermocapillary instability problems, as an 'of course' condition!

For the static motionless conduction state, when $\theta = \theta_S(z')$, the Davis condition (2.47) obtains, because in a conduction state, in a flat surface case, $\partial\theta/\partial n' \Rightarrow \mathrm{d}/\mathrm{d}z'$, $\mathrm{d}\theta_S/\mathrm{d}z' + 1 + B\theta_s = 0$, at $z' = 1$, and gives as solution $\theta_S(z') = 1 - z'$, in dimensionless form, and is independent of the Biot (in fact, conduction) number.

In a concise reply, as an answer (in February 27, 2003) to my interrogation concerning the above, *à la* Davis, derivation, Professor Stephen H. Davis wrote in a short letter:

> One is free to allow the heat-transfer coefficient, h, depending on a variety of things involving as much complexity as one wishes. The simplest case of constant h is satisfactory for many problems. If in a particular case the theory diverges from the experimental results, then one has a strong case to add 'new effects', and I have no objection to this. We chose the simplest case to analyze.

Unfortunately, this, 'rather trivial', answer has no relation to my above analysis, which shows that the problem is not linked with the constancy of the heat-transfer coefficient, h, in a convective regime or with any 'divergence' of the theory from the experimental results. The main mistake in Davis' [6] derivation, of his above condition (2.47), is mainly related with the assumption (in a 'hidden manner') that, conduction and convection heat-transfer coefficients are identical – which is, from the physics of the thin film problem, an untenable assertion!

For an arbitrary heat-transfer coefficient in a convection regime (q_{conv}), obviously different from the conduction heat-transfer coefficient (q_s), the correct upper, free- surface condition for θ is the above dimensionless condition (2.46) when we adopt the Davis derivation way correctly!

In a paper by Scheid et al., the reader can find some remarks (see [7, pp. 241–242]) concerning the vanishing (single) Biot number case considered, in particular, by Takashima in 1981 [8], in his linear theory. This vanishing (convection) Biot number case is a special case which deserves a serious critical approach.

In a different approach from that performed (*à la* Davis) in [6]), Pearson's [9] theoretical treatment was based on a linear stability analysis and is discussed more in detail in Section 4.4. In fact, we can consider a slight extension of Pearson's approach (without any linearization) in order to obtain an upper, free-surface, boundary condition for the dimensionless temperature (see (1.17c),

$$\Theta\left[=\frac{(T-T_A)}{(T_w-T_A)}\right]. \tag{2.48}$$

Namely, from the general (see Section 4.4) upper, free-surface condition (4.46a), when for the rate of heat loss $Q(T)$ from the free surface we write

$$Q(T)=Q_s+\left[\frac{\mathrm{d}Q(T)}{\mathrm{d}T}\right]_A(T-T_A), \tag{2.49}$$

where T_A is again the constant ambient motionless air temperature above the upper, free surface. As a consequence, with $Q_s=k(T_A)\beta_s$, where here for

β_s we have the relation (1.21b), we obtain the upper, free-surface boundary condition,

$$-k(T_d)\frac{\partial T}{\partial n} = \left[\frac{\mathrm{d}Q(T)}{\mathrm{d}T}\right]_A (T - T_A) \quad \text{at } z = d + ah(t, x, y), \tag{2.50}$$

or, when we take into account that $T = T_A + (T_w - T_A)\Theta$, from (2.48),

$$\frac{\partial \Theta}{\partial n'} + L\Theta = 0, \quad \text{at } z' = 1 + \eta h'(t', x', y'), \tag{2.51}$$

which is the upper condition used by Pearson (see his condition (17), but written at $z' = 1$ for the function $g(z')$ [9, p. 495] when $h' \equiv 0$ (the linear case).

In [9], L is assumed a constant (L is in fact a function of the constant air temperature T_A). But in [9] we can read, also, that

> The values of L encountered in practice would depend on the thickness of the film and for *very thin films would tend to zero*!

From the above it is clear that Pearson well understood that it is necessary to recognize a difference between the conduction and (variable) convection Biot effects! It is also concluded (in [9]) that surface tension forces are responsible for cellular motion in many such cases where the criteria given in terms of buoyancy forces do not allow for instability – the buoyancy mechanism has no chance of causing convection cells, while the surface tension mechanism is almost certain to do so and observations support this conclusion! Finally, according to Pearson [9, p. 499],

> An intimation that the instability theory based on buoyancy forces would not account for all of Bénard's results, [10], appears in a paper by Volkovisky in a 1939 Scientific and Technical Publication of the French Air Ministry [11].

In a recent paper by Ruyer-Quil et al. [12], this above condition (2.51) is, in fact, adopted – but unfortunately, again, a confusion between conduction and convection Biot numbers has arisen, despite my advice! Obviously, when we work with Θ, then it seems judicious (by analogy) to assume that the variation of surface tension with temperature is modeled by the following linear approximation (instead of, for example, (2.36d)),

$$\sigma(T) = \sigma(T_A) - \gamma_\sigma (T - T_A), \tag{2.52}$$

with

$$\gamma_\sigma = -\left[\frac{\mathrm{d}\sigma(T)}{\mathrm{d}T}\right]_A . \tag{2.53}$$

In Section 4.4, we again discuss this problem concerning the thermal upper, free-surface condition, but mainly for the dimensionless temperature Θ, and also the dimensionless modellling of the term (2.36c) expressing the thermocapillary stress in the free-surface jump condition (2.35).

2.6 Influence of Initial Conditions and Transient Behavior

For the NS–F systems derived above (see Sections 2.3 and 2.4), consisting of three evolution equations for u_i, T and p, because the partial derivatives in time t, $\mathrm{d}u_i/\mathrm{d}t$, $\mathrm{d}T/\mathrm{d}t$ and $\mathrm{d}p/\mathrm{d}t$ are present (see for instance (2.32a–c)), it is necessary to assume that three initial data at initial time u_i^0, T^0 and p^0, are given as functions of coordinates; namely, for $t = 0$, we write

$$\text{at } t = 0,\ u_i = u_i^0,\ T = T^0 \text{ and } p = p^0, \tag{2.54}$$

where T^0 and p^0 are positive known data and we observe that in various technological applications, often, it is essential to take into account these above three initial conditions (2.54).

However, in the framework of a rational analysis and asymptotic modelling of a weakly expansible liquid layer heated from below, our main purpose is the formulation (when the expansibility parameter, ε tends to zero) of leading-order, approximate, consistent models, for the considered convection problem, in accordance with various values of the square of a reference Froude, Fr_d^2, number based on the thickness of the liquid layer d. Unfortunately, the passage from the full exact starting equations with given initial conditions and associated to convection problem boundary conditions, to a limiting approximate convection model is, in general, singular.

This singular nature is mainly expressed by the fact that: often via the limiting passage, some partial derivatives in time (present in full exact starting equations) disappear in derived model equations and, as a consequence, it is not possible to apply all the given in start initial conditions at $t = 0$.

As a consequence, certainly, the asymptotically-derived, limiting, approximate model equations are not valid in the vicinity of the initial time, and a short-time-scale, local in time, rational analysis is necessary!

In the framework of an asymptotic modelling, the logical rational way for solving the associated local/short-time problem is the consideration of an initial time layer near time $= 0$. In this initial time layer a new, local, dimensionless, model of unsteady equations is derived where, in place of the non-dimensional (evolution) time t', a new (adjustment) short-time, τ', is introduced in local equations governing the associated adjustment problem. For example, if the approximate limiting model problem (significant outside of the singular initial time layer) is asymptotically derived via the limiting process

$$\varepsilon \text{ tends to zero,} \tag{2.55}$$

then the corresponding short time is

$$\tau' = \frac{t'}{\varepsilon} \tag{2.56}$$

Often, the new, local-in time, dimensionless, model problem, with all derivatives relative to τ', is an unsteady linear problem with all (starting) initial conditions. This local-in time problem is an unsteady adjustment problem and (matching) when τ' tends to infinity, we discover the initial conditions at $t' = 0$ for the limiting model evolution problem derived, for instance, according to (2.55).

In other words, close to initial time, $\tau' = 0$, an unsteady (relative to short time τ') local problem is considered with the given starting initial conditions, and then the limiting value, when $\tau' \to \infty$, of the solution of this local problem are adjusted to a set of new initial conditions at $t' = 0$ for the previously derived, limiting, simplified evolution model equations.

Otherwise:

$$\lim(\text{local when } \tau' \to \infty) = \lim(\text{model at } t' = 0). \tag{2.57}$$

A typical example is considered in a paper by Dandapat and Ray [14], where the flow of a thin liquid film over a cold/hot rotating disk is analyzed for a small Reynolds number

$$\mathrm{Re} = \frac{U_0 h_0}{\nu} \ll 1; \tag{2.58}$$

in this 'low Reynolds number' situation, the balance of centrifugical force and the viscous shear across the film defines a characteristic time (denoted by the authors in [14] by t_b):

$$t_b = \frac{\nu}{(h_0^2 \Omega^2)}. \tag{2.59}$$

The characteristic velocity scale, U_0, is defined as h_0/ν, where h_0 is the initial (at time $= 0$) film thickness, Ω is angular velocity and ν the kinematic viscosity. We return to this thin liquid film problem over a cold/hot rotating disk in Chapter 10 of this book.

Here we observe, only, that the problem considered in [14] is, in fact, an extension of an unsteady problem considered in [15], by Higgins (and also in my recent book [16], section 5.4 devoted to very low Reynolds number flows, where the main lines of the Higgins problem are exposed). In [17], Hwang and Ma studied the film thickness and its dependence on various parameters. In the paper [14], Dandapat and Ray reconsider the problem examined in [18] by taking into account the effect of the variation of the surface tension with temperature (Marangoni effect) and thermal stress on the free surface (Newton's law of cooling is taking into account – but, in fact, the heat transfer coefficient is assumed later to be 0).

The influence of initial conditions on transient thin-film flow, has been recently examined by Khayat and Kim [19]. This study investigates, theoretically, the influence of initial conditions on the development of early transients for pressure-driven planar flow of a thin film over a stationary substrate, emerging from a channel. The flow is governed by the thin film equations of boundary-layer type and the wave and flow structure are examined for various initial conditions of flow and film profile. It is found that, depending on the initial film profile and velocity distribution, the limiting steady state may or may not be reached (stable) – alternatively, the instability of the steady state is shown to be closely linked to the existence of a gradient catastrophe.

In various classical simplified model problems for a film and, in particular, in the case of the derivation of a lubrication equation, via the long-wave approximation theory and, also, when an averaged integral boundary-layer (IBL) approach is used, the initial conditions must be specified for the model evolution (in time) equations – but the number of these initial conditions for these model equations is, usually, less than that for the full dominant starting equations? Again, this is caused by the fact that, during limiting processes, some time derivatives disappear in derived limiting model equations and, obviously, in each case, it is necessary to put the following question: what initial conditions can be imposed to a derived approximate model problem and how are these initial conditions related to the initial given data for full dominant starting equations?

It is also important to note that, depending on the kind of convection problems, we may have mainly two kinds of behavior, for the solution of the unsteady adjustment local-in time process, when the rescaled time goes to infinity! Either one may have a tendency towards a limiting steady state (and in such a case the matching is ensured) or an undamped set of oscillations appear (and it is necessary to apply a multiscale asymptotic method).

Actually, unfortunately, in RB, BM and lubrication problems and also in the averaged integral boundary-layer, IBL, approach, the associated unsteady adjustment model (inner) problems, valid in the vicinity of the initial time, are often not regarded or even discussed? In spite of the fact that, obviously, for many technological problems, related in particular with thin films, such an inner/local in time approach is required for a truthful time evolution prediction, the transient behavior playing more often than not an important role! In Section 10.8 of we give an interesting, singular, example where a matching is realized.

2.7 The Hills and Roberts' (1990) Approach

Hills and Roberts [20] – see the book by Straughan [21, pp. 48–49]) – for the full system of equations (1.1a–c) consider, first, the entropy production inequality

$$\rho\left[TS-\frac{\mathrm{d}e}{\mathrm{d}t}\right]+T_{ji}d_{ji}-\left(\frac{q_i}{T}\right)\left[\frac{\partial T}{\partial x_i}\right]\geq 0. \qquad (2.60)$$

Their interest is in liquids whose density ρ can be changed mainly by variations in the temperature T, but not in the thermodynamic pressure P (when the Stokes relation is not taken into account), so they formulate the constitutive theory in terms of P and T. They argue that the natural thermodynamic potential is the Gibbs energy:

$$G=e-ST+\frac{P}{\rho} \qquad (2.61)$$

and (2.60) may thus rewritten as

$$-\rho\left(\frac{\mathrm{d}G}{\mathrm{d}t}+S\frac{\mathrm{d}T}{\mathrm{d}t}\right)+\frac{\mathrm{d}P}{\mathrm{d}t}+(T_{ji}+P\delta_{ji})d_{ij}-\left(\frac{q_i}{T}\right)\left[\frac{\partial T}{\partial x_i}\right]\geq 0. \qquad (2.62)$$

Namely, as constitutive theory, according to Hills and Roberts' paper [20], we have

$$G=G(T,P),\qquad S=S(T,P),\qquad \rho=\rho(T), \qquad (2.63)$$

$$T_{ij}=-p\delta_{ij}+\lambda d_{mm}\delta_{ij}+2\mu d_{ij}, \qquad (2.64)$$

$$q_i=-k\frac{\partial T}{\partial x_i}, \qquad (2.65)$$

where p is the mechanical pressure, and the two coefficients, λ and μ, of the viscosity and coefficients k of the thermal conductivity depend on P and T.

For the form of ρ in (2.63), the continuity equation becomes (since $\chi \equiv 0$) as usually:

$$\frac{\partial u_i}{\partial x_i} = \alpha \frac{\mathrm{d}T}{\mathrm{d}t} \quad \text{with} \quad \alpha = -\left(\frac{1}{\rho}\right)\frac{\mathrm{d}\rho}{\mathrm{d}t}, \tag{2.66}$$

which is regarded as a constraint and then included in (2.62) via a Lagrange multiplier Λ. By using the arbitrariness of the body force (and eventually heat supply, ρr, in the third equation (1.1c) of the system of equations (1.1a–c) governing the continuum regime) Hills and Roberts deduce from (2.62) that

$$S = -\left[\frac{\mathrm{d}G}{\mathrm{d}T} + \frac{\Lambda\alpha}{\rho}\right], \tag{2.67a}$$

$$\frac{\mathrm{d}G}{\mathrm{d}P} = \left(\frac{1}{\rho}\right), \tag{2.67b}$$

$$p = P + \Lambda, \quad \Lambda = \Lambda(T, P), \tag{2.67c}$$

$$\lambda + (2/3)\mu \geq 0, \quad \mu \geq 0, \quad k \geq 0. \tag{2.67d}$$

They then work with another modified Gibbs energy:

$$G^* = G + \left(\frac{\Lambda}{\rho}\right), \tag{2.68}$$

which allows them to replace G, P, Λ by G^*, p, for which:

$$G^* = G^*(T, p) = G_0(T) + \left(\frac{p}{\rho}\right); \tag{2.69a}$$

$$\frac{\partial G^*}{\partial T} = -S; \tag{2.69b}$$

$$\frac{\partial G^*}{\partial p} = \left(\frac{1}{\rho}\right), \tag{2.69c}$$

where now the liquid parameters depend on mechanical pressure p and absolute temperature T.

The governing equations for the expansible liquid (with $\rho = \rho(T)$, but also with (2.69a–c)) become in such a case,

$$\alpha \frac{\mathrm{d}T}{\mathrm{d}t} = \frac{\partial u_i}{\partial x_i}; \tag{2.70a}$$

$$\rho\frac{\mathrm{d}u_i}{\mathrm{d}t} = \rho f_i - \frac{\partial p}{\partial x_i} + \frac{\partial}{\partial x_j}[\lambda d_{mm}\delta_{ij} + 2\mu d_{ij}]; \tag{2.70b}$$

$$-\alpha T\frac{\mathrm{d}p}{\mathrm{d}t} + C_p\left(\frac{\rho}{\alpha}\right)\frac{\partial u_i}{\partial x_i} = (\lambda d_{ii})^2 + 2\mu d_{ij}d_{ij} + \frac{\partial}{\partial x_i}\left[k\frac{\partial T}{\partial x_i}\right], \tag{2.70c}$$

where $C_p = T(\partial S/\partial T)_p$ is again the specific heat at constant pressure in equation (2.70c), which is an evolution equation for the pressure p.

The system (2.70a–c) is slightly more general than our system (2.32a–c), derived in Section 4.4, because the Stokes relation ($\lambda = -(2/3)\mu$) is not assumed.

References

1. J. Serrin, Mathematical principles of classical fluid mechanics. In *Handbuch der Physik*, Vol. WIII/1, S. Flügge (Ed.). Springer, Berlin, pp. 125–263, 1959.
2. G.K. Batchelor, *An Introduction to Fluid Dynamics*. Cambridge University Press, Cambridge, 1988.
3. J.A. Dutton and G.H. Fichtl, *J. Atmosph. Sci.* **26**, 241, 1969.
4. P.-A. Bois, *Geophys. Astrophys. Fluid Dynam.* **58**, 45–55, 1991.
5. S. Pavithran and L.G. Redeekopp, *Stud. Appl. Math.* **93**, 209, 1994.
6. S.H. Davis, *Ann. Rev. Fluid Mech.* **19**, 403–435, 1987.
7. B. Scheid, C. Ruyer-Quil, S. Kalliadasis, M.G. Velarde and R.Kh. Zeytounian, Thermocapillary long waves in a liquid film flow. Part 2. Linear stability and nonlinear waves. *J. Fluid Mech.* **538**, 223–244, 2005.
8. M. Takashima, *J. Phys. Soc. Japan* **50**(8), 2745–2750 and 2751–2756, 1981.
9. J.R.A. Pearson, *J. Fluid Mech.* **4**, 489–500, 1958.
10. H. Bénard, *Revue Gén. Sci. Pures Appl.* **11**, 1261–1271 and 1309–1328, 1900. See also *Ann. Chim. Phys.* **23**, 62–144, 1901.
11. V. Volkovisky, *Publ. Sci. Tech., Ministère de l'Air* **151**, 1939.
12. C. Ruyer-Quil, B. Scheid, S. Kalliadasis, M.G. Velarde and R.Kh. Zeytounian, Thermocapillary long waves in a liquid film flow. Part 1. Low-dimensional formulation. *J. Fluid Mech.* **538**, 199–222, 2005.
13. M. Van Dyke, *Perturbation Methods in Fluid Mechanics*. Academic Press, New York, 1964.
14. B.S. Dandapat and P.C. Ray, *Int. J. Non-Linear Mech.* **28**(5), 489–501, 1993.
15. B.G. Higgins, *Phys. Fluids* **29**, 3522, 1986.
16. R.Kh. Zeytounian, *Theory and Applications of Viscous Fluid Flows*, Springer-Verlag, Berlin/Heidelberg, 2004.
17. J.H. Hwang and F. Ma, *J. Appl. Phys.* **66**, 388, 1989.
18. B.S. Dandapat and P.C. Ray, *Int. J. Non-Linear Mech.* **25**, 589–501, 1990.
19. R.E. Khayat and K.-T. Kim, *Phys. Fluids* **14**(12), 4448–4451, 2002.
20. R. Hills and P. Roberts, *Stab. Appl. Anal. Continuous Media* **1**, 205–212, 1991.
21. B. Straughan, *Mathematical Aspects of Penetrative Convection*. Longman, 1993.

Chapter 3
The Simple Rayleigh (1916) Thermal Convection Problem

3.1 Introduction

Lord Rayleigh, in his December 1916, pioneering paper [1] devoted to 'On Convection Currents in a Horizontal Layer of Fluid, when the Higher Temperature is on the Under Side' first wrote

> The present paper is an attempt to examine how far the interesting results obtained during the years 1900–19001 by Bénard [2] in his careful and skilful experiments can be explained theoretically. Bénard worked with very thin layers, only about 1 mm. deep, standing on a levelled metallic plate which was maintained at a uniform temperature. The upper surface was usually free, and being in contact with the air was at a lower temperature. Various liquids were employed.
>
> The layer rapidly resolves itself into a number of cells, the motion being an ascension at the middle of a cell and a descension at the common boundary between a cell and its neighbours.

And also

> M. Bénard does not appear to be acquainted with James Thomson's paper 'On a changing Tesselated Structure in certain Liquids' (*Proc. Glasgow Phil. Soc.* 1881–1882), where a like structure is described in much thicker layers of soapy water cooling from the surface.

In Lord Rayleigh's paper the calculations are based upon approximate equations formulated by Boussinesq in his 1903 book [3] – the special limitation which characterizes these so-called 'Boussinesq equations' is the neglect of variation of density, except in so far as they modify the action of gravity. According to Lord Rayleigh: 'Of course, such neglect can be justified only

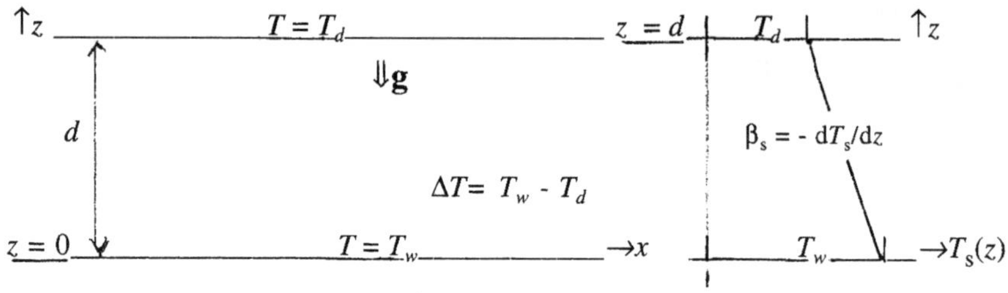

Fig. 3.1 Geometry of the Rayleigh simple convection problem.

under certain conditions, which Boussinesq has discussed'. In Zeytounian's 2003 paper [4], a hundred years later, the reader can find a rational/logical justification of this Boussinesq approximation and associated Boussinesq equations; the main lines of such a justification are presented briefly below.

In fact, the present chapter is an extended version of a paper written in 2006 for the 90 years of the above-mentioned Rayleigh's pioneering 1916 paper devoted to thermal convection, but … *unpublished*, for various reasons!

As equation of state in [1], Lord Rayleigh assumed, in fact, that

$$\rho = \rho(T), \tag{3.1a}$$

and in such a case

$$-\left(\frac{1}{\rho}\right)\frac{\mathrm{d}\rho}{\mathrm{d}T} = \alpha(T), \tag{3.1b}$$

where $\alpha(T)$ is the usual coefficient of volume/thermal expansion.

Lord Rayleigh [1] considered a particular, simple, 'Rayleigh Problem', when the fluid is supposed to be bounded by two infinite fixed planes, respectively, at $z = 0$ and $z = d$, where also the temperatures (respectively, T_w and T_d) are both maintained constant (see Figure 3.1).

In the case of this Rayleigh problem, we have a simple static, motionless conduction temperature state $T_s(z)$, such that

$$-\frac{\mathrm{d}Ts(z)}{\mathrm{d}z} \equiv \beta_s = \frac{(T_w - T_d)}{d} \equiv \frac{\Delta T}{d}, \quad T > 0, \tag{3.1c}$$

when the higher temperature T_w is below (at $z = 0$) and $T_s(z = d) = T_d$.

However, a little unexpectedly, it appears that the equilibrium may be thoroughly stable, if the coefficients of conductivity and viscosity are not too small – as the temperature gradient $\beta_s = (T_w - T_d)/d$ increases, and when $\beta_s = \Delta T/d$ exceeds a certain critical value β_{sC} (that is, when β_s just exceeds β_{sC}) instability enters which produces a permanent regime of regular

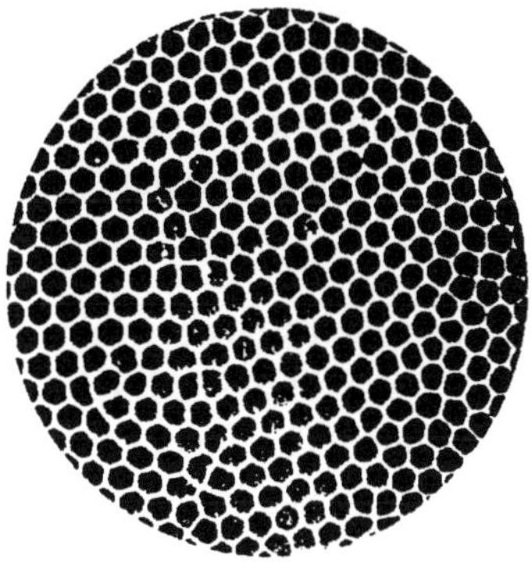

Fig. 3.2 Bénard cells in spermaceti. Reprinted with kind permission from [5].

hexagons – then these cells become equal and regular and align themselves (see, for instance Figure 3.2, which is a reproduction of one of Bénard's early photographs, from [5]). Because Lord Rayleigh [1] considered a liquid layer with a constant thickness d, the Marangoni and Biot effects are absent in the exact formulation of the thermal convection problem; the effect of the (constant) surface tension is also absent and the Weber number does not appear in the Rayleigh formulation of the problem. As a consequence the analytical problem considered by Rayleigh has no relation with the physical experimental problem considered by Bénard in his various experiments [2] – in Rayleigh's simple analytical problem the main driving force, which gives a bifurcation from a conduction motionless regime to convective motion, is the buoyancy (Achimedean) force. Nevertheless, Rayleigh's theoretical problem, leading to the famous 'Rayleigh–Bénard instability problem', is a typical problem in hydrodynamic instability and represents a transition to turbulence in a fluid system. In my recent book [6], the reader can find various aspects of this RB problem with recent references – here, in this chapter, our main purpose is to give a rational/asymptotic justification of the derivation of the RB model problem and show how it is possible to improve this leading-order RB model by a second-order consistent model which takes into account various non-Boussinesq effects!

Now we see that temperature-dependent surface tension forces, on the deformable free surface above a liquid layer, are sufficient to cause (Marangoni) instability and are responsible for many of the cellular patterns that have been observed in cooling fluid layers, with at least a free surface. In such a case, the driving force for this thermocapillary convective motion is provided by the flow of heat from the heated lower surface to the cooled upper, free surface. Obviously, in the case of the simple Rayleigh model problem, considered in the present chapter, the thermocapillary convective (Marangoni effect) is completely absent. Concerning the Bénard experiments, the reader can find in Chandrasekhar's book [5, §18] the following short description:

> Bénard carried out his experiments on very thin layers of fluid, about a millimetre in depth, or less, standing on a levelled metallic plane maintained at a constant temperature. The upper surface was usually free and being in contact with the air was at a lower temperature. He was particularly interested in the role of viscosity; and as liquid of high viscosity he used melted spermaceti and paraffin. In all cases, Bénard found that when the temperature of the lower surface was gradually increased, at a cerain instant, the layer became reticulated and revealed its dissection into cells. He further noticed that there were motions inside the cells: of ascension at the centre, and of descension at the boundaries with the adjoining cells.
>
> Bénard distinguished two phases in the succeeding development of cellular pattern: an initial phase of short duration in which the cells acquire a moderate degree of regularity and become convex polygons with four to seven sides and vertical walls; and a second phase of relative permanence in which the cells all become equal, hexagonal, and properly aligned.

In 1993, Koschmieder, who has been for decades a key figure in the experimental investigation of the Bénard problem, wrote a very valuable monograph [7] concerning the Bénard cells (and also Taylor vortices) and the reader can find in there many figures which are results of the Koschmieder experiences.

Although Bénard was aware of the role of surface tension and especially of the surface tension gradients (in particular in the case of a temperature-dependent surface tension) in his experiments, it took more than five decades to unambiguously assess, experimentally and theoretically (see for instance papers by Block [8] and Pearson [9]), that: 'Indeed the surface tension gradients rather than buoyancy was the main cause of Bénard cells in thin (weakly expansible) liquid films'.

Only in 1997 was this almost evident physical fact (see the book by Guyon et al. [10, pp. 459–462]) proved rigorously, through an asymptotic approach, in [11, 12] where is formulated an 'alternative' that is valid in an asymptotic significance in leading order:

Either the buoyancy is taken into account and in this case the free surface deformation effect is negligible and we have the possibility to take into account in the Rayleigh–Bénard model problem the Marangoni effect only partially or, the free surface deformation effect is taken into account and, in such a case, the buoyancy does not play a significant leading-order role in the Bénard–Marangoni full thermocapillary model problem.

Thanks to the above 'alternative' it has been possible to obtain various criteria for the validity of the leading order, Rayleigh–Bénard (RB), Bénard–Marangoni (BM), and Deep Convection (DB), model problems (see, for instance Chapter 8). These criteria have been derived thanks to our rational analysis and asymptotic approach, via different similarity rules.

Although Bénard initially assumed that surface tension at the free surface of the film was an important factor in cell formation, this idea was abandoned for some time as the result of the work of Rayleigh in 1916 [1] where he analyzed the buoyancy driven natural thermal convection of a layer of fluid heated from below. He found that if hexagonal cells formed, the ratio of the spacing to cell depth almost exactly equaled that measured by Bénard, an agreement which we now know to have been fortuitous! Rayleigh showed also that if the cells are to form, then the vertical adverse temperature gradient $\beta_s = \Delta T/d$, according to (3.1c), must be sufficiently large that a particular dimensionless parameter proportional to the magnitude of the gradient exceeds a critical value – we now call this parameter the *Rayleigh number*, Ra, defined in Chapter 1 by (1.9b), and rewritten here as

$$\mathrm{Ra} = \frac{\alpha(T_d)\beta_s g d^4}{\nu_d \kappa_d},$$

when we take into account (3.1c). This Rayleigh number is a characteristic ratio of the destabilizing effect of buoyancy to the stabilizing effects of diffusion and dissipation.

It was the experimental work (in 1956) of Block [8], cited above, which put to rest the confusion surrounding the interpretation of Bénard's experiments, and which demonstrated conclusively that Bénard's results were not a consequence of buoyancy but were (temperature-dependent) surface tension induced. Among other things, he showed that cellular convection took place for Rayleigh numbers more than an order of magnitude smaller than required by the Rayleigh theory.

Most importantly, if the cells are buoyancy induced, then if the thin film is cooled from below the density gradient and gravity will be in the same direction and the film will be stably stratified.

Finally, Block concluded that for thin films of thicknesses less than 1 mm, variations in surface tension due to temperature variations (Marangoni effect) were the cause of Bénard cell formation and not buoyancy as postulated by Rayleigh in his 1916 paper.

It is now generally agreed that for films smaller than about a few millimeters, surface tension is the controlling force, while for larger thicknesses buoyancy is the controlling force and there the Rayleigh mechanism delimits the stable and unstable regimes.

In Chapter 4, via a rational analysis, we quantitatively give a criterion for the separation of two model convection problems, specfically, Rayleigh–Bénard thermal convection from Marangoni–Bénard thermocapillary convection.

In spite of the fact that the Rayleigh interpretation of Bénard's experiments was erroneous, Rayleigh's 1916 pioneering paper is the foundation of scores of papers on thermal convection.

We observe also that Rayleighs model is in accord with experiments on layers of fluid with rigid boundaries (considered here below) and in thicker layers (considered in Chapter 5), because the importance of the variation of surface tension relative to that of buoyancy diminishes as the thickness of the layer increases – but for 'appreciably' thicker layers we have, in fact, a third form of (deep) convection, *à la Zeytounian*, considered in Chapter 6.

3.2 Formulation of the Starting *à la Rayleigh* Problem for Thermal Convection

Here we consider as an 'exact' starting simple problem for thermal convection, a so-called, '*á la* Rayleigh problem', the one governed by the equations formulated in Section 2.4. Namely:

$$\rho(T)\frac{\mathrm{d}\mathbf{u}}{\mathrm{d}t} + \nabla p + \rho(T)g\mathbf{k} = \mu_d[\nabla\mathbf{u} + (1/3)\nabla(\nabla\cdot\mathbf{u})], \tag{3.2a}$$

$$\nabla\cdot\mathbf{u} = \alpha(T)\frac{\mathrm{d}T}{\mathrm{d}t}, \tag{3.2b}$$

$$\rho(T)C(T)\frac{\mathrm{d}T}{\mathrm{d}t} + p(\nabla\cdot\mathbf{u}) = \Phi + k_d\Delta T, \tag{3.2c}$$

and we observe that ρ is not an unknown function but is given by the above relation, (3.1a) as a function of the temperature T only, $\rho = \rho(T)$.

Equation (3.2c) is, in fact, a direct consequence of the energy equation (2.5) in Section 2.2, when we take into account the relation (3.1a), for ρ, which shows that specific internal energy is also a function of temperature T only:

$$e = E(T) \tag{3.3a}$$

and

$$\frac{\mathrm{d}E}{\mathrm{d}t} = \left(\frac{\mathrm{d}E}{\mathrm{d}T}\right)\frac{\mathrm{d}T}{\mathrm{d}t},$$

where

$$C(T) \equiv \frac{\mathrm{d}E}{\mathrm{d}T} \tag{3.3b}$$

is our specific heat and the viscous dissipation function Φ is given by (2.22b) and here this function Φ is written as

$$\Phi = (1/2)\left[\frac{\mu_d}{\rho(T)}\right]\left\{\left[\frac{\partial u_i}{\partial x_j} + \frac{\partial u_j}{\partial x_i}\right]^2 - (1/3)(\nabla\cdot\mathbf{u})^2\right\}. \tag{3.4}$$

The constant coefficients μ_d and k_d, in (3.2a), (3.2c) and (3.4), are at constant temperature T_d which is the fixed (in the Rayleigh problem T_d is a given data) temperature of the upper fixed infinite (flat) plate $z = d$ (see Figure 3.1). For the three above starting, exact, *à la* Rayleigh equations (3.2a–c), governing our 'Rayleigh thermal convection problem', we write as simple (assuming a constant liquid layer of the thickness d) boundary conditions for the velocity vector $\mathbf{u}$ and temperature T,

$$\mathbf{u} = 0 \quad \text{and} \quad T = T_w \equiv T_d + \Delta T \quad \text{on } x_3 \equiv z = 0, \tag{3.5a}$$

$$\mathbf{u} = 0 \quad \text{and} \quad T = T_d \quad \text{on } x_3 \equiv z = d. \tag{3.5b}$$

The above formulated *à la* Rayleigh thermal convection problem, (3.2a–c)–(3.5a, b), is a 'typical problem' and makes it possible to explain very simply our asymptotic modelling approach, founded on a careful, rational non-dimensional analysis. This rational approach also makes it possible to derive in a consistent way from the considered starting, exact, simple Rayleigh thermal shallow convection (see Sections 3.5 and 3.6) an associated, approximate model problem with RB leading first order. This is a significant second-order approximate model problem that takes into account some non-Boussinesq effects neglected on the level of the RB model problem formulated in Section 3.4.

3.3 Dimensionless Dominant Rayleigh Problem and the Boussinesq Limiting Process

A dominant dimensionless form of the above Rayleigh, starting, thermal convection problem, (3.2a–c)–(3.5a, b), is derived when we use, at first, the non-dimensional quantities (denoted by a prime); this non-dimensionalization is a twice necessary first step in the rational approach given below. Namely, we write:

$$(x', y', z') = \left(\frac{x_1}{d}, \frac{x_2}{d}, \frac{x_3}{d}\right); \quad t' = \frac{t}{(d^2/\nu_d)}; \quad \nu_d = \frac{\mu_d}{\rho_d}; \tag{3.6a}$$

$$u_i' = \frac{u_i}{(\nu_d/d)}, \quad \nabla' = d\nabla; \quad \Delta' = d^2\Delta; \tag{3.6b}$$

$$\pi = \left(\frac{1}{\mathrm{Fr}_d^2}\right)\left\{\left[\frac{(p - p_d)}{gd\rho_d}\right] + z' - 1\right\}; \tag{3.6c}$$

$$\theta = \frac{(T - Td)}{\Delta T}, \tag{3.6d}$$

where π (unlike (1.25) because the upper surface is here the solid flat plane $z' = d$) and θ (as in (1.13)), are respectively, *dimensionless* pressure perturbation (when $z' = 1$, then $p = p_d$) and *dimensionless* temperature reckoned from the temperature/point where $T = T_d$.

We observe from the above relation (3.6d) for π that in dimensionless form the pressure is reckoned from the point where $p = p_d$ and is related, in fact, to the reference pressure $\rho_d(\nu_d/d)^2$ which is $\ll gd\rho_d$, when we assume that $\mathrm{Fr}_d^2 = (\nu_d/d)^2/gd \ll 1$, which is just the case of a thermal convection, where the main driving force is the buoyancy and the Boussinesq limiting process (3.22) is considered. Now with an error of ε^2, where the expansibity parameter (defined in Chapter 1 by (1.10a)):

$$\varepsilon = \alpha(T_d)\Delta T, \tag{3.7}$$

is our main small parameter, we can write the following approximate, leading-order equation of state (instead of $\rho = \rho(T)$),

$$\rho(T) = \rho(T_d + \Delta T\theta) \approx \rho_d[1 - \varepsilon\theta]. \tag{3.8}$$

On the other hand, in a system of the starting equations (3.2a–c), with $\rho(T)$, we have also two other functions of the temperature T, $\alpha(T)$ and $C(T)$. By analogy with (3.8) we write

$$\alpha(T) = \alpha_d[1 - \varepsilon A_d \theta], \tag{3.9a}$$

and

$$C(T) = C_d[1 - \varepsilon \Gamma_d \theta], \tag{3.9b}$$

where, respectively,

$$\Gamma_d = \left[\frac{(\mathrm{d}\log C/\mathrm{d}T)}{(\mathrm{d}\log \rho/\mathrm{d}T)}\right]_d \tag{3.10a}$$

$$A_d = \left[\frac{(\mathrm{d}\log \alpha/\mathrm{d}T)}{(\mathrm{d}\log \rho/\mathrm{d}T)}\right]_d. \tag{3.10b}$$

It seems judicious (and reasonable), if we have the ambition to derive a second-order [with the terms proportional to ε (see Section 3.5)] thermal convection model problem, associated with the classical/leading-order RB shallow thermal convection model problem (formulated in Section 3.4) to assume that the coefficients Γ_d and A_d, in (3.10a, b), are both not very small or not very large (in fact, we presuppose that Γ_d and A_d are both ≈ 1). In the framework of an asymptotic theory with first-order (RB model problem) and second-order (model problem with non-Boussinesq effects) approximate problems (instead of the full exact, starting, thermal convection problem, (3.2a–c)–(3.5a, b)), it seems that this is a very rational approach.

Now, with the above results, (3.8), (3.9a, b), and (3.10a, b), we can rewrite the vectorial equation (3.2a), of convection motion, for $\mathbf{u}'$, in the following dimensionless form, when we use the dimensionless quantities (3.6a–d).

$$[1 - \varepsilon\theta]\frac{\mathrm{d}\mathbf{u}'}{\mathrm{d}t'} + \nabla'\pi - \mathrm{Gr}\theta = \nabla'\mathbf{u}' + (1/3)\nabla'(\nabla' \cdot \mathbf{u}'), \tag{3.11}$$

where the Grashof number, Gr, according to (1.12), is the ratio of the two small parameters, namely, the expansibility parameter, $\varepsilon = \alpha(T_d)\Delta T$, to the squared Froude number $\mathrm{Fr}_d^2 = (\nu_d/d)^2/gd$:

$$\mathrm{Gr} = \frac{\varepsilon}{\mathrm{Fr}_d^2} \equiv \frac{g\alpha(T_d)d^3\Delta T}{\nu_d^2}. \tag{3.12}$$

This Grashof number (3.12), defined from a fluid dynamical point of view, is directly responsible for taking account of the buoyancy effect – the main driving, *Archimedean*, force in thermal convection – and, because $\varepsilon \ll 1$, it is also necessary to assume that $\mathrm{Fr}_d^2 \ll 1$. Associated to the Grashof number (3.12), the Rayleigh number is defined as (see (1.9b))

$$\mathrm{Ra} = \frac{g\alpha(T_d)d\Delta T}{\nu_d \kappa_d} \equiv \mathrm{PrGr} \tag{3.13a}$$

with, according to (1.9c),

$$\mathrm{Pr} = \frac{\nu_d}{\kappa_d}. \tag{3.13b}$$

Here, in fact, it is assumed that the Prandtl number is not very small or not very large (κ_d is the thermal diffusivity) – the cases of a small or large Pr require special attention, and has been considered by various authors (see, for instance, comments and references in Section 10.10).

Next, from the continuity equation (3.2b), with (3.9a), we derive (again with an error of ε^2) the following constraint for the dimensionless velocity vector $\mathbf{u}'$:

$$\nabla' \cdot \mathbf{u}' = \varepsilon \frac{\mathrm{d}\theta}{\mathrm{d}t'}. \tag{3.14}$$

Finally, the third dimensionless equation for θ, written with an error of ε^2, is derived from (3.2c), with (3.8) and (3.9b), taking into account (3.14). The result is the following equation:

$$\begin{aligned} &\{1 - \varepsilon(1 + \Gamma_d)\theta + \varepsilon\,\mathrm{Bo}[(p_d)' + \mathrm{Fr}_d^2\pi + 1 - z']\}\frac{\mathrm{d}\theta}{\mathrm{d}t'} \\ &= \left(\frac{1}{\mathrm{Pr}}\right)\Delta'\theta + (1/2\,\mathrm{Gr})\varepsilon\,\mathrm{Bo}\left[\frac{\partial u_i'}{\partial x_j'} + \frac{\partial u_j'}{\partial x_i'}\right]^2, \end{aligned} \tag{3.15}$$

where $(p_d)' = p_d/gd\rho_d$.

In the dimensionless equation (3.15) we have a new parameter (see (1.14)–(1.16) in Chapter 1) denoted by Bo:

$$\mathrm{Bo} = \frac{gd}{C_d \Delta T}, \tag{3.16a}$$

and, more precisely in (3.13),

$$\mathrm{Pr} = \frac{\mu_d C_d}{k_d} = \frac{\nu_d}{(k_d/C_d\rho_d)} \equiv \frac{\nu_d}{\kappa_d},$$

where the thermal diffusivity is

$$\kappa_d \equiv \frac{k_d}{C_d\rho_d}. \tag{3.16b}$$

We observe that, in the framework of the Bénard–Marangoni (BM) convection (considered in Chapter 7), usually

$$\mathrm{Bo} = \frac{\mathrm{Cr}}{\mathrm{Fr}_d^2}$$

is the classical Bond number which is related to the Weber, We ($\equiv 1/\mathrm{Cr}$), number defined by (1.18a), the parameter Cr ($= 1/\mathrm{We}$) being the crispation/capillary number. As in the present book we do not make use of the Bond, Bo, number; our notation Bo, as a ratio of two lengths, d and $C_d\Delta T/g$, a number similar to a Boussinesq number (used in derivation of the Boussinesq approximate equations for the various meso or local atmospheric motions, see Chapter 9), seems not to introduce any confusion!

The dimensionless equation (3.15), for the dimensionless temperature θ, shows explicitly the role of the dissipation number Di*, defined by (1.14) with (1.15) – namely:

$$\mathrm{Di}^* = (1/2)\left[\frac{\varepsilon\,\mathrm{Bo}}{\mathrm{Gr}}\right] \equiv (1/2)\mathrm{Bo}\mathrm{Fr}_d^2 = \frac{(\nu/d)^2}{2C_d\Delta T}, \tag{3.17}$$

as a measure for the viscous dissipation. If we assume that $\mathrm{Di}^* \approx 1$, then we obtain the following estimation for the thickness of the liquid film

$$d \approx \frac{\nu_d}{[2C_d\Delta T]^{1/2}}. \tag{3.18}$$

The condition

$$\mathrm{Di}^* \approx 1, \tag{3.18a}$$

which allows us, in the thermal convection model problem, to take into account the viscous dissipation, leads also to the following relation for the difference of the temperature $\Delta T = T_w - T_d$:

$$\Delta T \approx \frac{(\nu/d)^2}{2C_d}. \tag{3.19}$$

For the above dimensionless dominant equations (3.11), (3.14) and (3.15), for $\mathbf{u}'$, π and θ we have from (3.5a, b) the following dimensionless boundary conditions:

$$\mathbf{u}' = 0 \quad \text{and} \quad \theta = 1 \quad \text{on } z' = 0; \tag{3.20a}$$

$$\mathbf{u}' = 0 \quad \text{and} \quad \theta = 0 \quad \text{on } z' = 1. \tag{3.20b}$$

As a conclusion, we also observe that for the motionless conduction temperature $T_s(z) = T_w - \beta_s z$, as a conduction dimensionless temperature $\theta_s(z)$, associated with θ, we have

$$\theta_s(z) = \frac{[T_w - d\beta_s z' - T_d]}{(T_w - T_d)} = \left[1 - \left(\frac{d\beta_s}{\Delta T}\right) z'\right],$$

or, according to relation (1.19c) for β_s,

$$\theta_s(z) = 1 - z', \tag{3.21a}$$

and the companion dimensionless perturbation pressure, $\pi_s(z)$, is

$$\pi_s = \varepsilon[z' - (1/2)z'^2]. \tag{3.21b}$$

In the above system of dimensionless dominant equations (3.11), (3.14) and (3.15), the expansibility parameter, $\varepsilon = \alpha(T_d)\Delta T$, is our main small parameter because all the usual liquids are weakly expansible: $\alpha(T_d) \approx 5 \times 10^{-4}$ and for moderate ΔT, we have always $\varepsilon \ll 1$. On the other hand, in equation (3.11), for $\mathbf{u}'$ the term proportional to $\mathrm{Gr} = \varepsilon/\mathrm{Fr}_d^2$ is a ratio of ε and Fr_d^2, while in equation (3.15) for θ we have two terms proportional to $\varepsilon\mathrm{Bo}$! As a consequence

- First, if we want to take into account the buoyancy term $-[\varepsilon/\mathrm{Fr}_d^2]\theta\mathbf{k}$, in equation (3.17) for the convective motion (for $\mathbf{u}'$), then, obviously, it is necessary to consider the following, *à la* Boussinesq, limiting process:

$$\varepsilon \downarrow 0 \quad \text{and} \quad \mathrm{Fr}_d^2 \downarrow 0 \quad \text{such that } \mathrm{Gr} = \varepsilon/\mathrm{Fr}_d^2 = \mathrm{O}(1), \tag{3.22}$$

  Gr being a fixed driving parameter for the RB model problem.
- Then, it is necessary to consider two cases:

$$\mathrm{Bo} = \mathrm{O}(1), \quad \text{fixed}, \tag{3.23a}$$

  or

$$\mathrm{Bo} \gg 1, \quad \text{such that } \varepsilon\mathrm{Bo} \equiv B^* = \mathrm{O}(1), \text{ fixed}. \tag{3.23b}$$

We observe that the Boussinesq limiting process (3.22) is considered when all the time-space variables, t' and x', y', z', Prandtl number, Pr, Γ_d, and $(p_d)'$ are fixed and O(1).

3.4 The Rayleigh–Bénard Rigid-Rigid Problem as a Leading-Order Approximate Model

Obviously, now, the asymptotic derivation of the classical Rayleigh–Bénard, RB, problem for the shallow convection, when $\mathrm{Bo} = \mathrm{O}(1)$ as in (3.23a) is fixed, from the above dominant dimensionless equations (3.11), (3.14) and (3.15) via the Boussinesq limiting process (3.22), is a very easy, even elementary, task! Namely, we consider for $\mathbf{u}'$, π and θ the following three expansions relative to ε:

$$\mathbf{u}' = \mathbf{u}_{RB} + \varepsilon \mathbf{u}_1 + \cdots, \quad \theta = \theta_{RB} + \varepsilon \theta_1 + \cdots, \quad \pi = \pi_{RB} + \varepsilon \pi_1 + \cdots. \tag{3.24}$$

As a leading-order result we derive, from the dominant Rayleigh equations (3.11), (3.14), (3.15), via the Boussinesq limiting process (3.22), associated with the three asymptotic expansions (3.24), under the constraint (3.23a), the following Boussinesq, shallow convection, RB model, leading-order equations:

$$\frac{d\mathbf{u}_{RB}}{dt'} + \nabla' \pi_{RB} - \mathrm{Gr}\, \theta_{RB} \mathbf{k} = \mathbf{\Delta}' \mathbf{u}_{RB}, \tag{3.25a}$$

$$\nabla' \cdot \mathbf{u}_{RB} = 0, \tag{3.25b}$$

$$\frac{d\theta_{RB}}{dt'} = \left(\frac{1}{\mathrm{Pr}}\right) \mathbf{\Delta}' \theta_{RB}. \tag{3.25c}$$

As boundary conditions for these above RB model equations (3.25a–c) we write, according to (3.20):

$$\mathbf{u}_{RB} = 0 \text{ and } \theta_{RB} = 1 \quad \text{on } z' = 0; \quad \mathbf{u}_{RB} = 0 \text{ and } \theta_{RB} = 0 \quad \text{on } z' = 1. \tag{3.25d}$$

With (3.21a, b) it is possible to write the above RB model equations (3.25a–c) in a more usual form. Namely, for this we introduce, instead of $\mathbf{u}_{RB}$, π_{RB} and θ_{RB}, the following three new functions:

$$\mathbf{U}_{Sh} = \mathrm{Pr}\, \mathbf{u}_{RB}, \tag{3.26a}$$

$$\Theta_{Sh} = z' - 1 + \theta_{RB}, \tag{3.26b}$$

$$\Pi_{Sh} = \mathrm{Gr}\, z'[(z'/2) - 1] + \pi_{RB}. \tag{3.26c}$$

As a result, with the new functions (3.26a–c), instead of the equations (3.25a–c), we derive the following shallow convection – RB – equations for U_{Sh}, Θ_{Sh}, Π_{Sh}:

$$\frac{d\mathbf{U}_{Sh}}{dt'} + \mathrm{Pr}\, \nabla' \Pi_{Sh} - \mathrm{Ra}\, \Theta_{Sh} \mathbf{k} = \mathbf{\Delta}' \mathbf{U}_{Sh}, \tag{3.27a}$$

$$\nabla' \cdot \mathbf{U}_{Sh} = 0, \tag{3.27b}$$

$$\mathrm{Pr} \frac{d\Theta_{Sh}}{dt'} - W_{Sh} = \mathbf{\Delta}' \Theta_{Sh}. \tag{3.27c}$$

where $W_{Sh} = \mathbf{U}_{Sh} \cdot \mathbf{k}$ is the vertical component of the velocity $\mathbf{U}_{Sh}$ in the direction of z', and $\mathrm{Ra} = \mathrm{Pr}\,\mathrm{Gr}$ is the Rayleigh number defined by (3.13a). These equations (3.27a–c) with the homogeneous boundary conditions:

$$\mathbf{U}_{Sh} = \Theta_{Sh} = 0 \quad \text{at } z' = 0 \text{ and } z' = 1, \tag{3.27d}$$

govern the rigid-rigid RB problem.

Thus, we recover the classical RB, shallow, thermal convection model rigid-rigid problem, (3.27a–d), which is usually derived in an *ad hoc* manner – see, for example the useful books by Drazin and Reid [13] and by Chandrasekhar [5].

Usually, in classical hydrodynamic instability theory, a linearized approach is chosen. Because, for the above rigid-rigid RB shallow thermal convection model problem (3.27a–d), the basic motionless conduction state is characterized by the following 'zero' solution:

$$\mathbf{U}_{\mathrm{Sh}} = \Theta_{\mathrm{Sh}} = \Pi_{\mathrm{Sh}} = 0, \tag{3.28}$$

then, in the case of an usual linearization when $\mathrm{d}\mathbf{U}_{\mathrm{Sh}}/\mathrm{d}t'$ and $\mathrm{d}\Theta_{\mathrm{Sh}}/\,\mathrm{d}t'$ are replaced in (3.27a) and (3.27c), respectively, by $\partial\mathbf{U}_{\mathrm{LSh}}/\partial t'$ and $\partial\Theta_{\mathrm{LSh}}/\partial t'$, we derive a single linear equation for the 'vertical' – relative to z' – component of the velocity $\mathbf{U}_{\mathrm{LSh}}$,

$$\mathbf{U}_{\mathrm{LSh}} \cdot \mathbf{k} = W_{\mathrm{LSh}}(z')f(x', y')\exp[\sigma t']. \tag{3.29}$$

Namely, after some simple manipulations we obtain for $W_{\mathrm{LSh}}(z')$ the following linear differential equation in z':

$$D^2(D^2 - \sigma)(D^2 - \sigma\,\mathrm{Pr})W_{\mathrm{LSh}}(z') = -a^2\,\mathrm{Ra}\,W_{\mathrm{LSh}}(z'), \tag{3.30a}$$

with

$$\frac{\partial^2 f}{\partial x'^2} + \frac{\partial^2 f}{\partial y'^2} + a^2 f = 0, \tag{3.30b}$$

where a is the wave number and $D^2 = [\mathrm{d}^2/\mathrm{d}z'^2 - a^2]$.

The relevant boundary conditions, for the rigid-rigid, linear RB problem, for the function $W_{\mathrm{LSh}}(z')$, solution of the linear equation (3.30a), at the rigid flat surfaces $z' = 0$ and $z' = 1$, are:

$$W_{\mathrm{LSh}}(z') = 0, \tag{3.30c}$$

$$\frac{\mathrm{d}W_{\mathrm{LSh}}(z')}{\mathrm{d}z'} = 0, \tag{3.30d}$$

$$D^2(D^2 - \sigma)W_{\mathrm{LSh}}(z') = 0, \tag{3.30e}$$

Linear equation (3.30a) for $W_{\mathrm{LSh}}(z')$ with the boundary conditions (3.30c–e) determines a so-called 'self-adjoint eigenvalue problem' for the parameters Ra, a^2 and σ, when Pr is fixed.

First, it can be proven that:

When Ra is less than a certain critical value Ra_c, all small disturbances of the purely conductive basic motionless equilibrium (conduction) state decay in time (stability). Whereas, if Ra exceeds the critical value Ra_c, instability occurs in the form of a convection in cells of a polygonal platform.

These cells are called Bénard cells, discovered in 1900 thanks to his quantitative experiments (in [5, sec. 18], the reader can find an account of some of the experimental work, up to 1960, on the onset of thermal instability in fluids). The formation of Bénard cells in a weakly expansible liquid layer is one of the most remarkable examples of bifucation phenomena (the bifurcations in dissipative, dynamical systems, are, in particular, investigated in [14, chapter 10]. We observe that:

From a physical viewpoint, the fundamental process involved in RB instability is the transformation of the potential energy of the convective disturbance.

Here, we note only that, in 1940, Pellew and Southwell [15] made a comprehensive study of linearized Bénard convection, and they conclusively proved (principle of the exchange of stabilities) that when the basic conduction temperature decreases upward, the only type of disturbance that can appear corresponds to real σ, so that an amplifying wave motion is not possible. In other words, the 'principle of exchange of stabilities' to hold if, in a given system, the growth rate $\sigma = \sigma_r + i\sigma_i$ in solution (3.29), is such that

$$\sigma \in \mathbf{R} \quad \text{or} \quad \sigma_i \neq 0 \Rightarrow \sigma_r < 0,$$

the marginal states being characterized by $\sigma = 0$, when Ra is assumed to be > 0.

Since σ is real for all positive Rayleigh numbers, i.e. for all adverse conduction temperature gradients β_s, defined by (3.1c), it follows that the transition from stability to instability must occur only via a stationary state.

The equations governing the marginal state are therefore to be obtained by setting $\sigma = 0$, in the relevant linear equation (3.30a) and linear conditions (3.30c–e). In such a case, instead of the linear problem (3.30a–e) we obtain the following simplified problem for the stationary state:

$$D^6 W_{\mathrm{LSh}}(z') = -a^2 \,\mathrm{Ra}\, W_{\mathrm{LSh}}(z'), \tag{3.31a}$$

with, as boundary conditions for $z' = 0$ and $z' = 1$,

$$W_{\mathrm{LSh}}(z') = 0, \tag{3.31b}$$

$$\frac{\mathrm{d}W_{\mathrm{LSh}}(z')}{\mathrm{d}z'} = 0, \tag{3.31c}$$

$$D^4 W_{\mathrm{LSh}}(z') = 0. \tag{3.31d}$$

On the other hand, when $\sigma = 0$, the first variational principle of Pellew and Southwell [15] leads to an energy-balance relation which establishes a precise balance between the rate of supply of kinetic energy to the velocity field and the rate of dissipation of kinetic energy (see, for instance, [5, pp 27–31]). In section 13 of [5], there is a second variational principle of Pellew and Southwell [15], which shows that:

> the Rayleigh number, at which disturbances of an assigned wave number become unstable, is the minimum value which a certain ratio of two positive definite integrals can attain. (See, [5, p. 32, relation (169)])

Also a physical content (thermodynamic significance) of this second, Pellew and Southwell, variational principle is shown, namely:

> Instability occurs at the minimum temperature gradient at which a balance can be steadily maintained between the kinetic energy dissipated by viscosity and the internal energy released by the buoyancy force.

When the principle of exchange of stabilities holds, convection sets in as stationary convection. If, on the other hand, at the onset of instability, $\sigma = i\sigma_i$, with $\sigma_i \neq 0$, the convection mechanism is referred to as oscillatory convection. But it must be emphasized that the linearized theory only yields a boundary for the instability. Whenever $\mathrm{Ra} > \mathrm{Ra}_c$ the (linear) solution grows (with an evolution in time of the form $\exp[\sigma t']$) and is unstable – the linearized equations do not yield any information on nonlinear stability.

It is, in general, possible for the solution to become unstable at a value of Ra lower than Ra_c, and in this case, a sub-critical instability (bifurcation) is said to occur. But for the standard RB linear problem, (3.30a)–(3.30c–e), we prove by energy stability theory that sub-critical instability is not possible (see, for example, the result of Joseph [16]). More precisely for the full RB problem (3.25a–d) it holds that the linear instability boundary $\equiv$ to the nonlinear stability boundary, and so no sub-critical instabilities are possible.

In Chapter 5, we give various complementary analytical results concerning the above RB thermal shallow convection problem (3.25a–d) or (3.27a–d).

3.5 Second-Order Model Equations Associated with the RB Shallow Convection Equations (3.25a–c)

We return to the dominant thermal convection system of equations derived above in Section 3.4. Namely, we again write, first, the following Rayleigh dominant system of three equations (3.11), (3.14) and (3.15),

$$[1-\varepsilon\theta]\frac{\mathrm{d}u}{\mathrm{d}t'}+\nabla'\pi-\mathrm{Gr}\,\theta\mathbf{k}=\Delta\mathbf{u}'+(1/3)\nabla'(\nabla'\cdot\mathbf{u}'),$$

$$\nabla'\cdot\mathbf{u}'=\varepsilon\frac{\mathrm{d}\theta}{\mathrm{d}t'},$$

$$\left\{1-\varepsilon(1+\Gamma_d)\theta+\varepsilon\,\mathrm{Bo}[(p_d)'+\mathrm{Fr}_d^2\pi+1-z']\right\}\frac{\mathrm{d}\theta}{\mathrm{d}t'}=\left(\frac{1}{\mathrm{Pr}}\right)\Delta'\theta$$

$$+(1/2\mathrm{Gr})\varepsilon\,\mathrm{Bo}\left[\frac{\partial u_i'}{\partial x_j'}+\frac{\partial u_j'}{\partial x_i'}\right]^2.$$

Then we consider, again, the following three asymptotic expansions (3.24):

$$\mathbf{u}'=\mathbf{u}_{\mathrm{RB}}+\varepsilon\mathbf{u}_1+\cdots,\qquad \theta=\theta_{\mathrm{RB}}+\varepsilon\theta_1+\cdots,\qquad \pi=\pi_{\mathrm{RB}}+\varepsilon\pi_1+\cdots,$$

associated with the Boussinesq limiting process (3.22)

$$\varepsilon\downarrow 0\quad\text{and}\quad \mathrm{Fr}_d^2\downarrow 0\quad\text{such that } \mathrm{Gr}=\varepsilon/\mathrm{Fr}_d^2=\mathrm{O}(1)\text{ fixed.}$$

The above three equations, for $\mathbf{u}'$, θ and π (valid with an error of order ε^2), subject to three asymptotic expansions (relative to expansibility parameter ε) with the associated Boussinesq limiting process, give a rational framework for a rational, consistent, asymptotic derivation of a set of second-order model equations for the functions $\mathbf{u}_1$, θ_1 and π_1 in the above expansion.

Indeed, this rational method is the only one for the obtention of a significant set of companion, three second-order equations and boundary conditions, for the shallow leading-order RB model problem (3.25a–d).

We assume that Pr and Bo are fixed (and have 'moderate' values) when the above (3.22) Boussinesq limiting process is carried out. In such a case, for $\mathbf{u}_1$, θ_1 and π_1, we derive our set of consistent second-order equations associated with the RB model equations (3.25a–c), when we take into account the well-balanced terms proportional to ε. In our asymptotic rational and consistent approach, only these terms – proportional to ε – can be present, below, in second-order model equations (3.32a–c), which are associated to shallow convection RB equations (3.25a–c).

Namely, we obtain as second-order equations, for $\mathbf{u}_1$, θ_1 and π_1, when we take into acount that

$$\frac{\mathrm{d}}{\mathrm{d}t'} = \frac{\partial}{\partial t'} + (\mathbf{u}' \cdot \nabla)\mathbf{u}',$$

the following system of linear, but non-homogeneous, system of three dimensionless equations, with zero boundary conditions at $z' = 0$ and $z' = 1$, for $\mathbf{u}_1$ and θ_1:

$$\frac{\partial \mathbf{u}_1}{\partial t'} + (\mathbf{u}_{\mathrm{RB}} \cdot \nabla')\mathbf{u}_1 + (\mathbf{u}_1 \cdot \nabla')\mathbf{u}_{\mathrm{RB}} + \nabla'\pi_1 - \mathrm{Gr}\,\theta_1\mathbf{k} - \mathbf{\Delta}'\mathbf{u}_1$$
$$= \theta_{\mathrm{RB}}\frac{\mathrm{d}\mathbf{u}_{\mathrm{RB}}}{\mathrm{d}t'} + (1/3)\nabla'\left(\frac{\mathrm{d}\theta_{\mathrm{RB}}}{\mathrm{d}t'}\right); \tag{3.32a}$$

$$\nabla' \cdot \mathbf{u}_1 = \frac{\mathrm{d}\theta_{\mathrm{RB}}}{\mathrm{d}t'}; \tag{3.32b}$$

$$\frac{\partial \theta_1}{\partial t'} + \mathbf{u}_{\mathrm{RB}} \cdot \nabla'\theta_1 + \mathbf{u}_1 \cdot \nabla'\theta_{\mathrm{RB}} - \left(\frac{1}{\mathrm{Pr}}\right)\Delta'\theta_1$$
$$= \left\{(1+\Gamma_d)\theta_{\mathrm{RB}} - \mathrm{Bo}[(p_d)' + 1 - z']\right\}\frac{\mathrm{d}\theta_{\mathrm{RB}}}{\mathrm{d}t'}$$
$$+ (1/2)\left[\frac{\mathrm{Bo}}{\mathrm{Gr}}\right]\left[\frac{\partial u_{\mathrm{RB}i}}{\partial x'_j} + \frac{\partial u_{\mathrm{RB}j}}{\partial x'_i}\right]^2, \tag{3.32c}$$

with

$$\mathbf{u}_1 = 0 \quad \text{and} \quad \theta_1 = 0 \quad \text{at } z' = 0 \text{ and } z' = 1. \tag{3.32d}$$

The terms on the right-hand side of equations (3.32a–d) are given by the RB leading-order model equations (3.25a–c). The second-order system of equations (3.32a–c), with zero conditions (3.32d) for $\mathbf{u}_1$ and θ_1, associated with the RB leading-order model problem (3.25a–d) – which is the only consistent one – takes into account the low expansibility effects and viscous dissipation in a weakly expansible liquid – both these effects are ‘non-Boussinesq effects’.

It seems that the above second-order model problem (3.32a–d), associated with the leading-order RB classical problem (3.25a–d), has not been obtained before. The analysis of the second-order model problem (3.32a–d) is obviously interesting for a more realistic estimation of the results obtained via the usual RB problem. This second-order model problem (3.32a–d) will serve for both postgraduate research workers and young researchers in fluid dynamics as a research problem; but it will require some efforts to fully comprehend the basic material (and philosophy) presented in this book.

3.6 Second-Order Model Equations Following from the Hills and Roberts Equations (2.70a–c)

In this section, the starting equations (2.70a–c), are the ones derived by Hill and Roberts:

$$\alpha\frac{\mathrm{d}T}{\mathrm{d}t}=\frac{\partial u_i}{\partial x_i};$$

$$\rho\frac{\mathrm{d}u_i}{\mathrm{d}t}=\rho f_i-\frac{\partial p}{\partial x_i}+\frac{\partial}{\partial x_j}[\lambda d_{mm}\delta_{ij}+2\mu d_{ij}];$$

$$-\alpha T\frac{\mathrm{d}p}{\mathrm{d}t}+C_p\left(\frac{\rho}{\alpha}\right)\frac{\partial u_i}{\partial x_i}=(\lambda d_{ii})^2+2\mu d_{ij}d_{ij}+\frac{\partial}{\partial x_i}\left[k\frac{\partial T}{\partial x_i}\right],$$

In the equation of motion for the velocity component u_i (with $i = 1, 2$ and 3) we assume, on the one hand, that $f_1 = f_2 = 0$ and $f_3 = -g$. On the other hand, in this equation of motion for u_i, when both viscous coefficients λ and μ are assumed constant (respectively λ_d and μ_d, as functions of the constant temperature T_d), we write the viscous term $\partial/\partial x_j[\lambda d_{mm}\delta_{ij}+2\mu d_{ij}]$ on the right-hand side of the above second equation (for u_i), as $\mu_d\{\Delta u_i + [1+(\lambda_d/\mu_d)]\nabla(\partial u_i/\partial x_i)\}$.

First, by analogy with the non-dimensional analysis performed in Section 3.3, instead of the above three equations, we derive a dominant dimensionless system of equations, which replace equations (3.11), (3.14) and (3.15) of Section 3.3, and includes the terms proportional to ε – the terms proportional to ε^2 being neglected. Namely, for our $\mathbf{u}'$, π and θ, with our notations, we obtain the following system of three dimensionless dominant equations:

$$[1-\varepsilon\theta]\frac{\mathrm{d}\mathbf{u}'}{\mathrm{d}t'}+\nabla'\pi-\mathrm{Gr}\,\theta\mathbf{k}=\Delta'\mathbf{u}'+\left[1+\left(\frac{\lambda_d}{\mu_d}\right)\right]\nabla'(\nabla'\cdot\mathbf{u}'), \quad (3.33a)$$

$$\nabla'\cdot\mathbf{u}'=\varepsilon\frac{\mathrm{d}\theta}{\mathrm{d}t'}, \quad (3.33b)$$

$$[1-\varepsilon(1+\Gamma_{pd})\theta]\frac{\mathrm{d}\theta}{\mathrm{d}t'}-\varepsilon\,\mathrm{Bo}\left[\left(\frac{T_d}{\Delta T}\right)+\theta\right]\left[\mathrm{Fr}_d^2\frac{\mathrm{d}\pi}{\mathrm{d}t'}-\mathbf{u}'\cdot\mathbf{k}\right]$$

$$=\left(\frac{1}{\mathrm{Pr}}\right)\Delta'\theta+(1/2\mathrm{Gr})\varepsilon\,\mathrm{Bo}\left[\frac{\partial u_i'}{\partial x_j'}+\frac{\partial u_j'}{\partial x_i'}\right]^2, \tag{3.33c}$$

where, according to (3.9a, b), the following two relations:

$$\alpha(T)=\alpha_d[1-\varepsilon A_d\theta] \quad \text{and} \quad C_p(T)=C_{pd}[1-\varepsilon\Gamma_{pd}\theta],$$

have been used. The coefficient A_d is given by the relation (3.10b) and the coefficient

$$\Gamma_{pd}=\frac{[(1/C_p)\,\mathrm{d}C_p/\mathrm{d}T]_d}{\alpha(T_d)}$$

is given according to the relation (3.10a), but written as C_p instead of $C(T)$.

In reality, in equation (3.33c), the term

$$\varepsilon\,\mathrm{Bo}\,\mathrm{Fr}_d^2\left[\left(\frac{T_d}{\Delta T}\right)+\theta\right]\frac{\mathrm{d}\pi}{\mathrm{d}t'}$$

is an ε^2-order term, because $\mathrm{Fr}_d^2=\varepsilon/\mathrm{Gr}$ according to Boussinesq limiting process (3.22). With $\mathrm{Bo}=\mathrm{O}(1)$, Pr and Gr fixed, when $\varepsilon\downarrow 0$, we again recover, at the leading order, the RB shallow convection model equations (3.25a–c), as expected!

Then, a second-order companion system of equations to RB equations (3.25a–c), is derived from the above dominant equations (3.33a–c), with the following three asymptotic expansions:

$$\mathbf{u}'=\mathbf{u}_{\mathrm{RB}}+\varepsilon\mathbf{u}_1+\cdots,\qquad \theta=\theta_{\mathrm{RB}}+\varepsilon\,\theta_1+\cdots,\qquad \pi=\pi_{\mathrm{RB}}+\varepsilon\pi_1+\cdots,$$

relative to ε.

Namely, we obtain the following consistent system of three second-order equations for three functions $\mathbf{u}_1$, π_1 and θ_1:

$$\frac{\partial\mathbf{u}_1}{\partial t'}+(\mathbf{u}_{\mathrm{RB}}\cdot\nabla')\mathbf{u}_1+(\mathbf{u}_1\cdot\nabla')\mathbf{u}_{\mathrm{RB}}+\nabla'\pi_1-\mathrm{Gr}\,\theta_1\mathbf{k}-\Delta'\mathbf{u}_1$$

$$=\theta_{\mathrm{RB}}\frac{\mathrm{d}\mathbf{u}_{\mathrm{RB}}}{\mathrm{d}t'}+\left[1+\left(\frac{\lambda_d}{\mu_d}\right)\right]\nabla'\left(\frac{\mathrm{d}\theta_{\mathrm{RB}}}{\mathrm{d}t'}\right), \tag{3.34a}$$

$$\nabla'\cdot\mathbf{u}_1=\frac{\mathrm{d}\theta_{\mathrm{RB}}}{\mathrm{d}t'}, \tag{3.34b}$$

$$\frac{\partial \theta_1}{\partial t'} + \mathbf{u}_{\mathrm{RB}} \cdot \nabla' \theta_1 + \mathbf{u}_1 \cdot \nabla' \theta_{\mathrm{RB}} - \left(\frac{1}{\mathrm{Pr}}\right) \Delta' \theta_1$$
$$= [(1 + \Gamma_{pd})\theta_{\mathrm{RB}}] \frac{\mathrm{d}\theta_{\mathrm{RB}}}{\mathrm{d}t'} + \mathrm{Bo}\left[\left(\frac{T_d}{\Delta T}\right) + \theta_{\mathrm{RB}}\right](\mathbf{u}_{\mathrm{RB}} \cdot \mathbf{k})$$
$$+ (1/2)\left[\frac{\mathrm{Bo}}{\mathrm{Gr}}\right]\left[\frac{\partial u_{\mathrm{RB}i}}{\partial x'_j} + \frac{\partial u_{\mathrm{RB}j}}{\partial x'_i}\right]^2. \tag{3.34c}$$

Unfortunately, the system (unless non-dimensionalization holds) derived in an *ad hoc* manner by Hills and Roberts [17] in 1991 – see, for instance, [18, pp. 50, 51] – have nothing to do with the above second-order equations (3.34a–c)! The equations derived by Hills and Roberts [17], for a so-called fluid motion that is incompressible in a generalized sense and its relation to the Boussinesq approximation, are in fact, not consistent mainly as a consequence of their 'exotic' limit process: 'gravity g tends to infinity and $\alpha(T_d)$ tends to zero, such that their product, $g\alpha(T_d)$ remains finite', which is, in fact, a 'bastardized', non-formalized version of our *à la* Boussinesq limit process (3.22), coupled with the asymptotic expansions (3.24). Straughan writes [18, p. 50]:

> The key philosophy of the Hills and Roberts paper [17] is that typical acceleration promoted in the fluid by variations in the density are always much less than the acceleration of gravity. The resulting equations from the Boussinesq approximation, the so-called Oberbeck–Boussinesq (O–B) equations, arise by taking the simultaneous limits
>
> $$g \to \infty, \quad \alpha_d \to 0, \quad \text{with the restriction that } g\alpha_d \text{ remains finite.}$$

In [17], Hills and Roberts expand the pressure, velocity, and temperature fields, in their dimensional equations (2.70a–c), in $1/g$ ($\to 0$) such that

$$p = p^0 g + p^1 + \left(\frac{1}{g}\right) p^2 + \cdots,$$
$$u_i = u_i^1 + \left(\frac{1}{g}\right) u_i^2 + \cdots,$$
$$T - T_d = T^1 - T_d + \left(\frac{1}{g}\right)[T^2 - T_d] + \cdots$$

Then, they derive the O–B equations at the level O(1) with the limit

$$\varepsilon_{\text{H-R}} = \frac{g\alpha_d d}{C_{pd}} \to 0.$$

In fact, their small parameter, $\varepsilon_{\text{H-R}}$, is our above εBo!

Their derived equations are

$$\frac{\partial u_i}{\partial x_i} = 0,$$

$$\frac{\mathrm{d}u_i^1}{\mathrm{d}t} = -\mathrm{Ra}\, T^1\delta_{i3} - \frac{\partial p^1}{\partial x_i} + \Delta u_i^1,$$

$$\frac{\mathrm{d}T^1}{\mathrm{d}t} - \varepsilon_{\text{H-R}}(T_d + T^1)u_3^1 = \left(\frac{1}{\mathrm{Pr}}\right)\Delta T^1 + \left(\frac{2}{\mathrm{Ra}}\right)\varepsilon_{\text{H-R}} d_{ij}^1 d_{ij}^1.$$

However, these Hills and Roberts equations above contain (some) first-order effects of compressibility via the $\varepsilon_{\text{H-R}}$ terms – the various terms in these equations being very poorly balanced and, unfortunately, consistent (and give only our RB model equations) only when their $\varepsilon_{\text{H-R}} \to 0$.

We see that, even if an ad hoc derivation is often able to give a valuable result at the leading order, in spite of the fact that a deficient approach has been chosen, such an approach will in no way be able to derive consistently a rational second-order approximation with well balanced second-order ε terms. This strong observation is one of the main reasons for our present approach and for the publication of this book!

3.7 Some Comments

We have already observed that, for a rational formulation of the RB thermal convection model equations, the smallness of the Froude number is a requisite condition which makes it possible to take into account the buoyancy as a main (Archimedean) driving force in this shallow thermal convection model problem. But, this constraint has also an important consequence on the upper, free-surface, condition relative to pressure p!

Namely, this upper boundary condition, (2.42a) written in Section 5.5, where d_{ij} are given by (2.2) in Chapter 2, is

$$p = p_A + \mu_0[d_{ij}n_i n_j - (2/3)(\nabla \cdot \mathbf{u})] + \sigma(T)(\nabla_{\|} \cdot \mathbf{n}),$$
$$\text{at } z = d + ah(t, x, y).$$

With the non-dimensional quantities (3.6a–d) this above upper boundary condition for the pressure is rewritten relative to the dimensionless pressure,

$$\pi = \left(\frac{1}{\mathrm{Fr}_d^2}\right)\left[\frac{(p - p_A)}{gd\rho_d}\right] + z' - 1,$$

in the dimensionless form

$$\pi_{1+\eta h'} = \left(\frac{\eta}{\mathrm{Fr}_d^2}\right) h'(t', x', y') + \cdots . \tag{3.35}$$

We do not have write the above dimensionless, upper, free-surface condition (3.35) for π at $z' = 1 + \eta h'$ in detail (this is done in Chapter 4). For the moment, the important point here is just the first term in condition (3.35) for $\pi_{1+\eta h}$.

Indeed, because $\mathrm{Fr}_d^2 \ll 1$ for a rational derivation of the RB model equations, it is obvious that a necessary condition (in the framework of an asymptotic modeling of the shallow thermal convection) for a rational approach is the following:

$$\eta \ll 1, \tag{3.36a}$$

with the similarity rule

$$\frac{\eta}{\mathrm{Fr}_d^2} \equiv \eta^* \approx 1, \tag{3.36b}$$

when the free-surface amplitude η and square of the Froude number, Fr_d^2, both tend to zero. In this case, for the shallow thermal convection model limit, equations (3.25a–c), the upper boundary conditions are written (again) for a non-deformable free-surface, simulated by $z' = 1$.

As a consequence, in a shallow thermal convection problem, when buoyancy is the main (Archimedean) driving force, in the leading order, the free surface deformation has no influence.

On the other hand, for the dimensionless temperature θ, such that $T = T_d + (T_w - T_d)\theta$, at the undeformable free-surface $z' = 1$, we can write as an upper boundary condition (rigid-free problem) for the convection model equations (3.25a–c),

$$\left.\frac{\partial \theta_{\mathrm{RB}}}{\partial z'}\right|_{z'=1} = -1, \tag{3.37}$$

when the Biot effect is neglected. But, from two upper, free-surface, conditions (2.42b, c),

$$\mu_d d_{ij}(\mathbf{t}^{(1)})_i n_j = \left[\frac{\mathrm{d}\sigma(T)}{\mathrm{d}T}\right](\mathbf{t}^{(1)})_i\left(\frac{\partial T}{\partial x_i}\right);$$

$$\mu_d d_{ij}(\mathbf{t}^{(2)})_i n_j = \left[\frac{\mathrm{d}\sigma(T)}{\mathrm{d}T}\right](\mathbf{t}^{(2)})_i\left(\frac{\partial T}{\partial x_i}\right),$$

at $z = d + ah(t, x, y)$,

rewritten with the non-dimensional quantities (3.6a–d) when the Marangoni effect is neglected and (3.36a, b) is taken into account, we obtain from (2.44a–c), instead of these above two tangential conditions, at $z' = 1$:

$$\frac{\partial u_1'}{\partial z'} + \frac{\partial u_3'}{\partial x_2'} = 0,$$

$$\frac{\partial u_2'}{\partial z'} + \frac{\partial u_3'}{\partial x_1'} = 0,$$

and from the kinematic condition (2.38), again, with (3.36a, b), we have only

$$u_3' = 0, \quad \text{at } z' = 1. \tag{3.38a}$$

As a consequence, at $z' = 1$ we obtain two conditions:

$$\frac{\partial u_1'}{\partial z'} = 0, \tag{3.38b}$$

and

$$\frac{\partial u_2'}{\partial z'} = 0. \tag{3.38c}$$

Finally, from the second, divergence free condition, $\nabla' \cdot \mathbf{u}_{\mathrm{RB}} = 0$, for the velocity vector in the shallow convection model equation (3.25b), and condition (3.38a–c), we obtain the following two upper, free-surface conditions at $z' = 1$:

$$w_{\mathrm{RB}} = 0, \tag{3.39a}$$

and

$$\frac{\partial^2 w_{\mathrm{RB}}}{\partial z'^2} = 0 \tag{3.39b}$$

for the shallow convection model equations (3.25a–c).

The three conditions (3.37) for θ_{RB} and (3.39a, b) for w_{RB} at $z' = 1$, replace the condition $\mathbf{u}_{\mathrm{RB}} = 0$ and $\theta_{\mathrm{RB}} = 1$ at $z' = 1$, written in (3.25d), when we consider for the RB equations (3.25a–c) a rigid-free model problem. This rigid-free, RB, model problem is the only significant limiting approximate problem, emerging rationally from the full Bénard exact problem (heated from below) when we take into account a deformable free surface.

In Chapter 4, devoted to the full Bénard thermal convection problem, heated from below, we give a complete account of the Weber, Biot and Marangoni effects. We only observe here that an explicit and detailed account of boundary conditions – especially at the upper, deformable free-surface – is indeed a matter of prime importance. A rigid conducting surface behaves in a drastically different way from a free and insulating surface. The more significant (but difficult) case being, obviously, the thermal interaction of the fluid layer with the eventual boundary.

Concerning the Biot effect, the formal limit Biot $\downarrow 0$, roughly, corresponds to the extreme situation of a perfectly conducting boundary. In the case of an upper surface open to the air – a *free-surface* – from the thermal point of view the exchange of energy is affected by means of radiation, conduction, and convection. These three phenomena together can be accounted for by a so-called Robin condition that merely reduces to condition (2.46) or (2.47), or else (2.51). We observe that often the condition at a free surface for the temperature is written under the hypothesis, $T_d = T_A$, at the free surface – between the passive air and the liquid the continuity of the temperature distribution is assumed but this seems rather a non-realistic hypothesis! As in [9] the more realistic upper, free surface condition is linked with the continuity of the heat flux across the free surface.

Quantitative, as well as qualitative, differences in behavior are to be expected between the two extreme cases (Biot $\downarrow 0$ and Biot $\uparrow \infty$) of highly conducting and insulating boundaries – in the former a fluctuation of temperature carried to the boundary soon relaxes through the exterior ambient air and, for a discussion from the physicist's point of view, see the survey paper by Normand et al. [19, pp. 597–598].

On the other hand, the quantity β_s (> 0), which is defined as the negative of the vertical conduction temperature gradient, $-\mathrm{d}T_s(z)/\mathrm{d}z$, that would appear in a purely conductive state (since in the pure heat conducting state, the temperature at the upper surface is uniform), there is no ambiguity in determining experimentally

$$\Delta T = d\beta_s = T_w - T_d$$

and, according to (1.21b) with (1.21c), we have the following formula:

$$\beta_s = \frac{(T_w - T_A)}{[(k_d/q_s) + d]}. \tag{3.40}$$

For more details, the reader is referred to Koschmieder and Prahl's 1990 work [20]. We stress again that when the fluid (the expansible liquid) is set in motion, the conduction temperature gradient β_s is no longer the temperature gradient in the liquid layer since convection induces a non-zero mean perturbative temperature at the upper fluid surface.

As a consequence (as is pertinently observed in a paper by Parmentier et al. [21]) of this 'obvious fact', the dimensionless Marangoni and Rayleigh numbers (see definitions (1.19d) and (3.2)) must be experimentally evaluated with β_s as given by (3.40), where q_s is present. It is easily seen that the 'problem with two Biot numbers' (outlined in Chapter 2) should be questioned seriously (see also the discussion in Chapter 4), but a quantitative and accurate description of this problem requires specific and likely lengthy treatments, which are outside the scope of the present book.

In general, the density ρ as a (solely) function of T, according to equation of state $\rho = \rho(T)$, can be written approximately as (see (2.39)):

$$\rho = \rho(T) = \rho_d \Big\{ 1 - \alpha(T_d)(T - T_d) + (1/2)\left[\alpha^2(T_d) - \left(\frac{\partial \alpha(T)}{\partial T} \right)_{T_d} \right] (T - T_d)^2 + \cdots \Big\}, \quad (3.41)$$

when an expansion in a Taylor's series, about a constant temperature reference (fiducial), T_d, is performed. With (3.41), the main problem concerns the influence of the term proportional to $(T - T_d)^2$, when we want to derive a second-order approximate model. In a first naive approach we can write (3.41) in the following dimensionless form:

$$\frac{\rho}{\rho_d} \equiv \rho'(\theta) = 1 - \varepsilon\theta + (1/2)\left[1 - \left(\frac{1}{\alpha_d^2} \right) \left(\frac{\partial \alpha(T)}{\partial T} \right)_{T_d} \right] \varepsilon^2 \theta^2 + \cdots. \quad (3.42)$$

With (3.42), in the second-order set of equations (3.32a–c), various new terms appear. For example, first in equations (3.32a) for $\mathbf{u}_1$ on the right-hand side, we have the following complementary term:

$$-(1/2)\left[1 - \left(\frac{1}{\alpha_d^2} \right) \left(\frac{\partial \alpha(T)}{\partial T} \right)_{T_d} \right] \mathrm{Gr}\, \theta_{\mathrm{RB}}^2 \mathbf{k}, \quad (3.43a)$$

but it is not clear: what is the value of the coefficient:

$$\left(\frac{1}{\alpha_d^2} \right) \left(\frac{\partial \alpha(T)}{\partial T} \right)_{T_d}. \quad (3.43b)$$

in (3.42), for various liquids?

However, it seems (see the paper by Perez and Velarde [22]) that this coefficient may have an effect on the second-order complementary term (3.43a), in the equation for $\mathbf{u}_1$! On the other hand, in [23] Knightly and Sather consider in an *ad hoc* manner a quadratic term in θ, in the leading-order shallow thermal convection equations. For this, according to (3.42), it is obviously necessary that

$$\varepsilon^2\left(\frac{1}{\alpha_d^2}\right)\left(\frac{\partial\alpha(T)}{\partial T}\right)_{T_d} = \varepsilon\varphi_d \tag{3.44a}$$

with

$$\varphi_d \equiv \left(\frac{\Delta T}{\alpha_d}\right)\left(\frac{\partial\alpha(T)}{\partial T}\right)_{T_d} \tag{3.44b}$$

and in such a case, instead of the RB leading-order equation (3.25a) for $\mathbf{u}_{\mathrm{RB}}$, we obtain the following shallow thermal convection model equation for $\mathbf{u}_{\mathrm{Sh}}$:

$$\frac{d\mathbf{u}_{\mathrm{Sh}}}{dt'} + \nabla'\pi_{\mathrm{Sh}} - \mathrm{Gr}\left[\theta_{\mathrm{Sh}} + \left(\frac{\varphi_d}{2}\right)\theta_{\mathrm{Sh}}^2\right]\mathbf{k} = \mathbf{\Delta}'\mathbf{u}_{\mathrm{Sh}}, \tag{3.45}$$

which is a leading-order thermal shallow convection equation for the velocity $\mathbf{u}_{\mathrm{Sh}}$, analogous of one considered in [23] – but, obviously, more investigation into this way is necessary.

Concerning the various analyses for RB convection, the reader can find recent developments in a review paper by Boenschatz et al. [24]. In a short paper by Manneville [25] on the same subject and written 100 years later (in French) for the *Journée H. Bénard, ESPCI, 25/06/2001*, the reader can find also a digest concerning various facets of RB convection, such as linear and nonlinear convection from Rayleigh to Busse, a physicist approach, transition to chaos, the Newell–Whitehead–Segel-amplitude equations approach, with various references (in particular, [26–30]). Chapter 10 in [6], is devoted entirely to 'asymptotic modelling of thermal convection (RB model) and interfacial phenomena (Marangoni effect and BM model)'.

In Chapters 5 to 7 we return carefully to three main convection model problems: RB, BM and deep-*à la* Zeytounian, and these three model convection problems are also the main subject of the discussion in Chapter 8.

I shall close this chapter, by quoting a few lines extracted from a recent paper by Paul Germain [31], which concerns very directly our above approach related to the rational obtention of the second-order thermal convection model equations. According to Germain [31]:

> ... it seems of great importance that a rational approach be adopted to make sure, for example, that terms neglected really are much smaller

> than those retained. Until this is done, and even now it is possible in part, it will be difficult to convince the detached and possibly skeptical reader of their value as an aid to understanding.

On the other hand, from [6, p. xv] we quote:

> For some time the growth in capabilities of numerical simulation in fluid dynamics will be strongly dependent on, or at least closely related to, the development of the rational modelling.

References

1. Lord Rayleigh, On convection currents in horizontal layer of fluid when the higher temperature is on the under side. *Philos. Mag. Ser. 6* **32**(192), 529–546, 1916.
2. H. Bénard, Les tourbillons cellulaires dans une nappe liquide. *Revue Gén. Sci. Pures Appl.* **11**, 1261–1271 and 1309–1328, 1900. See also: Les tourbillons cellulaires dans une nappe liquide transportant de la chaleur par convection en régime permanent. *Ann. Chimie Phys.* **23**, 62–144, 1901.
3. J. Boussinesq, *Théorie analytique de la chaleur*, Vol. II. Gauthier-Villars, Paris, 1903.
4. R.Kh. Zeytounian, Joseph Boussinesq and his approximation: A contemporary view. *C.R. Mec.* **331**, 575–586, 2003.
5. S. Chandrasekhar, *Hydrodynamic and Hydromagnetic Stability*. Clarendon Press, Oxford, 1961. See also Dover Publications, New York, 1981.
6. R.Kh. Zeytounian, *Asymptotic Modelling of Fluid Flow Phenomena*. Fluid Mechanics and Its Applications, Vol. 64, Kluwer, Dordrecht, 2002.
7. E.L. Koschmieder, *Bénard Cells and Taylor Vortices*. Cambridge University Press, Cambridge, 1993.
8. M.J. Block, *Nature* **178**, 650–651, 1956.
9. J.R.A. Pearson, On convection cells induced by surface tension. *J. Fluid Mech.* **4**, 489, 1958.
10. E. Guyon, J-P. Hulin, L. Petit and C.D. Mitescu, *Physical Hydrodynamics*. Oxford University Press, Oxford, 2001.
11. R.Kh. Zeytounian, The Bénard–Marangoni thermocapillary instability problem: On the role of the buoyancy. *Int. J. Engrg. Sci.* **35**(5), 455–466, 1997.
12. R.Kh. Zeytounian, The Bénard–Marangoni thermocapillary-instability problem, *Phys. Uspekhi* **41**(3), 241–267, 1998 [English edition].
13. P.G. Drazin and W.H. Reid, *Hydrodynamic Instability*. Cambridge University Press, Cambridge, 1981.
14. R.Kh. Zeytounian, *Theory and Applications of Viscous Fluid Flows*. Springer-Verlag, Berlin/Heidelberg, 2004.
15. A. Pellew and R.V. Southwell, *Proc. Roy. Soc.* **A176**, 312–343, 1940.
16. D.D. Joseph, *Stability of Fluid Motions*, Vol. II. Springer, Heidelberg, 1976.
17. R. Hills and P. Roberts, *Stab. Appl. Anal. Continuous Media* **1**, 205–212, 1991.
18. B. Straughan, *The Energy Method, Stability, and Nonlinear Convection*. Appl. Math. Sci. Vol. 91. Springer-Verlag, New York, 1992.

19. C. Normand, Y. Pomeau and M.G. Velarde, Convective instability: A physicist's approach. *Rev. Modern Phys.* **48**(3), 581–624, 1977.
20. E.L. Koschmieder and S. Prahl, *J. Fluid Mech.* **215**, 571, 1990.
21. P.M. Parmentier, V.C. Regnier and G. Lebond, Nonlinear analysis of coupled gravitational and capillary thermoconvection in thin fluid layers. *Phys. Rev. E* **54**(1), 411–423, 1996.
22. R. Perez Cordon and M.G. Velarde, *J. Physique* **36**(7/8), 591–601, 1975.
23. G.H. Knightly and D. Sather, *Stability of Cellular Convection, Archive for Rational Mechanics and Analysis* **97**(4), 271–297, 1987.
24. E. Boenschatz, W. Pesch and G. Ahlers, *Ann. Rev. Fluid Mech.* **32**, 708–778, 2000.
25. P. Manneville, Convection de Rayleigh–Bénard. *Journée H. Bénard, ESPCI, Paris* **25/06**, 2001.
26. F.H. Busse, Transition to turbulence in thermal convection. In: *Convective Transport and Instability Phenomena*, J. Zierep, H. Oertel (Eds.). Braun, Karlsruhe, 1982.
27. A.C. Newell, Th. Passot and J. Legat, *Ann. Rev. Fluid Mech.* **25**, 399–453, 1993.
28. P. Bergé, *Nucl. Phys. B* (Proc. Suppl.) **2**, 247–258, 1987.
29. Y. Pomeau and P. Manneville, *J. Phys. Lettr.* **40**, L-610, 1979.
30. P. Manneville, *J. Physique*, 44, 759-765, 1983.
31. P. Germain, The 'new' mechanics of fluids of Ludwig Prandtl. In: *Ludwig Prandtl, ein Führer in der Strömungslehre*, G.E.A. Meier (Ed.), pp. 31–40. Vieweg, Braunschweig, 2000.

Chapter 4
The Bénard (1900, 1901) Convection Problem, Heated from Below

4.1 Introduction

In this chapter, we take into account the influence of a deformable free surface and, as a consequence, we revisit the mathematical formulation of the classical problem describing the Bénard instability of a horizontal layer of fluid, heated from below, and bounded by an upper deformable free surface. Because the deformation of the free surface, subject to a temperature-dependent surface tension, is taken into account in the full Bénard convection problem, heated from below, we have not specified this convection as being a 'thermal convection'.

Indeed, in the full starting Bénard problem, heated from below, when we take into account the influence of a deformable free surface, subject to a temperature-dependent surface tension, the fluid being an expansible liquid, it is necessary to take into account, simultaneously, four main effects. Namely:

(a) the conduction adverse temperature gradient (Bénard) effect in motionless steady-state conduction temperature,
(b) the temperature-dependent surface tension (Marangoni) effect,
(c) the heat flux across the upper, free surface (Biot) effect, and
(d) the buoyancy (Archimedean–Boussinesq) effect arising from the volume (gravity) forces.

Particular attention is also paid to the approximate form of the equation of state for a *weakly* expansible liquid.

In his two pioneering papers [1], Henri Bénard (1900, 1901) considered a very thin layer of fluid, about a millimeter in depth, or less, standing on a levelled metallic plate maintained at a constant temperature. The upper surface was usually free and, being in contact with the air, was at a lower temperature. Bénard experimented with several liquids of differing physical constants. He was particularly interested in the role of viscosity (in fact, liquid with a high viscosity and a low Reynolds number, which is linked with the lubrication approximation).

In all cases, Bénard found that when the temperature of the lower plate was gradually increased, at a certain instant the layer became reticulated and revealed its dissection into cells. Concerning the experiments on the onset of (thermal) instability in fluids, see [2, pp. 59–75], and also the more recent book by Koschmieder [3].

More precisely, the Bénard cells are primarily induced (in a very thin layer of expansible liquid) by the temperature-dependent surface tension gradients resulting from temperature variations on the deformable free surface. The corresponding instability phenomenon is usually known as the Bénard–Marangoni (BM) thermocapillary instability. The first scientist whose works enlightened the way to our understanding of surface tension gradient-driven flows was Carlo Marangoni (1865, 1871), in [4], who was known to have lively exchanges with Joseph Plateau (1849, 1873), see [5].

Unfortunately, the Bénard cells phenomenon was confused (over the course of many years) with the well-known Rayleigh–Bénard (RB) buoyancy driven instability according to Rayleigh's (1916) interpretation, via the so-called Boussinesq approximation (1903), and assuming (as in Chapter 3) that the fluid is confined between two planes, the influence of the deformation of a free surface being completely eliminated! Indeed, the RB instability appeared in situations when the liquid layer, with a substantial thickness, is bounded by a flat (non-deformable) upper, free surface, the buoyant volume (Archimedean) forces being the main driving operative effect. See in [6], *Zeytounian's alternative* which is quoted in Section 3.1.

The inappropriateness of Lord Rayleigh's (1916) model to Bénard's experiment was not adequately explained until Pearson in 1958 [12] showed (in an *ad hoc* linear theory) that, rather than being a buoyancy driven flow, Bénard cells are the consequence of a temperature-dependent surface tension. However, we observe that, two years before Pearson, in 1956, Block [8] in a short (2 pages) paper showed that: 'tension [is] the cause of Bénard cells and surface deformation in a liquid film' and 'when a liquid film (about 0.08 cm thick) with a free surface was cooled at its base, cellular patterns more regular than Bénard cells were observed'.

Of course in practice, usually, both the buoyancy effect (in the liquid layer) and the temperature-dependent surface tension effect (on a free deformable surface) are operative, so it is natural to ask: how are the two effects coupled?

Along this line, Nield [7] combined both mechanisms into a single (rather simple) analysis and found that: 'as the depth of the liquid layer decreases, the surface tension mechanism becomes more dominant and when the depth of the layer is less than 0.1 cm the buoyancy effect can safely be neglected for most liquids'.

In reality, all the above assertions are true only in the *leading order* for a weakly expansible liquid, in the framework of an *approximate* rational theory and *asymptotic* modelling.

This role of surface tension cannot be explained by existing theories in which the free surface is assumed to be flat – non-deformable – and only in 1997, via a coherent, asymptotic modelling [9] was I able to prove consistently my 'alternative' [6], which is based on the value of the squared Froude number

$$\mathrm{Fr}_d^2 = \frac{(\nu_d/d)^2}{gd}, \tag{4.1}$$

where d is the thickness of the fluid layer in the motionless state, ν_d the constant kinematic viscosity (dependent on the constant temperature T_d) and g the magnitude of the gravity force.

More precisely, when $\mathrm{Fr}_d^2 = \mathrm{O}(1)$, and in such a case $d \approx 1$ mm, at the leading order we derive the BM problem and, for $\mathrm{Fr}_d^2 \ll 1$, at the leading order, we derive the RB problem, with an upper bound for the thickness (d), when we do not take into account (at the leading order) the term with the viscous dissipation function in the full energy equation written for the dimensionless temperature θ of the liquid.

On the other hand, in our asymptotic rational analysis, the main small (expansibility) parameter (where $\Delta T = \mathrm{d}\beta_s$, with β_s the adverse conduction temperature gradient),

$$\varepsilon = \alpha_d \Delta T, \tag{4.2}$$

is linked with the weakly thermal expansion (α_d is the cubic dilatation at temperature T_d) of the liquid and the classical Grashof number,

$$\mathrm{Gr} = \frac{\varepsilon}{\mathrm{Fr}_d^2}, \tag{4.3}$$

a ratio of ε to Fr_d^2, is in fact a similarity parameter (the ratio of two small parameters). We note that in some works devoted to thermal convection the non-dimensional expansibility, small parameter ε is called a 'Boussinesq' number and

$$\mathrm{Ga} = \frac{\mathrm{Pr}}{\mathrm{Fr}_d^2} \equiv \frac{(\nu_d/\kappa_d)gd}{(\nu_d/d)^2} = \frac{gd^3}{\nu_d \kappa_d} \tag{4.4}$$

is a 'Galileo' number, where Pr is the Prandtl number – the ratio of kinematic viscosity ν_d and thermal diffusivity κ_d. In such a case the Rayleigh number

$$\mathrm{Ra} = \frac{\alpha_d \Delta T g d^3}{\nu_d \kappa_d} \equiv \varepsilon\,\mathrm{Ga} \tag{4.5}$$

appears as the product of a Boussinesq number $\alpha_d \Delta T$, with a Galileo number Ga, defined by (4.4).

As a consequence, if a situation is considered for which the above Rayleigh number (4.5) is of order unity or higher, the use of the Boussinesq approximation (and in such a case the driving force in thermal convection is the buoyancy term) implies $\mathrm{Ga} \gg 1$, because $\varepsilon \ll 1$.

Typical values of the Prandtl number Pr are the following:

- for water = 6; silicone oil = 5×10^4; mercury = 0.026; air = 0.7;
- for liquids used for experiments on Bénard instabilities = 5 or more, especially for highly viscous oils;
- for liquid metals = 10^{-2}–10^{-3};
- for gases ≈ 1.

For a consistent derivation of the shallow convection (Boussinesq) equations governing the RB problem (where, in fact, $\rho = \rho(T)$) it is necessary to take into account a second similarity relation (see (2.30)) between the small, expansibility parameter ε and the small isothermal compressibility coefficient $\ni$ (defined by (2.29)) such that

$$\frac{\varepsilon^2}{\ni} \equiv K_0 = \mathrm{O}(1). \tag{4.6}$$

Some aspects of the interfacial phenomena – Marangoni and Biot effects – have been discussed in [6] and, more recently, in [10].

From the above brief discussion, we see that the RB and BM instability model problems are both limiting cases of the full classical Bénard, heated from below, instability problem, when we assume that the liquid is weakly expansible, the Froude number being small or O(1). This point of view allows us to formulate both these convection problems in a coherent way and, if necessary, to derive the corresponding second-order model approximate problems which take into account the influence of a weak expansibility and viscous dissipation of the liquid – a *third type of convection*, named 'deep convection' (DC) in [17], emerges also (see Chapter 6) from this full classical Bénard instability problem, heated from below. We observe, for instance, that our consistent asymptotic approach gives (unexpectedly) the possibility to obtain, in the framework of the RB problem, a partial differential equation for deformation of the free surface (which depends on a Weber number assumed to be large and linked with the constant part of the surface tension, see (1.28a, b)) via the dimensionless pressure at an upper non-deformable free surface which is a known function, when the RB model problem is resolved.

In practice, the first fundamental effect in the full Bénard thermal instability problem is strongly related to the definition of the adverse conduction temperature gradient β_s in motionless steady-state conduction temperature $T_s(z)$, where z is the vertical (to $z = 0$, lower heated horizontal solid plane) coordinate in the direction of the unit vector $\mathbf{k}$ (see Section 4.2). For the definition of β_s, via the data of the Bénard problem, it seems necessary (as realized in Chapter 1, see (1.20)–(1.21)) to introduce a constant conduction heat transfer coefficient q_s and write the corresponding Newton's cooling law (see equation (4.12b)) for the motionless conduction-steady (function only of z) temperature, $T_s(z) = T_w - \beta_s z$, at $z = d$, which places the mean, flat, position of the free surface in a steady-state motionless conduction regime. In Section 4.4, we consider another scenario related to the dimensionless temperature (1.17c)/(2.48),

$$\Theta = \frac{(T - T_A)}{(T_w - T_A)},$$

the temperature-dependent surface tension being modeled by the linear approximation (2.52) with (2.53)

$$\sigma(T) = \sigma(TA) - \gamma_\sigma(T - T_A), \quad \text{with} \quad \gamma_\sigma = -\left[\frac{d\sigma(T)}{dT}\right] \quad \text{at } T = T_A,$$

in a convective regime; this model for temperature-dependent surface tension has been used in various recent papers.

The lower heated plate temperature $T_w = T_s(0)$ being a given data, in this case, the adverse conduction temperature gradient β_s appears as a known function of $(T_w - T_A)$, where T_A $(< T_w)$ is the known temperature of the passive air far above the free surface when the conduction constant Biot number

$$\mathrm{Bi}_s = \frac{d q_s(T_d)}{k_d}, \tag{4.7}$$

with a constant thermal conductivity k_d, and the constant $q_s(T_d)$, is known (see in Chapter 1, (1.21b) and (1.22)). Here, we use the reference temperature

$$T_s(z = d) \equiv T_d = T_w - \beta_s d, \tag{4.8}$$

and it is clear that β_s is always different from zero in the framework of the Bénard instability problem heated from below, and as a consequence in what follows it is necessary to have in mind that, always, the conduction Biot number, defined by (4.7), is also different from zero: $\mathrm{Bi}_s \neq 0$!

This trivial remark has, in fact, an important consequence and shows (again) that it is necessary, without fail, to work with the two Biot numbers – the first being the above constant conduction Biot number Bi_s, defined by (4.7). The second (in general variable) convective Biot number, being strongly linked with the formulation of the thermocapillary BM convective model problem, where again Newton's cooling law is used to obtain a boundary condition on a free deformable surface $z = H(t, x, y)$ for the dimensionless temperature (see Section 2.5)

$$\theta = \frac{(T - T_d)}{\Delta T}, \quad \text{with } T_w - T_d \equiv \Delta T, \tag{4.9}$$

(or Θ), but with a (variable, second) heat transfer convection coefficient, q_{conv}; see for instance (1.23b).

We again observe that in the framework of an approach *à la* Davis [11], but with two Biot numbers, the *correct* upper boundary condition for the dimensionless temperature θ (defined by the relation (4.9)) is the dimensionless condition (2.46), where $\mathrm{Bi}_{\text{conv}}$ is a non-constant convective Biot number! Only (2.46) is the correct condition for θ, contrary to the Davis condition (2.47) where Bi_s and $\mathrm{Bi}_{\text{conv}} = dq_{\text{conv}}/k_d$ are replaced by a single B (surface Biot number, with a unit thermal surface conductance h instead of our q_{conv}).

Obviously, as a first tentative approach, we can assume that $\mathrm{Bi}_{\text{conv}}$ is a function of temperature T of the liquid or else a function of the full thickness $H(t, x, y)$ of the deformable thin film. In such a case, it seems judicious to assume that the associated constant conduction Biot number Bi_s is (when the deformation of the free surface is absent) respectively, a function of T_d, $\mathrm{Bi}_s(T_d)$, or else a function of d, $\mathrm{Bi}_s(d)$. As has been noted in Section 2.5, the assumption concerning the necessity to introduce a variable convective heat transfer coefficient is present in the pioneering paper by Pearson [12], where a small disturbance analysis is carried out. On the other hand, for a thin layer with strong surface deformation (nonlinear case) and, especially, in the framework of the derivation of the lubrication equation, via the long-wave approximation (see Sections 7.3 and 7.4), a dependence of q_{conv} on the full thickness $H(t, x, y)$ seems reasonable (see, for instance, the pertinent paper by Vanhook et al. [13]). It is also important to observe that we have assumed above that the conduction heat transfer coefficient $q_s = \text{const}$ and, as a consequence, $\mathrm{Bi}_s = \text{const}$, because the reference temperature $T_d = \text{const}$ is uniform along the flat free surface, $z = d$, in the conduction regime linked with the motionless steady-state temperature $T_s(z) = T_w - \beta_s z$. Obviously, as has been stressed in Section 3.7, this is no longer true in thermocapillary convective instability because the dimensionless temperature θ, at the upper deformable free surface $z = H(t, x, y)$, varies from point to point. The heat transfer convective coefficient q_{conv} or its dimensionless expression, the *convection Biot*, Bi_{conv}, number, is then not a constant (see, again, the discussion in [14,15]). Therefore, it seems necessary to work with these two Biot numbers, Bi_s and Bi_{conv}, according to the upper, free-surface condition (2.46) for θ.

This leads to some modifications in the formulation of the BM model problem and its various applications – but, when we identify Bi_{conv} with Bi_s (as is the case in [11]) we recover again Davis' condition, derived in 1987, and from our results it is possible to obtain again some usual results.

These common results, obtained with a single (conduction) Biot number in the linear theory, are questionable (at least from a logical point of view) for the case of a zero Biot (convection) number. The conduction (different from zero) Biot number allows us (in the case of a Bénard convection in a thin liquid layer with an upper deformable free surface) to define β_s according to (1.21b).

I do not claim that our approach is the more effective approach or that it resolves the 'two Biot numbers paradox', which deserves obviously further careful attention. But I observe that our derived boundary condition (2.46) with two Biot numbers, seems to me more appropriate and justified. In any case, our condition (2.46) is the only correct one when we use Davis's [11] 'imaginative, 1987, approach' – but without Davis's confusion, which identifies Bi_{conv} *with* Bi_s*. From another point of view, it is also possible to replace Davis's (derived in [11]) upper condition (2.47) for* θ*, by the à la Pearson condition (2.51) for dimensionless temperature* Θ*. In this upper, free surface, at* $z' = 1 + \eta h'(t', x', y')$*, dimensionless condition, the parameter* L *is, in fact, a convective Biot number,*

$$L = \left(\frac{\mathrm{d}}{k_d}\right)\left[\frac{\mathrm{d}Q(T)}{\mathrm{d}T}\right]_A. \tag{4.10}$$

Finally, as the upper, free-surface condition for Θ*, it is also adequate to use directly the condition (2.37)!*

We now devote Section 4.2 to a detailed dimensionless mathematical formulation of the full Bénard, starting, exact convection problem heated from below. In this mathematical dimensionless, starting formulation, the buoyancy (Rayleigh–Bénard) and thermocapillary (Marangoni) are the two main effects taken into account. In this starting dimensionless problem the Weber and Biot numbers are also present, as well as the Boussinesq number linked with the viscous dissipation term; the upper, free surface, separating the passive air from the liquid layer, is assumed deformable.

In Section 4.3 we will show the fundamental role played by the Froude number, in the case of a weakly expansible liquid, for a consistent derivation of two main approximate rational models: first, in the buoyancy driven shallow thermal convection model RB equations and associated upper boundary conditions, then in the thermocapillary convection BM model problem with a deformable free surface.

The role of the Boussinesq number, Bo (defined by (1.16)), in the case of the deep thermal convection model problem with viscous dissipation term, is also considered in Section 4.3. However, the case of ultra-thin films is not considered in this book (see the references at the end of Chapter 10).

Finally, in Section 4.4 the reader can find some complements and concluding remarks. First, we consider the upper, free-surface, boundary condition (at $z' = 1 + \eta h'(t'x'_1, x'_2)$) for the dimensionless temperature, $\Theta = (T - T_A)/(T_w - T_A)$, instead of θ. Then, a discussion relative to long-wave approximation used in lubrication theory is given. Finally, we consider briefly some film flows in various geometries, for example, down an inclined plane and a vertical plate, down inside a vertical circular tube, coating of a liquid film, over a substrate with topography, liquid hanging below a solid ceiling, etc.

4.2 Bénard Problem Formulation, Heated from Below

We consider the horizontal one-layer classical Bénard problem, heated from below, consisting of a (weakly) expansible viscous and heat conductor liquid bounded below by a rigid horizontal flat surface (a plate) and above by a passive gas (an ambient air having negligible density, viscosity and known constant pressure (p_A) and temperature (T_A) far from the free surface) separated by a deformable free surface. This free surface, separating the liquid layer from the passive ambient air, in conduction motionless steady conduction state coincides with the flat plate $z = d$, the thickness of the liquid layer in conduction state being d. The rigid plate $z = 0$ is a perfect heat conductor fixed at temperature T_w and the free surface (non-dimensional equation), in the convection process is

$$\frac{z}{d} \equiv z' = 1 + \eta h'(t', x'_1, x'_2), \tag{4.11}$$

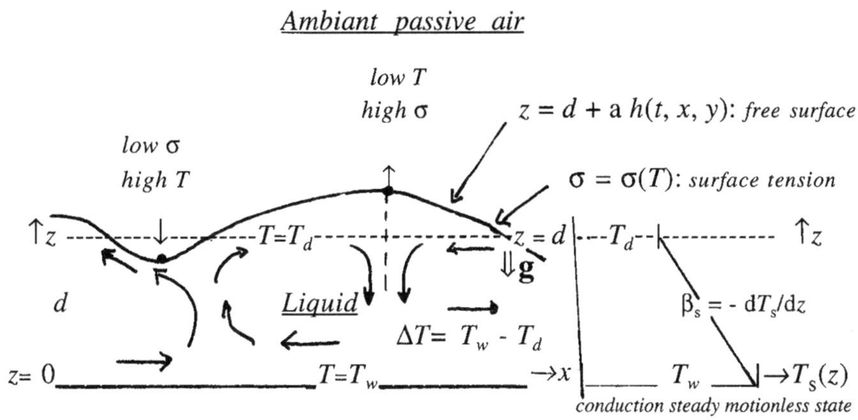

Fig. 4.1 Geometry of the full Bénard thermal convection problem.

with $\eta \equiv a/d$, the amplitude parameter of the deformable (with as a dimensionless free surface deformation function, $h'(t', x'_1, x'_2)$. The geometry of the full Bénard thermal convection problem is sketched in Figure 4.1.

The various non-dimensional quantities (denoted by a prime were introduced in Section 3.3 by (3.6a–d)). In the case of a strong deformation of the free surface, in the nonlinear case, we assume that $\eta = \mathrm{O}(1)$. In a steady-state motionless conduction state, when the temperature is $T_s(z)$, we obtain (with $k \equiv k_d = \text{const}$) the following simple conduction problem (see, for instance, Section 4.3):

$$\frac{\mathrm{d}^2 T_s(z)}{\mathrm{d}z^2} = 0, \quad \text{with } T_s(0) = T_w, \tag{4.12a}$$

and

$$k_d \frac{\mathrm{d}T_s(z)}{\mathrm{d}z} + q_s[T_s(z) - T_A] = 0, \quad \text{at } z = d. \tag{4.12b}$$

The condition at $z = d$ being the usual Newton's cooling law, but written for the steady-state temperature $T_s(z)$ in a conduction motionless regime, when the free surface is flat, $z = d$. Thanks to Newton's cooling law (4.12b) we are able to determine the adverse temperature gradient β_s, in motionless conduction steady state, via the constant (at constant temperature T_d) heat transfer, conduction coefficient q_s. Namely, the solution of (4.12a) with (4.12b) is obviously of the form (mentioned in the Introduction, Chapter 1, see (1.19a) and (1.21b)),

$$T_s(z) = T_w - \beta_s z, \quad \text{with} \quad \beta_s = \left[\frac{\mathrm{Bi}_s}{(1+\mathrm{Bi}_s)}\right]\left[\frac{(T_w - T_A)}{d}\right], \tag{4.12c}$$

where the conduction Biot number Bi_s is defined by (4.7). We observe also that if, in the starting Bénard problem, the given data are respectively d, q_s, k_d (or Bi_s), T_w and T_A, then when β_s is defined by the above relation (valid in the conduction regime) we determine also the reference temperature:

$$T_s(z = d) = T_d = T_w - \beta_s d,$$

and we have also

$$\beta_s = \frac{(T_w - T_d)}{d} \equiv \frac{\Delta T}{d}, \quad \text{or} \quad \beta_s = \mathrm{Bi}_s(T_d)\left[\frac{(T_d - T_A)}{d}\right],$$

or else (4.12c) for β_s.

Since Bi_s is different from zero, in a similar manner β_s is always also different from zero. The reference temperature T_d is obviously assumed different from the air temperature T_A and in such a case at the flat surface $z = d$ a discrete jump in temperature is realized! Indeed, the conduction Biot number, $\mathrm{Bi}_s \neq 0$, defined above plays an important role in the mathematical formulation of the Bénard dimensionless problem heated from below, and appears as a Bénard conduction effect which is always operating in convection model problems – this point is rarely noted, as if the authors of various papers on film problems do not wish to raise doubts concerning this Biot problem?

Finally, in the conduction steady motionless state, instead of (4.9), we obtain for the dimensionless conduction temperature:

$$\theta_s(z') \equiv \frac{[T_s(z') - T_d]}{(T_w - T_d)} = 1 - z'. \tag{4.13}$$

Usually it is assumed that for the temperature-dependent surface tension $\sigma = \sigma(T)$, we have the following approximate equation of state (according to (1.17a, b)):

$$\sigma(T) = \sigma(T_d) - \gamma_\sigma(T - T_d), \quad \text{where} \quad \gamma_\sigma = -\left.\frac{\mathrm{d}\sigma(T)}{\mathrm{d}T}\right|_{T=T_d} = \text{const} > 0, \tag{4.14}$$

where T_d is present, so that there is surface flow from the hot end toward the cold end and, since the bulk fluids are viscous, they are dragged along. As a consequence bulk-fluid motion results from interfacial temperature gradient, a so-called Marangoni thermocapillary effect. With the references constant density ρ_d, viscosity ν_d, constant tension $\sigma(T_d)$ and constant tension gradient γ_σ, for our viscous liquid, we define the Marangoni and Weber numbers as

$$\mathrm{Ma} = \frac{\gamma_\sigma \beta_s d^2}{\rho_d \nu_d^2}, \qquad \mathrm{We} = \frac{\sigma(T_d) d}{\rho_d \nu_d^2}. \tag{4.15}$$

And instead of (4.14), we write

$$\frac{\sigma(T)}{\sigma(T_d)} = 1 - (\mathrm{Ma}/\mathrm{We})\theta. \tag{4.16}$$

Instead of the Weber number, according to the physicists usually a so-called crispation number (slightly 'modified', see also Section 3.3), the following formula is introduced:

$$\mathrm{Cr} = \frac{1}{\mathrm{We\,Pr}} = \frac{\rho_d \nu_d \kappa_d}{\sigma(T_d) d} \tag{4.17a}$$

where $\mathrm{Pr} = \nu_d/\kappa_d$ is the Prandtl number with κ_d the constant thermal diffusivity of the heat conductor liquid. For most liquids in contact with air, Cr is very small and as a consequence the thermocapillary effect is significant (according to (4.16)) only for a relatively large modified Marangoni number

$$Ma = \mathrm{Pr\,Ma} = \frac{\gamma_\sigma \beta_s d^2}{\rho_d \nu_d \kappa_d}. \tag{4.17b}$$

and see, for instance in [16, p. 2748], the parameter range of this crispation, Cr parameter. In fact, the modified Marangoni number, Ma, is the ratio of a driving force, due to change in the surface tension, to viscous frictional force. This driving force for the Marangoni effect acts only at the free surface of the fluid layer and, as a consequence, in the upper, free-surface, boundary condition, at $z = H(t, x, y)$, we have a complementary term (see (2.36c)):

$$\nabla_{\|}\sigma(T) = \left[\frac{\mathrm{d}\sigma(T)}{\mathrm{d}T}\right]\nabla_{\|}T, \tag{4.18}$$

where $\nabla_{\|}$ is a surface gradient defined only in the free surface by the relation (2.36a). Finally, it is important to note that the Marangoni effect, characterized by the Marangoni number Ma, is operative only when Bi_s is different from zero, or when the Bénard conduction effect is really taken into account (when at the flat surface, $z = d$, a discrete jump in temperature is realized); this follows from the relation (4.15) for Ma where β_s is present.

Below, the system of three equations (2.32a–c), derived in Section 2.4, are our starting exact full equations, with dimensional quantities. Namely:

$$\frac{\mathrm{d}\rho}{\mathrm{d}t} + \rho(\nabla \cdot \mathbf{u}) = 0, \tag{4.19a}$$

$$\rho \frac{\mathrm{d}u}{\mathrm{d}t} + \nabla p = \rho \mathbf{f} + \mu_d[\Delta \mathbf{u} + (1/3)\nabla(\nabla \cdot \mathbf{u})], \tag{4.19b}$$

$$\rho C_p \frac{\mathrm{d}T}{\mathrm{d}t} - \alpha T \frac{\mathrm{d}p}{\mathrm{d}t} = \Phi + k_d \Delta T, \tag{4.19c}$$

with for density ρ, according to (2-32d), the following approximate equation of state is used:

$$\rho = \rho_d \left\{ 1 - \varepsilon \left[\frac{(T - T_d)}{\Delta T} \right] + \left(\frac{1}{K_0} \right) \varepsilon^2 \left[\frac{(p - p_A)}{g d \rho_d} \right] \right\}, \tag{4.19d}$$

As upper boundary conditions (with dimensional quantities), at the free surface, we have the following set of four conditions, according to (2.38)–(2.42a–c); also see Section 2.5. Namely, at $z = d + ah(t, x_1, x_2) \equiv H(t, x_1, x_2)$ the kinematic condition is

$$\frac{\mathrm{d}}{\mathrm{d}t}[z - H(t, x, y)] = 0, \tag{4.20a}$$

and the three jump conditions for the stress tensor are:

$$p = p_A + \mu_d \left[\left(\frac{\partial u_i}{\partial x_j} + \frac{\partial u_j}{\partial x_i} \right) n_i n_j - (2/3)(\nabla \cdot \mathbf{u}) \right] + \sigma(T)(\nabla_{\parallel} \cdot \mathbf{n}); \tag{4.20b}$$

$$\mu_d \left(\frac{\partial u_i}{\partial x_j} + \frac{\partial u_j}{\partial x_i} \right) \mathbf{t}_i^{(1)} n_j = \left[\frac{\mathrm{d}\sigma(T)}{\mathrm{d}T} \right] \mathbf{t}_i^{(1)} \left(\frac{\partial T}{\partial x_i} \right); \tag{4.20c}$$

$$\mu_d \left(\frac{\partial u_i}{\partial x_j} + \frac{\partial u_j}{\partial x_i} \right) \mathbf{t}_i^{(2)} n_j = \left[\frac{\mathrm{d}\sigma(T)}{\mathrm{d}T} \right] \mathbf{t}_i^{(2)} \left(\frac{\partial T}{\partial x_i} \right). \tag{4.20d}$$

For the non-dimensional temperature θ, given by (4.9), we write (see (2.46)) our upper dimensionless boundary condition at $z' = 1 + \eta h'(t', x', y') \equiv H'(t', x', y')$, as

$$\frac{\partial \theta}{\partial n'} + \left\{ \frac{\mathrm{Bi}_{\mathrm{conv}}}{\mathrm{Bi}_s(T_d)} \right\} [1 + \mathrm{Bi}_s(T_d)\theta] = 0, \tag{4.20e}$$

where the convective, $\mathrm{Bi}_{\mathrm{conv}}$, Biot number is, in general, a non-constant parameter and $Q_0 = 0$.

In equation (4.19c) the viscous dissipation term Φ is given by

$$\Phi = 2\mu_d\{\mathbf{D}(\mathbf{u}) : \mathbf{D}(\mathbf{u}) - (1/3)(\nabla \cdot \mathbf{u})^2\}, \tag{4.21a}$$

and the term $\rho \mathbf{f} \equiv -\rho g \mathbf{k}$, the single body force being the gravity force. On the other hand, the kinematic condition (4.20a), written at $z = H(t, x, y)$, is rewritten as

$$\mathbf{u}\cdot\mathbf{k} = \left[\frac{\partial}{\partial t} + u_1\frac{\partial}{\partial x_1} + u_2\frac{\partial}{\partial x_2}\right] H(t, x_1, x_2). \qquad (4.21b)$$

In upper boundary condition (4.20b), for the pressure difference, $p - p_A$, the term, $\nabla_{\|}\cdot\mathbf{n}$ is given by (2.43a), with (2.43b–d). The outward normal unit vector $\mathbf{n}$ is given by (2.44c) and the two unit tangent vectors $\mathbf{t}^{(1)}$ and $\mathbf{t}^{(2)}$ in above (4.20c, d), parallel to the upper, free surface $z = H(t, x, y)$, are given by (2.44a, b).

We observe also that, below as in Section 3.3, the dimensionless quantities (see (3.6a, b)) are denoted by a prime, and on the other hand the dimensional Cartesian coordinates are, in various parts of this book, designated by

$$x_1 \equiv x, \quad x_2 \equiv y \quad \text{and} \quad x_3 \equiv z.$$

In, upper, free-surface condition (4.20e) for θ, we have (see (3.6a, b) and (2.43b)), in dimensionless form:

$$\frac{\partial\theta}{\partial n'} = \nabla'\theta\cdot\mathbf{n}' = \left(\frac{1}{N'}\right)^{1/2}\left\{\frac{\partial\theta}{\partial z'} - \eta(\mathbf{D}'\theta\cdot\mathbf{D}'h')\right\}, \qquad (4.22a)$$

where, in dimensionless form,

$$N' = 1 + \eta^2\mathbf{D}'^2h', \quad \mathbf{n}' = \left(\frac{1}{N'}\right)^{1/2}\left(-\eta\frac{\partial h'}{\partial x'}; -\eta\frac{\partial h'}{\partial y'}; +1\right), \qquad (4.22b)$$

$$\nabla' = \left(\frac{\partial}{\partial z'}\right)\mathbf{k} + \mathbf{D}', \quad \text{with} \quad \mathbf{D}' = \left(\frac{\partial}{\partial x'}; \frac{\partial}{\partial y'}\right). \qquad (4.22c)$$

In a linear theory when

$$\theta = 1 - z' + \eta\theta'(t', x', y', z') + \cdots, \qquad (4.23a)$$

and, when $\mathrm{Bi}_{\mathrm{conv}}$ is assumed a function of $H'(t', x', y') \equiv 1 + \eta h'(t', x', y')$, in a convective regime, we have the approximate relation

$$\mathrm{Bi}_{\mathrm{conv}} \equiv \mathrm{Bi}(H'(t', x', y')) = \mathrm{Bi}(1) + \eta\Lambda(H' \equiv 1)h', \qquad (4.23b)$$

where

$$\Lambda = \frac{\mathrm{dBi}}{\mathrm{d}H'}, \qquad (4.23c)$$

and with $\mathrm{Bi}(1) \equiv \mathrm{Bi}_s$, we derive, instead of a full, upper, free-surface condition (4.20e) for θ, the following linear (at the order $\mathrm{O}(\eta)$ and written at $z' = 1$), upper, free-surface condition for θ' in the convective regime

$$\frac{\partial \theta'}{\partial z'} + \mathrm{Bi}_s(\theta' - h') + \left(\frac{1}{\mathrm{Bi}_s}\right)\Lambda(H' \equiv 1)h' = 0, \quad \text{at } z' = 1. \tag{4.23d}$$

In the above linearized, free-surface, boundary condition for θ', (4.23d), we have again Bi_s, but not a Biot number related to the convection regime, in front of the term with $(\theta' - h')$, as this is the case usually (see, for instance, [16, p. 2747]). On the other hand, we have a complementary term proportional to h' which emerges from the variability of the convection Biot number! An accurate linearization shows easily that, in fact, it is not possible to work with two constant Biot numbers, at least with the choice $Q_0 = 0$.

Indeed, if $\mathrm{Bi}_{\mathrm{conv}} \equiv \mathrm{Bi}^0_{\mathrm{const}}$ is a constant, then the linearization is possible only when we assume that

$$Q_0 = k\beta_S \left[1 - \left(\frac{\mathrm{Bi}^0_{\mathrm{conv}}}{\mathrm{Bi}_s}\right)\right], \tag{4.23e}$$

and in such a case, and only for this case, we recover a linear, upper, free-surface condition for θ', *à la* Takashima [16],

$$\frac{\partial \theta'}{\partial z'} + \mathrm{Bi}^0_{\mathrm{conv}}(\theta' - h') = 0, \quad \text{at } z' = 1, \tag{4.23f}$$

where, as a constant coefficient, in front of $(\theta' - h')$ the constant convective Biot number $\mathrm{Bi}^0_{\mathrm{conv}}$ appears – obviously, in such a case, the results of Takashima [16] are consistent when $\mathrm{Bi}^0_{\mathrm{conv}} = 0$, but not with $\mathrm{Bi}_s = 0$!

Obviously, with the Davis' approach [11], the classical linear theory, *à la* Takashima [18], with a (single conduction) Biot number equal to zero seems questionable. Our approach above, which gives (4.23f), explains clearly the 'zero (convective) Biot number case' in linear theory.

It is necessary (in a simple case, for instance) to assume that the convection (in a convection regime, with a deformable free surface) Biot number, $\mathrm{Bi}_{\mathrm{conv}}$, is a function of the full thickness of the liquid layer, $H' \equiv 1 + \eta h'(t', x', y')$, or a function of dimensionless temperature, $\theta(t', x', y', z')$, or else (at least) that $\mathrm{Bi}_{\mathrm{conv}}$ is a function of the small parameter η, $\mathrm{Bi}_{\mathrm{conv}} = B(\eta)$ – especially in the linear theory. It is important to note, also we are not concerned here with a thorough analysis of the mechanism of heat transfer to and from the liquid layer, though these matters become relevant in the investigations of any particular physical phenomenon. It must be made clear that the evaluation of $\mathrm{Bi}(H')$ in condition (4.23b) or $\Lambda(H' \equiv 1)$ in linear condition (4.23d), in any physically observed circumstances is not necessarily easy – it is, however, a separate problem.

As observed by Pearson [12], the introduction of a different heat transfer coefficient, at least in the classical form of Newton's cooling law, is of crucial importance and by means of a suitable choice many physical interfacial phenomena may be very reasonably idealized. Obviously, the aim in this account is not to provide an exhaustive description of these phenomena and their relevant idealizations, but rather to provide a general treatment that illustrates the fundamental surface tension mechanism and comprehends its many realizations.

In particular, the evaluation of the role of the second variable, $\mathrm{Bi}(H')$, convective Biot number on the lubrication equation (see, for instance, Section 4.4) is an interesting problem.

Finally, another advantage of the introduction of a second variable convective Biot number is the clear distinction between the Biot and Marangoni effects in the thermocapillary convection problem.

For me it is clear that it is necessary to strictly observe, first, at least in a fundamental modelization of a physical phenomena (such as convection in fluids), the rigor and consistency of the elaborated theory and, above all, not to 'adapt' this theory for an eventual approximate evaluation of the considered physical phenomena – this is obviously often a difficult challenge but so fruitful in consequences!

It is necessary to add to the above equations (4.19a–c), with (4.19d) and (4.21a), and upper conditions (4.20a–e), the following condition for $\mathbf{u}$ and θ at $z' = 0$:

$$\mathbf{u} = 0 \quad \text{and} \quad \theta = 1. \tag{4.24}$$

We observe also that, in the approximate equation of state (4.19d), for the density ρ, the constant $K_0 = \mathrm{O}(1)$ is given by the following relation (when we use the relation (2.27), and definitions (1.10a), (2.29)):

$$K_0 = \frac{C_{vd}(\gamma - 1)(\Delta T)^2}{g d T_d}. \tag{4.25a}$$

As a consequence, we obtain the following constraint for ΔT:

$$\Delta T \approx \left[\frac{g d T_d}{(\gamma - 1) C_{vd}}\right]^{1/2}, \tag{4.25b}$$

when we assume that

$$K_0 = \mathrm{O}(1).$$

For the derivation of not only leading-order approximate equations, from the above starting exact system for the full Bénard convection problem, heated from below, but also companion second-order approximate equations, a careful inspection shows that it is sufficient to use in exact equation (4.19c) for C_p and α the relations (3.9a, b) with (3.10a, b).

Below, in the so-called 'dominant' dimensionless Bénard problem, we take into account only the terms which are necessary for a rational and consistent derivation of these leading and second-order equations. The dimensionless quantities (with the prime) are given by (3.6a–d), but in relation (3.6c), for π, instead of $p - p_d$ we write obviously $p - p_A$, because we take the existence of a deformable upper, free surface into account.

First, with the approximate equation of state (4.19d), instead of the continuity equation (4.19a) we can write the following *dominant dimensionless equation of continuity*:

$$\nabla' \cdot \mathbf{u}' = \varepsilon \frac{\mathrm{d}\theta}{\mathrm{d}t'}, \tag{4.26a}$$

the term $(\varepsilon^2/K_0)[\mathrm{Fr}_d^2 \mathrm{d}\pi/\mathrm{d}t' - u_3']$, on the right-hand side (see (4.19d)) being neglected as a higher term, even for a second-order approximate continuity equation.

Then, instead of equation (4.19b) for $\mathbf{u}$, we obtain for the non-dimensional velocity vector $\mathbf{u}'$, as a dominant dimensionless equation:

$$[1 - \varepsilon\theta]\frac{\mathrm{d}\mathbf{u}'}{\mathrm{d}t'} + \nabla'\pi - \left(\frac{\varepsilon}{\mathrm{Fr}_d^2}\right)\left[\theta - \left(\frac{1}{K_0}\right)\varepsilon(1 - z')\right]\mathbf{k}$$
$$= \mathbf{\Delta}'\mathbf{u}' + (1/3)\varepsilon\nabla'\left(\frac{\mathrm{d}\theta}{\mathrm{d}t'}\right), \tag{4.26b}$$

after the cancellation of the term $(1/K_0)\varepsilon^2\pi\mathbf{k}$ (proportional to ε^2) and when we take into account that, in particular, for the derivation of the RB thermal shallow convection model problem (considered in Chapter 5) it is necessary to take into account the limiting process *à la* Boussinesq (3.22) with a fixed Grashof number $\mathrm{Gr} = (\varepsilon/\mathrm{Fr}_d^2) = \mathrm{O}(1)$.

Indeed, in the case of an RB thermal convection model, the buoyancy being the main driving force, the term

$$\left(\frac{\mathrm{Gr}}{K_0}\right)[1 - z']\mathbf{k}, \tag{4.26c}$$

appears necessarily in a second-order (terms proportional to ε) system of equations associated with the leading-order RB model. This, second-order term (4.26c) is, in fact, a trace of the influence of the pressure (in the equation of state) according to (4.19d) when instead of (3.6c) we write

$$\frac{(p - p_A)}{g d \rho_d} = \mathrm{Fr}_d^2 \pi + 1 - z'.$$

Obviously, the influence of this second-order term (4.26c) has been possible to detect, but, thanks to our rational approach, from an *ad hoc* manner one cannot reveal such a second-order term.

Before the derivation, from the 'exact' equation (4.19c), of the associated dominant (with an error proportional to ε^2) dimensionless equation for the temperature θ, we observe that the first term $\rho C_p\, \mathrm{d}T/\mathrm{d}t$ on the left-hand side of (4.19c) can be written as:

$$\left(\frac{\nu_d}{d^2}\right) \rho_d \Delta T C_{pd} (1 - \varepsilon B_d \theta) \frac{\mathrm{d}\theta}{\mathrm{d}t'},$$

where $B_d \equiv 1 + \Gamma_{pd} = \text{const}$, when we take into account the relations (4.19d), for ρ, and (3.9b) for C_p. In such a case, for the term α, in the front of the second term on the left-hand side of (4.19c), we use (3.9a) and the dominant dimensionless equation for θ associated to (4.19c) has the following form which includes terms proportional to ε:

$$[1 - \varepsilon B_d \theta] \frac{\mathrm{d}\theta}{\mathrm{d}t'} - \varepsilon\, \mathrm{Bo} \left[\left(\frac{T_d}{\Delta T} \right) + \theta \right] \left[\mathrm{Fr}_d^2 \frac{\mathrm{d}\pi}{\mathrm{d}t'} - u_3' \right]$$

$$= \left(\frac{1}{\mathrm{Pr}}\right) \Delta' \theta + (1/2\,\mathrm{Gr}) \varepsilon\, \mathrm{Bo} \left[\frac{\partial u_i'}{\partial x_j'} + \frac{\partial u_j'}{\partial x_i'} \right]^2, \qquad (4.26\mathrm{d})$$

where the term $-(2/3\,\mathrm{Gr})\varepsilon^3\, \mathrm{Bo}[\mathrm{d}\theta/\mathrm{d}t']^2$ has been neglected as a high-order term (of order ε^3, when $\mathrm{Bo} = \mathrm{O}(1)$, and of order ε^2, when $Bo \gg 1$ such that $\varepsilon\, \mathrm{Bo} = \mathrm{O}(1)$). The above system (4.26a, b, d), of three dominant dimensionless equations for the velocity $\mathbf{u}' = (u_1', u_2', u_3')$, pressure π and temperature fields θ, is very significant (with an error of $\mathrm{O}(\varepsilon^2)$) for a rational analysis and an asymptotic modelling of the Bénard full convection problem heated from below, when we assume that the considered liquid is weakly expansible, such that the expansibility parameter is the main small parameter $\varepsilon \ll 1$.

This system, (4.26a, b, d), of three dominant dimensionless equations, with ε-order terms, allows us without any doubt to derive RB thermal convection model equations and also their companion-associated second-order model equations.

However, in system (4.26a, b, d), besides this main small parameter ε, we have also three other dimensionless parameters. First, the square of the Froude number:

$$\mathrm{Fr}_d^2 = \frac{(\nu_d/d)^2}{gd},$$

which is, in particular, a function of thickness d, as $\mathrm{Fr}_d^2 = (\nu_d^2/g)/d^3$.

The second parameter is our 'Boussinesq' parameter:

$$\mathrm{Bo} = \frac{gd}{C_{pd}\Delta T},$$

and this parameter plays a decisive role in taking account of the viscous dissipation term – the last term on the right-hand side of equation (4.26d) for θ proportional to $(1/2)\mathrm{Gr}\,\varepsilon\,\mathrm{Bo}$.

Finally, the third parameter is the Prandtl number, $\mathrm{Pr} = \nu_d/\kappa_d$, which governs the relative role of the viscous (by ν_d) and thermal diffusivity (by κ_d) effects – for the various liquids it is necessary to consider the cases when $\mathrm{Pr} \ll 1$ (dominant thermal diffusivity effect) or else $\mathrm{Pr} \gg 1$ (dominant viscous effect); in Section 10.10, the reader can find some information concerning these two limit cases.

Concerning the role of Fr_d^2, for the present we observe only that it is necessary to analyze three main cases:

1. the *Boussinesq thermal case*, for the RB model problem:

$$\mathrm{Fr}_d^2 \ll 1 \quad \text{with} \quad \varepsilon \ll 1, \quad \text{such that } \mathrm{Gr} = \varepsilon/\mathrm{Fr}_d^2 = \mathrm{O}(1); \tag{4.27a}$$

2. the *incompressible thin layer case*, for the BM model problem:

$$\mathrm{Fr}_d^2 \approx 1 \Rightarrow \mathrm{Gr} \approx \varepsilon; \tag{4.27b}$$

3. the *deep layer dissipative case*, for the DC model problem

$$\mathrm{Fr}_d^2 \ll 1, \quad \varepsilon \ll 1 \quad \text{and} \quad \mathrm{Bo} \gg 1, \quad \text{with } \varepsilon\,\mathrm{Bo} = \mathrm{O}(1), \tag{4.27c}$$

but also the *ultra-thin film case*,

$$\mathrm{Fr}_d^2 \gg 1! \tag{4.27d}$$

The case linked with (4.27a) is analyzed in detail in Chapter 5; then the case with the constraint (4.27b), but also with the Marangoni, Weber and Biot effects associated with the, upper, free-surface, dimensionless boundary conditions (see below) is considered at length in Chapter 7; the case (4.27c) is analyzed in Chapter 6. The last case, with (4.27d), deserves further consideration.

A more complicated (but mainly technical) problem is the derivation of significant dominant, dimensionless, upper boundary conditions on the free surface, for leading and second-order approximate equations, from (4.21b), (4.20b–d) and (4.20e). For this, we take into account the results of Section 2.5.

First, from kinematic condition (4.21b), at an upper, free surface, we obtain for the vertical (along axis Oz') component, u'_3, of the dimensionless velocity, the following dimensionless condition:

$$u'_3 = \frac{\partial H'}{\partial t'} + u'_1 \frac{\partial H'}{\partial x'} + u'_2 \frac{\partial H'}{\partial y'}, \quad \text{on } z' = H'(t', x'_1, x'_2), \tag{4.28a}$$

where $H'(t', x'_1, x'_2) = H/d \equiv 1 + \eta h'$. Then instead of the jump condition for the difference of the pressure, $p - p_A$, (4.20b), with (4.26a) and (4.16), when we replace, $p - p_A$ by $gd\rho_d[\mathrm{Fr}_d^2 \pi + 1 - z']$, we obtain for π the following dominant, dimensionless, upper boundary condition:

$$\begin{aligned}\pi \;=\; & \frac{[H'(t', x'_1, x'_2) - 1]}{\mathrm{Fr}_d^2} + \left(\frac{\partial u'_i}{\partial x'_j} + \frac{\partial u'_j}{\partial x'_i} \right) n'_i n'_j \\ & + [\mathrm{We} - \mathrm{Ma}\,\theta](\nabla'_{\|} \cdot \mathbf{n}') - (2/3)\varepsilon \frac{\mathrm{d}\theta}{\mathrm{d}t'},\end{aligned} \tag{4.28b}$$

and then, instead of (4.20c–d), we derive the following two tangential conditions:

$$\left(\frac{\partial u'_i}{\partial x'_j} + \frac{\partial u'_j}{\partial x'_i} \right) t_i^{\prime(k)} n'_j + \mathrm{Ma}\, t_i^{\prime(k)} \frac{\partial \theta}{\partial x'_i} = 0, \tag{4-28c,d}$$

with $k = 1$ and 2.

Finally we add, for θ, the free-surface, dimensionless condition, derived from (4.20e) and written as

$$\nabla' \cdot \mathbf{n}' + \left\{ \frac{\mathrm{Bi}_{\mathrm{conv}}}{\mathrm{Bi}_s(T_d)} \right\} [1 + \mathrm{Bi}_s(T_d)\theta] = 0. \tag{4.28e}$$

All the above upper conditions (4.28b–e) are (as is the condition (4.28a)) written in the free surface, $z' = H'(t', x'_1, x'_2) \equiv 1 + \eta h'(t', x'_1, x'_2)$.

In the upper, free-surface, dimensionless conditions above we have three new parameters, We, Ma and $\mathrm{Bi}_{\mathrm{conv}}$, and usually for liquids the Weber number is a large parameter, $\mathrm{We} \gg 1$. For $(\nabla'_{\|} \cdot \mathbf{n}')$ and normal and tangential unit vectors we have the relations (2.43a), with (2.43b–d), and (2.44a–c).

The above dimensionless, dominant (with an error of $O(\varepsilon^2)$) Bénard convection problem, heated from below, (4.26a, b, d) with (4.28a–e) and the conditions, at the lower flat solid plate, $z' = 0$,

$$\mathbf{u}' = 0 \quad \text{and} \quad \theta = 1, \tag{4.29}$$

is a very complicated nonlinear problem even for a numerical computation, thus a preliminary rational analysis and asymptotic modelling will obviously be helpful!

In the next section we give some information concerning rational ways for a consistent simplification of this above formulated Bénard dominant problem and, in particular, the role played by the squared Froude (Fr_d^2) number and Boussinesq (Bo) number, in the derivation of simplified approximate models.

4.3 Rational Analysis and Asymptotic Modelling

A first important observation (see also our short discussion in Chapter 1, linked with the summary of Chapter 4) concerns the appearance of Fr_d^2 in the first term on the right-hand side of (4.28b) as denominator, the numerator ε being a small parameter! As a consequence, in the above-mentioned Boussinesq thermal convection case (4.27a), for the RB shallow convection, when $\mathrm{Fr}_d^2 \ll 1$, a singularity appears in the upper condition (4.28b) for π?

This singularity in the upper condition for π is removed only if we assume that

$$H'(t', x_1', x_2') - 1 = \eta h' \ll 1,$$

and, in such a case, it is necessary that

$$\eta \ll 1, \quad \text{because } h'(t', x_1', x_2') = O(1). \tag{4.30a}$$

In fact, the following similarity rule is assumed:

$$\frac{\eta}{\mathrm{Fr}_d^2} = \eta^* = O(1), \quad \text{when } \eta \downarrow 0 \text{ and } \mathrm{Fr}_d^2 \downarrow 0. \tag{4.30b}$$

Therefore, the RB model equations governing the thermal shallow convection problem, driven by the buoyancy force, are a *consistent limiting leading-order system* of equations, only if we assume that the deformation of the upper, free surface is negligible!

Under the constraints (4.30a,b), the above upper, free-surface, boundary conditions (4.28a–e) are written, at the leading order in the 'Boussinesq thermal shallow convection case', (4.27a), at a flat 'free surface': $z' = 1$. In addition, in the case (4.27a), the upper boundary conditions (4.28a–e) are strongly simplified. Namely, with $H' = 1 + \eta h'$, we have obviously, first, instead of (4.28a), at the leading order (when $\eta \downarrow 0$):

$$u'_3 = 0 \quad \text{at } z' = 1. \tag{4.31a}$$

On the other hand, in (4.28b), for the term $(\partial u'_i/\partial x'_j + \partial u'_j/\partial x'_i)n'_i n'_j$, we can write, with $H' = 1 + \eta h'$ and the amplitude parameter, $\eta \ll 1$:

$$\left\{\frac{2}{1+\eta^2[(\frac{\partial h'}{\partial x'_1})^2 + (\frac{\partial h'}{\partial x'_2})^2]}\right\}\left\{\frac{\partial u'_3}{\partial x'_3} - \eta\left[\left(\frac{\partial u'_1}{\partial x'_3}\right.\right.\right.$$
$$\left. + \frac{\partial u'_3}{\partial x'_1}\right)\frac{\partial h'}{\partial x'_1} + \left(\frac{\partial u'_2}{\partial x'_3} + \frac{\partial u'_3}{\partial x'_2}\right)\frac{\partial h'}{\partial x'_2}\Bigg]$$
$$+ \eta^2\left[\frac{\partial u'_1}{\partial x'_1}\left(\frac{\partial h'}{\partial x'_1}\right)^2 + \frac{\partial u'_2}{\partial x'_2}\left(\frac{\partial h'}{\partial x_2}\right)^2 + \left(\frac{\partial u'_1}{\partial x'_2}\right.\right.$$
$$\left.\left.\left. + \frac{\partial u'_2}{\partial x'_1}\right)\frac{\partial h'}{\partial x'_1}\frac{\partial h'}{\partial x'_2}\right]\right\}, \tag{4.32}$$

and when $\eta \to 0$, from (4.32) there remains only ($x'_3 \equiv z'$) the term

$$2\left(\frac{\partial u'_3}{\partial z'}\right). \tag{4.31b}$$

After that, from (4.28c), we obtain for $(\partial u'_i/\partial x'_j + \partial u'_j/\partial x'_i)t_i^{\prime(1)} n'_j$, when $\eta \to 0$, only the following two terms:

$$-(1/2)\left[\frac{\partial u'_1}{\partial z'} + \frac{\partial u'_3}{\partial x'_1}\right] \tag{4.31c}$$

and from (4.28d), we obtain for $(\partial u'_i/\partial x'_j + \partial u'_j/\partial x'_i)t_i^{\prime(2)} n'_j$, when $\eta \to 0$,

$$-(1/2)\left[\frac{\partial u'_2}{\partial z'} + \frac{\partial u'_3}{\partial x'_2}\right]. \tag{4.31d}$$

On the other hand, in (4.28c, d) we obtain also, when $\eta \to 0$ for the right-hand side the two limiting relations

$$\mathrm{Ma}\, t_i^{\prime(1)} \frac{\partial \theta}{\partial x'_i} \approx \left(\frac{\mathrm{Ma}}{2}\right)\frac{\partial \theta}{\partial x'_1} \tag{4.31e}$$

and

$$\mathrm{Ma}\, t_i'^{(2)} \frac{\partial \theta}{\partial x_i'} \approx \left(\frac{\mathrm{Ma}}{2}\right) \frac{\partial \theta}{\partial x_2'}. \tag{4.31f}$$

Finally, from (2.43a), when $\eta \to 0$, we derive in (4.28b) for the term proportional to $[\mathrm{We} - \mathrm{Ma}\,\theta]$:

$$\nabla_{\parallel}' \cdot \mathbf{n}' = -\eta \left[\frac{\partial^2 h'}{\partial x_1'^2} + \frac{\partial^2 h'}{\partial x_2'^2}\right], \tag{4.31g}$$

and instead of the convective, upper, free surface condition (4.28e), for θ we write the following dimensionless condition, with an error of ε^3 according to (4.30b) and (4.27a):

$$\frac{\partial \theta}{\partial z'} + \left\{\frac{\mathrm{Bi}_{\mathrm{conv}}}{\mathrm{Bi}_s}\right\} [1 + \mathrm{Bi}_s \theta] = 0, \quad \text{at } z' = 1, \tag{4.31h}$$

when we use (4.22a–c).

In Chapter 5, devoted to rational derivation of the model equations and upper, free-surface, boundary conditions for the shallow thermal Rayleigh–Bénard convection, when the main driving force is the buoyancy force, we take into account the above two relations (4.30a, b), which give (4.31a–h). But we work (see Section 4.4) with the dimensionless temperature Θ (first introduced in (1.17c)) instead of θ.

Now if we consider the case (4.27b) – the Marangoni, thermocapillary convection case, when $\mathrm{Fr}_d^2 \approx 1$ – then (4.30a) and (4.30b) are superfluous, because $\mathrm{Gr} \approx \varepsilon \to 0$, with $\varepsilon \to 0$. *In leading order the term with the buoyancy plays no role in the Marangoni case.* In this case, because

$$\eta = \mathrm{O}(1) \quad \text{in } H'(t', x_1', x_2') = 1 + \eta h'(t', x_1', x_2'),$$

it seems better to work with the dimensionless thickness $H'(t', x_1', x_2')$.

Thus, from the upper, free-surface, boundary condition (4.28b) we write at the leading order, when $\varepsilon \to 0$, first

$$\begin{aligned}
\pi = {} & \left(\frac{1}{\mathrm{Fr}_d^2}\right)(H' - 1) + \left(\frac{2}{N'}\right)\left\{\left(\frac{\partial u_1'}{\partial x_1'}\right)\left(\frac{\partial H'}{\partial x_1'}\right)^2 + \left(\frac{\partial u_2'}{\partial x_2'}\right)\left(\frac{\partial H'}{\partial x_2'}\right)^2\right. \\
& + \frac{\partial u_3'}{\partial x_3'} + \left(\frac{\partial u_1'}{\partial x_2'} + \frac{\partial u_2'}{\partial x_1'}\right)\left(\frac{\partial H'}{\partial x_1'}\right)\left(\frac{\partial H'}{\partial x_2'}\right) \\
& \left. - \left(\frac{\partial u_1'}{\partial x_3'} + \frac{\partial u_3'}{\partial x_1'}\right)\left(\frac{\partial H'}{\partial x_1'}\right) - \left(\frac{\partial u_2'}{\partial x_3'} + \frac{\partial u_3'}{\partial x_2'}\right)\left(\frac{\partial H'}{\partial x_2'}\right)\right\}
\end{aligned}$$

$$
-\left(\frac{1}{N'}\right)^{3/2}[\mathrm{We}-\mathrm{Ma}\,\theta]N_2'\left[\left(\frac{\partial^2 H'}{\partial x_1'^2}\right)\right.
$$

$$
\left.-2\left(\frac{\partial H'}{\partial x_1'}\right)\left(\frac{\partial H'}{\partial x_2'}\right)\left(\frac{\partial^2 H'}{\partial x_1'\partial x_2'}\right)+N_1'\left(\frac{\partial^2 H'}{\partial x_2'^2}\right)\right], \tag{4.32a}
$$

where

$$
\begin{aligned}
N' &= 1+\left(\frac{\partial H'}{\partial x_1'}\right)^2+\left(\frac{\partial H'}{\partial x_2'}\right)^2,\\
N_1' &= N'-\left(\frac{\partial H'}{\partial x_2'}\right)^2,\\
N_2' &= N'-\left(\frac{\partial H'}{\partial x_1'}\right)^2.
\end{aligned}
$$

Then, from (4.28c, d), first taking into account formula (2.44a) for the components of $t_i'^{(1)}$, we obtain

$$
\begin{aligned}
&\left(\frac{\partial u_1'}{\partial x_1'}-\frac{\partial u_3'}{\partial x_3'}\right)\left(\frac{\partial H'}{\partial x_1'}\right)+(1/2)\left(\frac{\partial u_1'}{\partial x_2'}+\frac{\partial u_2'}{\partial x_1'}\right)\left(\frac{\partial H'}{\partial x_2'}\right)\\
&\quad+(1/2)\left(\frac{\partial u_2'}{\partial x_3'}+\frac{\partial u_3'}{\partial x_2'}\right)\left(\frac{\partial H'}{\partial x_1'}\right)\left(\frac{\partial H'}{\partial x_2'}\right)\\
&\quad-(1/2)\left[1-\left(\frac{\partial H'}{\partial x_1'}\right)^2\right]\left(\frac{\partial u_3'}{\partial x_1'}+\frac{\partial u_1'}{\partial x_3'}\right)\\
&=\left(\frac{N'^{1/2}}{2}\right)\mathrm{Ma}\left[\frac{\partial\theta}{\partial x_1'}+\left(\frac{\partial H'}{\partial x_1'}\right)\frac{\partial\theta}{\partial x_3'}\right];
\end{aligned} \tag{4.32b}
$$

and, with the formula (2.44b), for the components of $t_i'^{(2)}$, which are more complicated (see [6, pp. 244, 245], where the formula (4.32c) was first used), we have

$$
\begin{aligned}
&\left(\frac{\partial u_1'}{\partial x_1'}-\frac{\partial u_2'}{\partial x_2'}\right)\left(\frac{\partial H'}{\partial x_2'}\right)\left(\frac{\partial H'}{\partial x_1'}\right)^2+\left(\frac{\partial u_2'}{\partial x_2'}-\frac{\partial u_3'}{\partial x_3'}\right)\left(\frac{\partial H'}{\partial x_2'}\right)\\
&\quad+\left(\frac{\partial u_1'}{\partial x_3'}+\frac{\partial u_3'}{\partial x_1'}\right)\left(\frac{\partial H}{\partial x_1'}\right)\left(\frac{\partial H}{\partial x_2'}\right)\\
&\quad+(1/2)\left[1+\left(\frac{\partial H'}{\partial x_1'}\right)^2-\left(\frac{\partial H'}{\partial x_2'}\right)^2\right]\left(\frac{\partial u_1}{\partial x_2'}+\frac{\partial u_2'}{\partial x_1'}\right)\left(\frac{\partial H'}{\partial x_1'}\right)
\end{aligned}
$$

$$
- (1/2)\left[1 + \left(\frac{\partial H'}{\partial x'_1}\right)^2 - \left(\frac{\partial H'}{\partial x'_2}\right)^2\right]\left(\frac{\partial u'_2}{\partial x'_3} + \frac{\partial u'_3}{\partial x'_2}\right)
$$

$$
= \left(\frac{N'^{1/2}}{2}\right)\mathrm{Ma}\left\{-\left(\frac{\partial H'}{\partial x'_1}\right)\left(\frac{\partial H'}{\partial x'_2}\right)\frac{\partial \theta}{\partial x'_1}\right.
$$

$$
\left. + \left[1 + \left(\frac{\partial H'}{\partial x'_1}\right)^2\right]\frac{\partial \theta}{\partial x'_2} + \left(\frac{\partial H'}{\partial x'_2}\right)\frac{\partial \theta}{\partial x'_3}\right\}. \tag{4.32c}
$$

Finally, instead of (4.28e), we derive for θ the following upper, free-surface, boundary condition (thanks to relation (4.22a)):

$$
\frac{\partial \theta}{\partial z'} = \frac{\partial \theta}{\partial x'_1}\left(\frac{\partial H'}{\partial x'_1}\right) + \frac{\partial \theta}{\partial x'_2}\left(\frac{\partial H'}{\partial x'_2}\right) - N'^{1/2}\left\{\frac{\mathrm{Bi}_{\mathrm{conv}}}{\mathrm{Bi}_s}\right\}[1 + \mathrm{Bi}_s \theta] \tag{4.32d}
$$

The conditions (4.32a–d) are written on the upper, deformable free surface:

$$
z' = H'(t', x'_1, x'_2) \equiv 1 + \eta h'(t', x', y'), \tag{4.33}
$$

and in Chapter 7 these conditions (4.32a–c) are used in the framework of a theory for the interfacial-thermocapillary phenomena, mainly linked with the Marangoni (Ma) convection.

Concerning the condition (4.32d), instead of θ, as in Section 4.4, we prefer to use the dimensionless temperature Θ with again a $\mathrm{Bi}_{\mathrm{conv}}$ different from $\mathrm{Bi}_s = \mathrm{const}$. In fact, $\mathrm{Bi}_s = \mathrm{const}$ allows us, from (1.21b), to determine only the temperature gradient β_s in purely static, motionless, basic conduction (subscript 's') state.

The third case, when we consider a deep liquid layer with dissipative effect, according to (4.27c), is also interesting and is considered in Chapter 6. In this 'deep' case the parameter $\mathrm{Bo} \gg 1$, such that $\varepsilon\,\mathrm{Bo} = \mathrm{O}(1)$, and in the dominant equation (4.26d) for dimensionless temperature θ, when for the thickness d of the layer we have the estimate (1.32)

$$
d = d_{\mathrm{depth}} \approx \frac{C_{vd}}{g\alpha_d}, \tag{4.34}
$$

two new terms appear – one coupled with $[T_d/\Delta T + \theta]u'_3$ and the second with the viscous dissipation $(1/2\mathrm{Gr})[\partial u'_i/\partial x'_j + \partial u'_j/\partial x'_i]^2$. The convection, in a deep layer with viscous dissipation, was first considered by Zeytounian in 1989 and in [17] a simple model, for a constant layer of thickness, d_{depth}, of a weakly expansible viscous dissipative liquid was developed.

The last case concerns an ultra-thin film when we have the constraint (4.27d). The main approach used in continuum theory of (free) ultra-thin films (10–100 nm) is to take into account the details of long-range intermolecular interactions within the film – and in such a case, mainly, an additional term may then appear in the equations of motion – the gradient of the van der Waals potential (a *disjoining pressure* is often used instead of the van der Waals potential) that models the long-range molecular forces. In reality, it is also necessary to take into account a second additional term which is the divergence of the Maxwell stress tensor that represents the electric double-layer repulsion. Usually, if the van der Waals attraction dominates the double-layer repulsion, the film is unstable; the instability leads to rupture of the film! But because of the thinness of the ultra-thin films ($\mathrm{Fr}_d^2 \gg 1$) we wish to consider also the influence of the Marangoni effect, the buoyancy effect, in the leading order, being neglected. In the paper by Idea and Miksis [18] the reader can find various pertinent references concerning the dynamics of thin films subject to van der Waals forces, surface tension and surfactants. The elucidation of the role of the Marangoni effect on the stability of a free ultra-thin film, subject to attractive van der Waals forces (as an extra body force in momentum equations with the Hamaker constant) and surface tension (via the Weber number) is a challanging problem (see Section 10.10).

Obviously, it is necessary to write also initial conditions for $\mathbf{u}'$ and θ at $t' = 0$, for equations (4.26b) and (4.26d)? Both of these initial data characterize the physical nature of the above derived dominant dimensionless Bénard problem, (4.26a, b, d) with (4.28a–e), (4.29). But strictly speaking, for instance, the given starting physical Bénard data for density are not necessarily adequate to approximate equation of state (4.19d)!

Unfortunately, the problem of initial data for the above dominant dimensionless Bénard problem is very poorly investigated, and certainly the above dominant Bénard (in fact, outer relative to time) problem is not significantly close to initial time, because the partial time derivative of the density is lost in this dominant dimensionless problem.

As a consequence, it is necessary to derive, close to initial time, a local dominant dimensionless Bénard problem (with partial short time derivative terms) and then to consider a so-called, unsteady adjustment (inner) problem. At the end of this adjustment process, when the short time tends to infinity, by matching, we have in principle the possibility to obtain well-defined data for the above dominant dimensionless Bénard (outer) evolution in time problem.

A pertinent initial boundary value problem for the development of nonlinear waves on the surface of a horizontally rotating thin film (but without Marangoni and Biot effects) was considered in 1987 by Needham and Merkin [19] and more recently (in 1995) by Bailly [20]. In these two works, the incompressible viscous liquid is injected onto the disk at a specified flow rate through a small gap of height a at the bottom of a cylindrical reservoir of radius l situated at the center of the disk. With $a \ll l$, a long-wave unsteady theory is considered with a thin 'inlet' region and also a region for very small time in which rapid adjustment to initial conditions occurs.

Through matching, these two local regions provide appropriate 'boundary' and 'initial' conditions for the leading-order (outer) evolution problem in the main region (far from local regions). Without doubt the asymptotic approach of Needham and Merkin [19] and Bailly [20], can be used for various thermal and thermocapillary instability convection model problems, which are usually non-valid close to initial time, and this obviously deserves further careful investigations.

Finally, we note that the unsteady adjustment problem is mainly a problem of acoustics, significantly close to initial time where the compressibility/expansibility effect is 'missed', in the framework of the asymptotic modelling of the full compressible NS-F problem, for a weakly expansible thin liquid film problem, this modelling process being singular close to initial time.

I think that the investigations linked with the local-in-time unsteady problem (for liquid films) which ensure the correct asymptotic derivation of the consistent initial conditions at $t' = 0$ for the model approximate equations, are very relevant and will allow young researchers to work on new challenging problems!

4.4 Some Complements and Concluding Remarks

First we consider again the problem concerning the upper, free-surface, boundary condition for the temperature. While the continuity of temperature across the boundary is realistic in most situations (especially for solid boundaries) and is allowed by various authors (for instance, in the book by Joseph and Renardy [21], for a boundary with a negligible thermal resistance), it is not valid when the boundary possesses a non-negligible thermal resistance. One could reject this assumption (which is probably rather realistic at a solid-solid interface) and on the contrary suppose that the temperature is dicontinuous at the interface. The heat flux across the interface is then related to the difference between the temperatures (at the interface/free surface) in the fluid and in surrounding air (usually assumed at constant temperature T_A). This continuity of the heat flux law is often written as 'Newton's law of cooling' (as in (1.23), with dimensional quantities):

$$-k(T_A)\frac{\partial T}{\partial n} = q_{\text{conv}}[T - T_A], \quad \text{on } z = H(t, x, y), \tag{4.35}$$

in the absence of radiation.

The temperature T being the temperaure of the considered liquid layer, T_A is the constant temperature on the other side of the free surface (the ambient temperature in an infinite layer of air at a large distance from the free surface); in fact, the details of what happens very close to the free surface need not be specified and are hidden in a phenomenological coefficient q_{conv}. The equation/condition (4.35) is phenomenological in the sense that it defines, rather, the heat transfer coefficient q_{conv}. Its validity thus depends on the particular situation considered. In reality, with (4.35), it seems more adequate to work (in particular, in the case of a thermocapillary/Marangoni convection and see, for instance the recent paper by Ruyer-Quil et al. [22]) with the following two characteristic temperatures; namely: T_w (at rigid lower plane) and T_A ($< T_w$ – with T_A as the temperature of the ambient gas/air phase), both being constant. The non-dimensional temperature is then (as in (4.10)

$$\Theta = \frac{(T - T_A)}{(T_w - T_A)} \Rightarrow T = T_A + (T_w - T_A)\Theta,$$

so that the dimensionless wall and air temperatures are

$$\Theta = 1 \quad \text{and} \quad \Theta = 0, \quad \text{respectively.} \tag{4.36a}$$

We know that the thermocapillary/Marangoni effect accounts for the emergence of interfacial shear stresses, owing to the variation of surface tension, $\sigma = \sigma(T)$, with temperature of the weakly expansible liquid T. The function $\sigma(T)$ is modeled again by a linear approximation as, instead of (4.14),

$$\sigma(T) = \sigma(T_A) - \left[-\frac{\mathrm{d}\sigma(T)}{\mathrm{d}T}\right]_A (T - T_A) \tag{4.36b}$$

and, in such a case, the formula (4.15) for the Marangoni and Weber number are replaced by (see, for instance, [22])

$$\text{Ma} = \left[-\frac{\mathrm{d}\sigma(T)}{\mathrm{d}T}\right]_A \frac{d(T_w - T_A)}{\rho_A \nu_A^2} \tag{4.37a}$$

and

$$\text{We} = \frac{\sigma_A d}{\rho_A \nu_A^2}, \tag{4.37b}$$

where the subscript A is relative to value, $T = T_A$.

With the above dimensionless temperature Θ, from Newton's cooling law (4.35), we obtain the following upper, free-surface, boundary condition:

$$\frac{\partial \Theta}{\partial n'} + \mathrm{Bi}_{\mathrm{conv}}\Theta = 0, \quad \text{at } z' = 1 + \eta h'(t', x', y'), \tag{4.38a}$$

where

$$\mathrm{Bi}_{\mathrm{conv}} = \frac{dq_{\mathrm{conv}}}{k(T_A)} \tag{4.38b}$$

and $\mathrm{Bi}_{\mathrm{conv}}$ is constituted with a variable convective heat transfer coefficient q_{conv}, different from the constant conduction heat transfer coefficient q_s in

$$\mathrm{Bi}_s = \frac{dq_s}{k(T_A)}. \tag{4.38c}$$

But, above, in both Biot numbers, $\mathrm{Bi}_{\mathrm{conv}}$ and Bi_s, the thermal conductivity has been assumed constant (at $T = T_A$).

We observe that in [23], by Oron, Davis and Bankoff, exactly this (4.38a) condition is also considered (10 years after the 'ambiguous', 1987 paper by the same Davis [11], devoted to 'thermocapillary instabilities') in [23, p. 943]. Such an approach is very pertinent and removes the necessity to use the relation (1.24a) (valid only in a conduction regime).

The convective Biot number, $\mathrm{Bi}_{\mathrm{conv}}$, given by (4.38b), is not a constant but a very complicated function, and we observe again that the conduction Biot number Bi_s plays a role only in the determination of the value of the purely static basic temperature gradient β_s; namely we have the relation

$$\beta_s = \left[\frac{\mathrm{Bi}_s}{1+\mathrm{Bi}_s}\right]\left[\frac{(T_w - T_A)}{d}\right]. \tag{4.39}$$

In a recent paper by Ruyer-Quil et al. [22], just this condition (4.38a) at a free surface has been used for the dimensional temperature Θ, defined above for the case of a film falling down a uniformly heated inclined plane. Unfortunately, in [22] again only a single constant conduction Biot number Bi_s appears in the considered convective problem, the same Biot number Bi_s used in (4.39) for the determination of β_s? During the derivation of the above, upper, free-surface condition (4.38a) for Θ, trying not to complicate the derivation of this condition, we have assumed the thermal conductivity as a constant, $k = k(T_A)$; obviously it is easy to assume that k is a function of Θ, such that

$$k = k(T_A)[1 - \varepsilon D_A \Theta], \tag{4.40a}$$

where, by analogy with for instance (3.9a),

$$D_A = \left[\frac{(\mathrm{d}\log k/\mathrm{d}T)}{(\mathrm{d}\log\rho/\mathrm{d}T)}\right]_A \tag{4.40b}$$

can be assumed fixed (when $\varepsilon \to 0$, and $D_A = \mathrm{O}(1)$).

With (4.40a, b) instead of the condition (4.38a), we obtain the following, upper, free-surface condition for Θ:

$$\frac{\partial\Theta}{\partial n'} + \mathrm{Bi}_{\mathrm{conv}}\Theta = \varepsilon D_A \Theta \frac{\partial\Theta}{\partial n'}, \quad \text{at } z' = 1 + \eta h'(t', x', y'). \tag{4.41a}$$

In accordance with (4.40a) instead of equation (4.26d), written for θ, the following equation for the dimensionless temperature Θ is derived:

$$\begin{aligned}
&[1 - \varepsilon B_d\Theta]\frac{\mathrm{d}\Theta}{\mathrm{d}t'} - \varepsilon\,\mathrm{Bo}'\left[\frac{T_d}{T_w - T_A} + \Theta\right]\left[\mathrm{Fr}_d^2\frac{\mathrm{d}\pi}{\mathrm{d}t'} - u_3'\right] \\
&= \left(\frac{1}{\mathrm{Pr}}\right)\Delta'\Theta + (1/2\,\mathrm{Gr})\varepsilon\,\mathrm{Bo}'\left[\frac{\partial u_i'}{\partial x_j'} + \frac{\partial u_j'}{\partial x_j'}\right]^2 \\
&- \varepsilon\left(\frac{1}{\mathrm{Pr}}\right) D_A \frac{\partial}{\partial x_j'}\left[\Theta\frac{\partial\Theta}{\partial x_j'}\right].
\end{aligned} \tag{4.41b}$$

In equation (4.41b) for the dimensionless temperature Θ, the 'modified' Boussinesq number Bo' is

$$\mathrm{Bo}' = \frac{gd}{(T_w - T_A)C_{pA}}. \tag{4.41c}$$

When we use the dimensionless temperature Θ, then for the density ρ, instead of (4.19d) we write

$$\rho = \rho_A\left\{1 - \varepsilon\left[\frac{(T - T_A)}{(T_w - T_A)}\right] + \left(\frac{1}{K_0}\right)\varepsilon^2\left[\frac{(p - p_A)}{gd\rho_d}\right]\right\}, \tag{4.42}$$

and our main small parameter (instead of $\varepsilon = \alpha(T_d)\Delta T$ with $\Delta T = T_w - T_d$) is

$$\varepsilon' = \alpha(T_A)(T_w - T_A). \tag{4.43}$$

Obviously with $T_s(z) = T_w - \beta_s z$, the corresponding function Θ for a conduction regime is now

$$\Theta_s(z') = 1 - \left[\frac{\mathrm{Bi}_s}{1 + \mathrm{Bi}_s}\right]z', \tag{4.44a}$$

when we take into account the above expression (4.39) for β_s.

With (4.44a) the above boundary condition (4.38a) is automatically satisfied at $z' = 1$, when in the conduction regime $\mathrm{Bi}_{\mathrm{conv}} \equiv \mathrm{Bi}_s$. Namely, we obtain

$$\left.\frac{\partial \Theta_s}{\partial z'}\right|_{z'=1} + \mathrm{Bi}_s \Theta_{s|z'=1} \equiv 0,$$

because

$$-\left[\frac{\mathrm{Bi}_s}{1+\mathrm{Bi}_s}\right] + \mathrm{Bi}_s\left[1 - \left[\frac{\mathrm{Bi}_s}{1+\mathrm{Bi}_s}\right]\right] = 0.$$

In a simple linear case, when $\mathrm{Bi}_{\mathrm{conv}} = B(\eta)$ – a function of the free-surface (simulated by the equation $z' = 1 + \eta h'(t', x', y')$) deformation amplitude parameter η – we can write

$$B(\eta) = B(0) + \eta\left[\frac{\mathrm{d}B(\eta)}{\mathrm{d}\eta}\right]_0; \quad B(0) \equiv \mathrm{Bi}_s, \tag{4.44b}$$

and assume that

$$\Theta = 1 - \left[\frac{\mathrm{Bi}_s}{1+\mathrm{Bi}_s}\right] z' + \eta\Theta' + \cdots. \tag{4.44c}$$

In such a case, from the boundary condition (4.38a) with an error of $O(\eta^3)$, see (4.22a–c), we derive the following linearized condition (at $z' = 1$) at the order η:

$$\left.\frac{\partial \Theta'}{\partial z'}\right|_{z'=1} + \mathrm{Bi}_s\left\{\Theta'|_{z'=1} - \left[\frac{\mathrm{Bi}_s}{1+\mathrm{Bi}_s}\right]h'\right\} + \left[\frac{1}{1+\mathrm{Bi}_s}\right]\left(\frac{\mathrm{d}B(\eta)}{\mathrm{d}\eta}\right)_{\eta=0} = 0. \tag{4.44d}$$

Once more, in this above linearized (4.44d) boundary condition (at $z' = 1$), we see that only Bi_s is present, when $\mathrm{Bi}_{\mathrm{conv}}$ is different from Bi_s! We observe that if we assume $(\mathrm{d}B(\eta)/\mathrm{d}\eta)_{\eta=0} = 0$, then $\mathrm{Bi}_{\mathrm{conv}}$ to be in such a case a constant and, automatically, we have $\mathrm{Bi}_{\mathrm{conv}} \equiv \mathrm{Bi}_s$.

It seems also that the validity of Newton's law with a constant heat transfer coefficient, in a convection regime, is not guaranteed, as this is obvious in various particular examples.

In Pearson's approach [12] the unperturbed rate of heat loss per unit area (heat flux) from the upper, free surface at the plane $z = d$, is written as

$$Q_s = k(T_d)\beta_s, \tag{4.45}$$

with $k(T_d)$ denoting the (constant) thermal conductivity of the considered liquid. This relation (4.45) is derived (see (1.20)) from the condition that the rate of heat supply to the free surface from the liquid must equal the rate of loss of heat from the surface to the air above. The magnitude of Q_s (via Newton's cooling law (1.20), written for the conduction state $T_s(z)$) is then defined by the free-surface temperature $T_s(z = d) = T_d$ and the cooling by the air above the free surface.

At the lower rigid surface, $z = 0$ (a plate), of the liquid layer where the conductivity is large compared to the liquid, it is $T_s(z = 0) = T_w$, which corresponds to a fixed temperature at the rigid plate $z = 0$. In a convection regime, at the upper, free surface, it is assumed that the boundary condition for temperature of the liquid T is well modeled by the balance between heat supply to and heat loss from the upper, free surface, i.e., with dimensional quantities:

$$-k(T_d)\frac{\partial T}{\partial n} = Q(T) + k(T_d)\beta_s \quad \text{at } z = d + ah(t, x, y), \tag{4.46a}$$

and as in [12], the rate of heat loss $Q(T)$ per unit area from the upper, free surface is a function of liquid temperature, T. In Pearson's linear theory [12], a perturbation temperature T', such that $T = T_s(z) + T'$, is considered in balance condition (4.46a) – but obviously without the term, $k(T_d)\beta_s$ – and the rate of heat loss $Q(T)$ from the free surface is defined as:[1]

$$Q(T) = Q_s + \left[\frac{\mathrm{d}Q(T)}{\mathrm{d}T}\right]_d T'_d, \tag{4.46b}$$

where $[\mathrm{d}Q(T)/\mathrm{d}T]_d$ is the value of $\mathrm{d}Q(T)/\mathrm{d}T$ at $T = T_d$ (the temperature at the flat free surface, $z = d$), and represents the rate of change with temperature of the rate of loss of heat from the upper, free surface to its upper environment. The term Q_s is defined, as before, by (4.45). The coefficient $[\mathrm{d}Q(T)/\mathrm{d}T]_d$ plays the role of a 'free-surface heat transfer coefficient', that is, the rate of change with respect to temperature of the heat flux from the free surface to the air – it is likely to be affected in a complicated way by the surface environment relations. As this is very well noted in Pearson's 1958 paper (see the footnote in [12, p. 492]):

> The boundary conditions [(including our (4.14) and (4.46a)] are of crucial importance; by means of a suitable choice for these, many physical phenomena may be very reasonably idealized. The aim in this account

[1] Pearson writes T_d for the value of T' at the undeformable free surface $z = d$.

> is not to provide an exhaustive description of these phenomena and their relevant idealizations, but rather to provide a general treatment that illustrates the fundamental surface tension mechanism and comprehends its many realizations.

On the other hand, Pearson [9, pp. 493, 494] also wrote:

> If, for simplicity, we consider a discrete jump in temperature as occurring at the free surface, then this jump may be small or large compared with the drop in temperature across the liquid layer, depending on the efficiency of the process for removing heat from the surface. Whatever the process, the equality
>
> $$-k(T_d)\frac{\partial T'}{\partial z} = \left[\frac{\mathrm{d}Q(T)}{\mathrm{d}T}\right]_d T', \quad \text{must hold at } z = d, \tag{4.46c}$$
>
> using the relation (4.45) and the reasons given to justify (at least in an *ad hoc* manner) the equality (4.46c).

In Pearson's paper [12, p. 495], the parameter

$$L = \left(\frac{d}{k(T_d)}\right)\left[\frac{\mathrm{d}Q(T)}{\mathrm{d}T}\right]_d \tag{4.46d}$$

plays obviously the role of a convection Biot number. It must be made clear, again, that the evaluation of this (convective Biot) Pearson parameter L, in any physically observed circumstances, is not necessarily easy; it is however a separate problem!

The limiting case $L = 0$, for the insulating[2] boundary condition, $\partial T'/\partial z = 0$, is particular and gives for the modified (physicist) Marangoni number $(= \gamma_\sigma \beta_s d^2/\rho_d \nu_d \kappa_d)$, as critical value, 48.

The arguments are not altered greatly, while the surface tension mechanism is almost certain to be, and observations support this. Since the choice of $L = 0$ was not critical, an exact analysis of the heat transfer at the free surface is not necessary to sustain the above argument.

In general, larger positive values of L lead to greater stability. Really, the values of L encountered in practice would depend on the thickness of the film and for very thin film would tend to zero. It is also important to observe that the buoyancy mechanism has no chance (at least, in a leading order in an asymptotic approach, for the weakly expansible liquids) of causing

[2] 'Insulating' as regards the perturbation temperature T', according to (4.46c), which corresponds to the case of a uniform heat flow.

convection cells: 'for a thin liquid layer, when the thickness d is as small as 1 mm, the squared Froude number (based on d) is unity and the Grashof number Gr tends to zero with the expansibility parameter ε'.

After this rather long digression on the thermal, free-surface, boundary conditions (for instance, see also the discussion in the book by Platten and Legros [24]), we see that the case of a zero convective Biot number poses a problem in linear theory.

When we consider this zero convective Biot number case,

$$\mathrm{Bi}_{\mathrm{conv}} = \frac{dq_{\mathrm{conv}}}{k(T_A)} = 0, \quad \text{because } q_{\mathrm{conv}} = 0, \text{ but } q_s = \text{const} \neq 0,$$

then it seems more judicious to consider, before the process of the linearization, a 'truncated' free surface condition,

$$\frac{\partial T}{\partial n} = 0, \quad \text{on } z = H(t, x, y), \tag{4.47}$$

instead of (4.35). At the end of [24], the reader can find some arguments concerning this zero (convective) Biot number case.

Our second discussion below is related to the *long-scale evolution of the thin liquid films* (see, for instance, the very pertinent review paper [23] by Oron, Davis and Bankoff), which gives a unified approach, taking into account the disparity of the length scales. Indeed, it is often very judicious to take advantage of the disparity of the length scales in view of an asymptotic procedure of reduction of the full set of governing equations (4.26a–c) and boundary conditions (4.28a–e) and (4.29) 'derived in Section 4.2 and discussed in Section 4.3 – to a simplified, but highly nonlinear evolution equation (a so-called 'lubrication' equation) or to a reduced set of (two) equations. As a result of this long-wave theory, a model problem is derived that does not have the full mathematical complexity of the model problem established (set up) in Section 4.2, but does preserve (via a rational analysis) many of the important features of its physics!

Below the basis of the long-wave theory is explained for the case of a two-dimensional problem and, in particular, the problem concerning the coupling of the buoyancy (RB model) and Marangoni (BM model) effects is considered. In [22, 23], the reader can find various references concerning applications of the long-wave theory to evolution of thin liquids films.

In long-wave theory, the reference Reynolds number

$$\mathrm{Re} = \frac{dU_0}{\nu_0}, \tag{4.48}$$

(with as characteristic velocity U_0) plays an important role for the derivation of a lubrication model evolution equation for a free surface. In [23] the reader can find such an evolution, first-order in time, partial differential equation (for the various particular cases) for a liquid film layer bounded below by a horizontal solid undeformable plate and above by an upper, free surface, separating the liquid and passive atmospheric air – the starting equations being the Navier incompressible equations for the velocity vector and pressure with an energy (for the temperature) equation in order to incorporate the thermocapillary effect without buoyancy.

We note that the long-scale approximations have their origins in the lubrication theory of viscous fluids and can be most simply illustrated by considering a fluid-lubricated slipper bearing – a machine part in which viscous fluid is forced into a converging channel. Many details related to Reynolds's and others' work can be found in [26] and the reader can in Schlichting's classical book [27] on boundary-layer theory also find the very reduced simplified incompressible model equations:

$$\frac{\partial u}{\partial x} + \frac{\partial w}{\partial z} = 0 \quad \text{and} \quad \frac{\partial p}{\partial x} = \mu_0 \frac{\partial^2 u}{\partial z^2}, \quad \text{with} \quad \frac{\partial p}{\partial z} = 0,$$

the boundary conditions below the bearing, $0 < x < L$, being

$$u(0) = U_0, \quad w(0) = 0,$$

and

$$u(h) = 0, \quad \text{at } z = h(x).$$

The lower boundary of the bearing, located at $z = h(x)$, is static and tilted at small angle α – the above equations for $\partial p/\partial x$ tell us that since α is (very) small, the flow is locally parallel. Beyond the bearing, $x < 0$ and $x > L$, the pressure is atmospheric, and in particular,

$$p(0) = p(L) = p_A.$$

When p depends on x only, one can solve the above problem (see [23, p. 936]).

The length scale in the x direction is defined by wavelength λ on a film of mean thickness of the liquid layer d. We consider the distortions to be of long scale if

$$\delta = \frac{d}{\lambda} \ll 1. \tag{4.49}$$

It seems natural to scale (as before) to d and the dimensionless z is here

$$Z = \frac{z}{d}, \tag{4.50a}$$

and x to λ, or equivalently d/δ. Then the dimensionless x-coordinate is given by

$$X = \delta\left(\frac{x}{d}\right). \tag{4.50b}$$

Time is scaled to $\lambda/U_0 = \delta(U_0/d)$, so that the dimensionless time is

$$T = \delta\left(\frac{U_0}{d}\right)t. \tag{4.50c}$$

Likewise if there are no rapid variations expected, relative to new time-space variables T, X, Z, as

$$\delta \to 0, \quad \text{with time } T \text{ and space variables } Z \text{ and } X \text{ fixed,} \tag{4.51a}$$

such that

$$\frac{\partial}{\partial T}, \quad \frac{\partial}{\partial X} \quad \text{and} \quad \frac{\partial}{\partial Z} \quad \text{are O}(1). \tag{4.51b}$$

On the other hand, if $u = \mathrm{O}(1)$, the dimensionless horizontal (in the X direction) is

$$U = \frac{u}{U_0} \tag{4.52a}$$

and then, for a consistent (not degenerate) limiting continuity equation (see below (4.53a),

$$W = \left(\frac{1}{\delta}\right)\left(\frac{w}{U_0}\right). \tag{4.52b}$$

Finally for the pressure p, notice that 'pressures' are large due to the lubrication effect, we choose as dimensionless pressure:

$$P = \frac{p}{(\mu_A U_0/\delta d)}. \tag{4.52c}$$

The density ρ is a function of the temperature only, with ρ_A as reference density, and is given below by (4.54a). For this temperature, the function Θ is the dimensionless temperature We obtain first, for the material derivative $\mathrm{d}/\mathrm{d}t = \partial/\partial t + u\partial/\partial x + w\partial/\partial z$, the dimensionless relation

$$\frac{\mathrm{d}}{\mathrm{d}t} = \left(\frac{\delta U_0}{d}\right)\left[\frac{\partial}{\partial T} + U\frac{\partial}{\partial X} + W\frac{\partial}{\partial Z}\right] = \left(\frac{\delta U_0}{d}\right)\frac{\mathrm{d}}{\mathrm{d}T}. \tag{4.52d}$$

These dimensionless variables (4.50a–c), functions (4.52a–c), and the relation (4.52d), for the material derivative, are substituted into the starting (with dimensions) equations, (4.53a–d), written below and governing the convection in a thin liquid heated from below, horizontal one-layer on a plane $z = 0$ (classical Bénard problem). Namely, with the Stokes hypothesis, when the second constant coefficient of viscosity, $\mu'_A \equiv \lambda_A + (2/3)\mu_A = 0$, the starting 2D NS–F equations (with dimensional quantities) are:

$$\frac{\mathrm{d}\log\rho(T)}{\mathrm{d}t} + \frac{\partial u}{\partial x} + \frac{\partial w}{\partial z} = 0, \tag{4.53a}$$

$$\rho(T)\frac{\mathrm{d}u}{\mathrm{d}t} + \frac{\partial p}{\partial x} = \mu_A\left[\frac{\partial^2 u}{\partial x^2} + \frac{\partial^2 u}{\partial z^2} + (1/3)\frac{\partial}{\partial x}\left(\frac{\partial u}{\partial x} + \frac{\partial w}{\partial z}\right)\right], \tag{4.53b}$$

$$\rho(T)\frac{\mathrm{d}w}{\mathrm{d}t} + \frac{\partial p}{\partial z} + \rho g$$
$$= \mu_A\left[\frac{\partial^2 w}{\partial x^2} + \frac{\partial^2 w}{\partial z^2} + (1/3)\frac{\partial}{\partial z}\left(\frac{\partial u}{\partial x} + \frac{\partial w}{\partial z}\right)\right], \tag{4.53c}$$

$$\rho(T)C(T)\frac{\mathrm{d}T}{\mathrm{d}t} + p\left(\frac{\partial u}{\partial x} + \frac{\partial w}{\partial z}\right)$$
$$= k_A\left[\frac{\partial^2 T}{\partial x^2} + \frac{\partial^2 T}{\partial z^2}\right] + \mu_A\left[2\left(\frac{\partial u}{\partial x}\right)^2 + 2\left(\frac{\partial w}{\partial z}\right)^2\right.$$
$$\left. + \left(\frac{\partial w}{\partial x} + \frac{\partial u}{\partial z}\right)^2 - (2/3)\left(\frac{\partial u}{\partial x} + \frac{\partial w}{\partial z}\right)^2\right]. \tag{4.53d}$$

With Θ, we use for ρ, as approximate equation of state

$$\rho = \rho_A[1 - \varepsilon'\Theta], \tag{4.54a}$$

where

$$\varepsilon' = -\left(\frac{1}{\rho_A}\right)\left[\frac{\mathrm{d}\rho}{\mathrm{d}T}\right]_A (T_w - T_A) \tag{4.54b}$$

is the thermal expansion, small expansibility, non-dimensional parameter, introduced (see (4.43)) by analogy with the small parameter ε defined in (1.10a), when instead of Θ defined by (1.17c), we have θ defined by (1.13) – see Chapter 1.

A challenging problem is to elucidate the relation between two small non-dimensional parameters δ and ε'. In other words, the question is:

whether or not it is possible to take into account, at the leading order, simultaneously in long-wave approximation the buoyancy and Marangoni effects, when the limiting process (4.51a), with (4.51b), is performed.

It seems that the answer is negative! Indeed, first from (4.53c), the dimensionless equation for W (given by (4.52b)) is:

$$\delta^3 \operatorname{Re}(1-\varepsilon'\Theta)\left[\frac{\partial W}{\partial T}+U\frac{\partial W}{\partial X}+W\frac{\partial W}{\partial Z}\right]+\frac{\partial P}{\partial Z}+\left(\frac{\delta \operatorname{Re}}{F^2}\right)[1-\varepsilon'\Theta]$$
$$=\delta^2\left[\frac{\partial^2 W}{\partial Z^2}+\delta^2\left(\frac{\partial^2 W}{\partial x^2}\right)+(1/3)\varepsilon'\frac{\partial}{\partial Z}\left(\frac{\mathrm{d}\Theta}{\mathrm{d}T}\right)\right], \tag{4.55}$$

and when the long-wave approximation (4.51a, b) is realized, assuming that Re given by (4.48) is $O(1)$ and fixed, instead of (4.55), we derive the following truncated limit equation, at the leading order (superscript '0') in an expansion in powers of δ,

$$* \qquad \frac{\partial P^0}{\partial Z}+\operatorname{Re}\left(\frac{\delta}{F^2}\right)[1-\varepsilon'\Theta^0]=0. \tag{4.56a}$$

If the Froude number (defined with U_0)

$$F^2=\frac{U_0^2}{gd}, \tag{4.56b}$$

is such that, Re being O(1),

$$\frac{\delta}{F^2}\approx 1 \Rightarrow \lambda \approx \frac{gd^2}{U_0^2}, \tag{4.56c}$$

then the above limit equation (4.56a) is reduced (at the leading order) to

$$\frac{\partial P^0}{\partial Z}=-\operatorname{Re}, \tag{4.57}$$

because, for a usual liquid, ε' is a *small parameter* (of the order 10^{-3}).

On the other hand, we observe that, with (4.52d) and (4.54a), from (4.53a) the dimensionless equation of the continuity is written as

$$** \qquad \frac{\partial U}{\partial X}+\frac{\partial W}{\partial Z}=\varepsilon'\frac{\mathrm{d}\Theta}{\mathrm{d}T}, \tag{4.58a}$$

and, again, for a usual liquid we have the possibility (at the leading order, when $\varepsilon' \to 0$) to use the incompressibility constraint

$$\frac{\partial U^0}{\partial X} + \frac{\partial W^0}{\partial Z} = 0. \tag{4.58b}$$

Then from (4.53d) we derive a dimensionless equation for the dimensionless temperature Θ – namely, neglecting the very small term proportional to ε'^2, we obtain the following equation:

$$\delta \,\mathrm{Re}\,\mathrm{Pr}\,[1 - \varepsilon'(1 + \Gamma_0)\Theta]\frac{\mathrm{d}\Theta}{\mathrm{d}T} + \varepsilon'\,\mathrm{Pr}\,\mathrm{Bo}'\,F^2 P\frac{\mathrm{d}\Theta}{\mathrm{d}T}$$
$$= \left[\frac{\partial^2\Theta}{\partial Z^2} + \delta^2\left(\frac{\partial^2\Theta}{\partial x^2}\right)\right] + 2\delta^2\,\mathrm{Pr}\,\mathrm{Bo}'\,F^2\left[\left(\frac{\partial U}{\partial X}\right)^2 + \left(\frac{\partial W}{\partial Z}\right)^2\right]$$
$$+\,\mathrm{Pr}\,\mathrm{Bo}'\,F^2\left(\frac{\partial U}{\partial z} + \delta^2\frac{\partial W}{\partial X}\right)^2, \tag{4.59a}$$

where Pr is the usual Prandtl number and

$$\mathrm{Bo}' = \frac{gd}{C_A(T_w - T_A)} \tag{4.59b}$$

is a non-dimensional parameter similar (modified, see (4.41c)) to Bo defined by (3.16a). In derivation of the above dimensionless equation (4.59a), for Θ we have used, by analogy with (3.9b) and (3.10a), the relation

$$C(T) = C_A[1 - \varepsilon\Gamma_A\Theta] \quad \text{with} \quad \Gamma_A = \left[\frac{(\mathrm{d}\log C/\mathrm{d}T)}{\mathrm{d}\log\rho/\mathrm{d}T)}\right]_A. \tag{4.60}$$

From the above dimensionless equation (4.59a), for Θ, at the leading order (subscript '0') in an expansion in powers of δ, we rigorously derive, in the long-wave approximation (4.51a, b), the following truncated limit equation for Θ^0:

$$*** \qquad \frac{\partial^2\Theta^0}{\partial Z^2} = \mathrm{Pr}\,\mathrm{Bo}'\,\mathrm{Fr}^2\left[\varepsilon'\frac{\mathrm{d}\Theta^0}{\mathrm{d}T} - \left(\frac{\partial U^0}{\partial z}\right)^2\right]. \tag{4.61a}$$

However, when we assume that the similarity rule (4.56c) between Fr^2 and δ remains valid, we have the possibility to derive the usual (in long-wave approximation theory, see [23, p. 944]) the following, strongly truncated, limiting equation for Θ^0:

$$\frac{\partial^2\Theta^0}{\partial Z^2} = 0. \tag{4.61b}$$

Next we consider equation (4.53b) – for U in dimensionless form we obtain:

$$\delta\,\mathrm{Re}(1-\varepsilon'\Theta)\left[\frac{\partial U}{\partial T}+U\frac{\partial U}{\partial X}+W\frac{\partial U}{\partial Z}\right]+\frac{\partial P}{\partial X}$$
$$=\frac{\partial^2 U}{\partial Z^2}+\delta^2\left(\frac{\partial^2 U}{\partial x^2}\right)+(1/3)\varepsilon'\delta^2\frac{\partial}{\partial X}\left(\frac{\mathrm{d}\Theta}{\mathrm{d}T}\right), \qquad (4.62)$$

and with the incompressible limit, $\varepsilon' \to 0$, we obtain as a truncated equation:

$$\delta\,\mathrm{Re}\left[\frac{\partial U}{\partial T}+U\frac{\partial U}{\partial X}+W\frac{\partial U}{\partial Z}\right]+\frac{\partial P}{\partial X}=\frac{\partial^2 U}{\partial Z^2}+\delta^2\left(\frac{\partial^2 U}{\partial x^2}\right). \qquad (4.62a)$$

The limiting process (4.51a, b) leads, from (4.62a), to the corresponding reduced equation:

$$\frac{\partial P^0}{\partial X}-\frac{\partial^2 U^0}{\partial Z^2}=0. \qquad (4.62b)$$

Finally, according to the above rational analysis, in long-wave approximation theory, we have the following leading-order system of four equations:

$$\frac{\partial P^0}{\partial Z}=-\mathrm{Re};$$
$$\frac{\partial P^0}{\partial X}-\frac{\partial^2 U^0}{\partial Z^2}=0;$$
$$\frac{\partial U^0}{\partial X}+\frac{\partial W^0}{\partial Z}=0;$$
$$\frac{\partial^2 \Theta^0}{\partial Z^2}=0. \qquad (4.63)$$

Concerning the dimensionless boundary conditions for the above long-wave system, (4.63), we write first at the solid lower plate (no slip and no penetration):

$$U^0=0,\quad W^0=0,\quad \Theta^0=1 \quad \text{at } Z=0. \qquad (4.64a)$$

On the free surface, $z = h(t, x)$ we write first the following dimensionless kinematic condition – namely, if

$$Z=\left(\frac{1}{d}\right)h\left(\frac{T}{(\lambda/U^0)},\lambda X\right)=H(T,X)$$

is the dimensionless thickness of the liquid layer, then the dimensionless kinematic condition on an upper, deformable free surface is

$$W^0 = \frac{\partial H}{\partial T} + U^0 \frac{\partial H}{\partial z}, \quad \text{on } Z = H(T, X). \tag{4.64b}$$

With (4-64b) for W^0 we have the possibility to integrate the reduced continuity equation (the third equation in system (4.63)) in Z, from 0 to $H(T, X)$, using integration by parts and also the second condition (4.64a) – the results are:

$$\frac{\partial H}{\partial T} + \frac{\partial}{\partial X}\left(\int_0^{H(T,X)} U^0 \, \mathrm{d}z\right) = 0; \tag{4.65}$$

this evolution equation, in time T, replaces in fact the reduced continuity equation and kinematic condition at an upper, free surface – it ensures conservation of mass on a domain with a deflecting upper boundary.

Now, as a second condition, for Θ^0 on an upper, deformable free surface we write, according to (4.35),

$$\frac{\partial \Theta^0}{\partial z} + \mathrm{Bi}_{\mathrm{conv}} \Theta^0 = 0, \quad \text{on } Z = H(T, X), \tag{4.66}$$

because in dimensionless form ($\mathbf{n}$ is the unit outward vector normal to an upper, free surface) we can write

$$\begin{aligned} \frac{\partial T}{\partial n} &= \nabla T \cdot \mathbf{n} \\ &= \left[1 + \delta^2 \left(\frac{\partial H}{\partial X}\right)^2\right]^{-1/2} \left[\frac{(T_w - T_A)}{d}\right] \left[\frac{\partial \theta}{\partial Z} - \delta^2 \left(\frac{\partial \theta}{\partial X}\right)\left(\frac{\partial H}{\partial X}\right)\right]. \end{aligned}$$

The reader may want to be convinced that it is possible to derive a more general lubrication model equation, when a variable convection Biot number, function of the dimensionless thickness of the liquid layer $H(T, X)$ – $\mathrm{Bi}_{\mathrm{conv}} = B(H)$ – is taken into account. However, here we assume simply that $\mathrm{Bi}_{\mathrm{conv}} = B_0$ is a constant.

The solution of the problem for Θ^0,

$$\frac{\partial^2 \Theta^0}{\partial Z^2} = 0; \quad \Theta^0 = 1 \quad \text{at } Z = 0;$$

$$\frac{\partial \Theta^0}{\partial z} + B_0 \Theta^0 = 0, \quad \text{on } Z = H(T, X),$$

is then

$$\Theta^0(H, Z) = 1 - \left[\frac{B_0}{1 + B_0 H}\right] Z. \tag{4.67}$$

Next, the continuity of the stress tensor of the liquid at the upper, deformable free surface, according to (4.20b, c, d) with (2.43a–d) and (2.44a–c), written for the two-dimensional case, gives the following two dimensionless conditions on $Z = H(T, X)$:

$$P - P_A = \left[1 + \delta^2\left(\frac{\partial H}{\partial X}\right)^2\right]^{-3/2}\left\{\delta^3\,\mathrm{Re}[\mathrm{We} - \mathrm{Ma}\,\Theta]\frac{\partial^2 H}{\partial X^2}\right\}$$
$$- 2\delta^2\left[1 + \delta^2\left(\frac{\partial H}{\partial X}\right)^2\right]^{-1}\left\{\left[\frac{\partial U}{\partial Z} + \delta^2\frac{\partial W}{\partial X}\right]\frac{\partial H}{\partial X}\right.$$
$$\left. + \left[1 - \delta^2\left(\frac{\partial H}{\partial X}\right)^2\right]\frac{\partial W}{\partial Z}\right\}$$
$$+ \delta^2\left\{(2/3) - 2\delta^2\left[1 + \delta^2\left(\frac{\partial H}{\partial X}\right)^2\right]^{-1}\right\}\varepsilon'\frac{\mathrm{d}\Theta}{\mathrm{d}T}, \tag{4.68a}$$

where We and Ma are defined by (4.37a, b), and

$$\left[1 + \delta^2\left(\frac{\partial H}{\partial X}\right)^2\right]\left[\frac{\partial U}{\partial Z} + \delta^2\frac{\partial W}{\partial X}\right] = -4\delta^2\frac{\partial H}{\partial X}\frac{\partial W}{\partial Z}$$
$$+ \delta\,\mathrm{Re}\,\mathrm{Ma}\left[1 + \delta^2\left(\frac{\partial H}{\partial X}\right)^2\right]^{1/2}\left[\frac{\partial \Theta}{\partial X} + \left(\frac{\partial H}{\partial X}\right)\frac{\partial \Theta}{\partial Z}\right]$$
$$- 2\delta^2\left(\frac{\partial H}{\partial X}\right)\varepsilon'\frac{\mathrm{d}\Theta}{\mathrm{d}T}. \tag{4.68b}$$

An examination of the above two dimensionless conditions (4.68a, b) poses a problem concerning the Weber and Marangoni effects via We and Ma? Obviously, if the Marangoni effect is taken into account then it is necessary that for $\mathrm{Re} = \mathrm{O}(1)$ in (4.68b), we have

$$\delta\,\mathrm{Ma} = \mathrm{Ma}^* = \mathrm{O}(1), \tag{4.69a}$$

and in such a case, in (4.68a), the Weber effect is taken into account if

$$\delta^3\,\mathrm{We} = \mathrm{We}^* = \mathrm{O}(1). \tag{4.69b}$$

Finally, in long-wave approximation theory, according to (4.68a, b) with (4.69a, b), under the limiting process (4.51a, b), we obtain the following, two leading-order upper, free-surface conditions on $Z = H(T, X)$:

$$P^0 = P_A - \mathrm{Re}\,\mathrm{We}^* \frac{\partial^2 H}{\partial X^2}, \tag{4.70a}$$

$$\frac{\partial U^0}{\partial Z} = \mathrm{Re}\,\mathrm{Ma}^* \left[\frac{\partial \Theta^0}{\partial X} + \left(\frac{\partial H}{\partial X} \right) \frac{\partial \Theta^0}{\partial Z} \right]. \tag{4.70b}$$

The reduced first two equations of the system (4.63), for P^0 and U^0, with the first condition (4.64a) for U^0, two conditions (4.70a, b) and the solution (4.67) for Θ^0, give together the possibility to derive a single evolution lubrication equation for the thickness $H(T, X)$ of the film from (4.65).

We consider this the 'lubrication problem' in Section 7.4 devoted to the Bénard–Marangoni thin film problem.

The Bénard classical problem – heated from below – is relative to a liquid layer on a lower heated horizontal solid plate.

But many convection problems are relative to a thin liquid film falling down a uniformly heated inclined plane with inclination angle β with respect to the horizontal direction, and Figure 4.2 below sketches the flow situation in a Cartesian coordinate system with x the streamwise coordinate in the flow direction and y the coordinate normal to the substrate.

In a recent paper [22] by Ruyer-Quil et al. (2005), a two-dimensional incompressible (*à la Navier*, with an equation for the temperature) problem is considered.

Consider a thin layer flowing down a plane inclined to the horizontal by angle β as shown in Figure 4.2. The starting dimensional (Navier) equations are consistent with a uniform film of depth h_N in parallel flow with profile

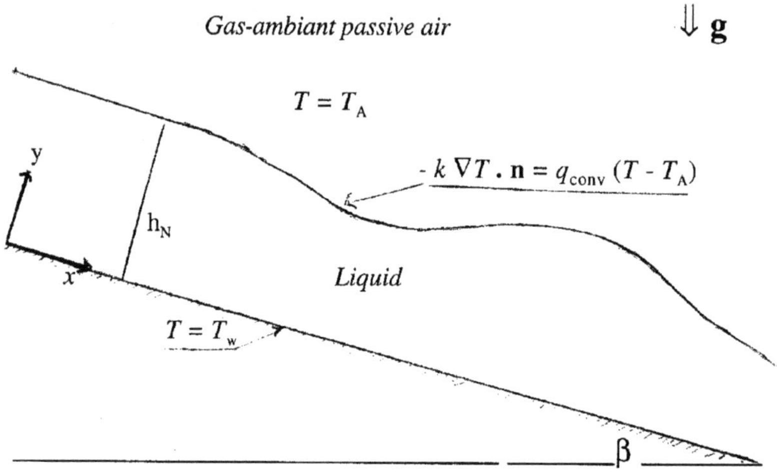

Fig. 4.2 Film falling down a substrate where h_N is the Nusselt flat film thickness.

$$U(z) = \nu_0 g \sin\beta[h_N z - (1/2)z^2], \tag{4.71a}$$

and hydrostaric pressure distribution

$$P(z) = p_A + pg\cos\beta[h_N - z]. \tag{4.71b}$$

This layer is susceptible to long-surface-wave instabilities as discovered by Yih [28] and Benjamin [29] using linear stability theory.

In [22, 25], the length and time scales are obtained from the streamwise gravitational acceleration $g\sin\beta$ and the constant kinematic viscosity $\nu_0 = \mu_0/\rho_0$, which yields

$$l_0 = \nu_0^{2/3}(g\sin\beta)^{-1/3} \tag{4.72a}$$

and

$$t_0 = \nu_0^{1/3}(g\sin\beta)^{-2/3} \tag{4.72b}$$

so that the velocity and pressure scales are

$$U_0 = l_0 t_0^{-1} = (\nu_0 g\sin\beta)^{1/3} \tag{4.72c}$$

and

$$P_0 = \rho_0(\nu_0 g\sin\beta)^{2/3}. \tag{4.72d}$$

The above scales express the importance of viscous and gravitational forces in the considered problem, $\sin\beta$ being of order unity and film flows of thickness h_N of the order of the length scale l_0. In this case the dimensionless equations, for velocity $\mathbf{u}$, pressure p and temperature Θ are:

$$\nabla\cdot\mathbf{u} = 0, \tag{4.73a}$$

$$\frac{\partial\mathbf{u}}{\partial t} + (\mathbf{u}\cdot\nabla)\mathbf{u} = -\nabla p + \mathbf{i} - \cot\beta + \nabla^2\mathbf{u}, \tag{4.73b}$$

$$\mathrm{Pr}\left(\frac{\partial\Theta}{\partial t} + \mathbf{u}\cdot\nabla\Theta\right) = \nabla^2\Theta, \tag{4.73c}$$

and the dimensionless conditions are, at the lower solid plane $y = 0$:

$$\mathbf{u}|_{y=0} = 0, \tag{4.74a}$$

$$\Theta|_{y=0} = 1; \tag{4.74b}$$

at the upper, free surface, $y = h$:

$$\left(\frac{\partial}{\partial t} + \mathbf{u}\cdot\nabla\right)[h - y] = 0; \tag{4.75a}$$

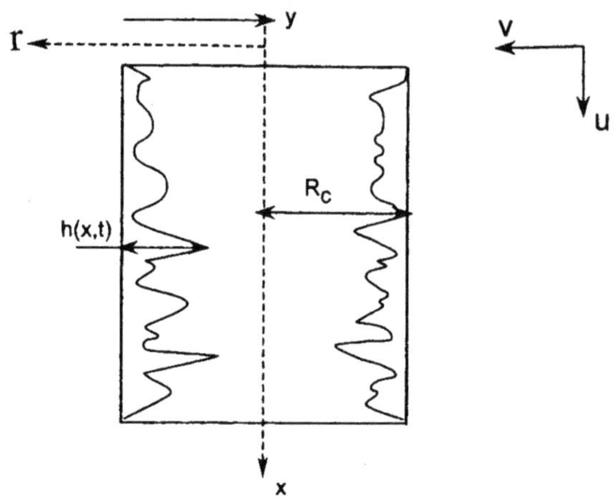

Fig. 4.3 Model of the flow geometry. Reprinted with kind permission from [30].

$$-p\mathbf{n} + (\nabla\mathbf{u} + \nabla\mathbf{u}^t)\cdot\mathbf{n} = -(\Gamma - \mathrm{Ma}\,\Theta)\nabla\cdot\mathbf{n}$$

$$- \mathrm{Ma}\,(\mathbf{I} - \mathbf{n}\otimes\mathbf{n})\cdot\nabla\Theta\cdot(\mathbf{I} - \mathbf{n}\otimes\mathbf{n}); \qquad (4.75\mathrm{b})$$

$$-\nabla\Theta|_{y=h}\cdot\mathbf{n} = \mathrm{Bi}\,\Theta|_{y=h}. \qquad (4.75\mathrm{c})$$

In continuity of stress at the free surface (4.75b), a Kapitza number is present:

$$\Gamma = \frac{\sigma(T_A)}{\rho_0 l_0^2 g \sin\beta}, \qquad (4.76\mathrm{a})$$

and the corresponding Marangoni number is here:

$$\mathrm{Ma} = \left[\frac{\Gamma}{\sigma(T_A)}\right]\left[-\left.\frac{\mathrm{d}\sigma(T)}{\mathrm{d}T}\right|_{T=T_A}(T_w - T_A)\right]. \qquad (4.76\mathrm{b})$$

A trivial solution of the above problem, (4.73a)–(4.75c), is a flat film of dimensionless thickness h_N with a parabolic velocity distribution and a linear temperature distribution; we obtain

$$u_b = h_N y - (1/2)y^2; \quad \Theta_b = 1 - \left[\frac{\mathrm{Bi}_b}{(1 + \mathrm{Bi}_b h_N)}\right]_y. \qquad (4.77)$$

An interesting case of a film falling down an inclined plane, is the case of a vertical plate, when $\beta = \pi/2$ and $\cot\beta = 0$! A somewhat complicated problem is related to the modeling of a liquid film flowing down inside a vertical circular tube (Figure 4.3) – see, for instance [30], where the case of

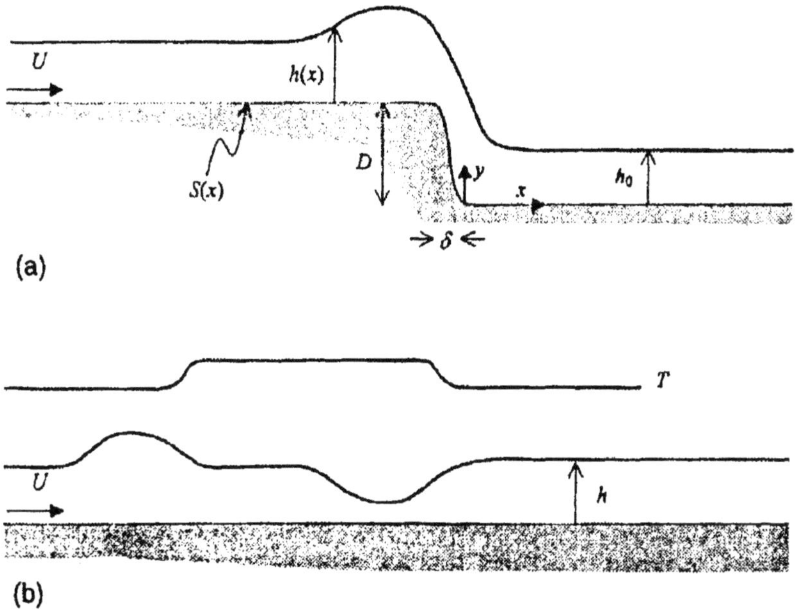

Fig. 4.4 Steady flow over a single sharp step down in topography. Reprinted with kind permission from [31].

a high Reynolds number Re is considered, in order to make a more realistic comparison with the experimental data. In [30], δ is the ratio between the film thickness and the characteristic length related to $\partial/\partial t \sim \partial x \sim \delta \ll 1$, and it is assumed that $\delta\,\mathrm{Re} = \mathrm{O}(1)$. Unfortunately, the liquid is assumed incompressible and the Marangoni effect is absent, only the (large) effect of the Weber number ($\delta^2\,\mathrm{We} = \mathrm{O}(1)$) is taken into account. On the other hand, a large number of applications require coating of a liquid film over a substrate with topography (see [31]). Steady one-dimensional flow over coating of a liquid film over a substrate with a one-dimensional feature such as a trench [32] is a prototype problem for more complicated situations such as coating over two-dimensional features and coating flows driven by complex body forces. Kalliadasis et al. [33] performed a parametric study of this problem, based only on lubrication theory. The combined influence of the topography and the surface tension results in a capillary ridge upstream of the step as sketched in Figure 4.4a. Flow over a planar substrate, with a portion of the surface heated by the temperature profil $T(x)$, showing a dip and a ridge as sketched in Figure 4.4b.

In a short note by Limat [34], instability of a liquid hanging below a solid ceiling (see Figure 4.5) is considered. According to the author, depending

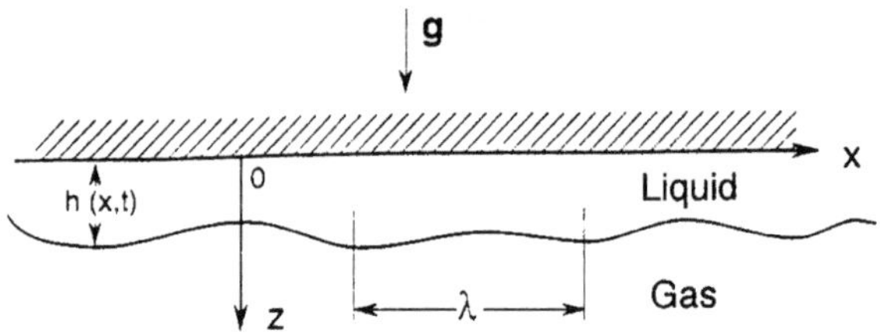

Fig. 4.5 Studied geometry. Reprinted with kind permission from [23].

on the value of the ratio l_c/l_v ($l_c = (\sigma/\rho g)^{1/2}$ and $l_v = (\mu^2/g\rho^2)^{1/3}$ two different behaviors (inviscid or viscous) can be observed.

For a given liquid layer, varying the (finite) thickness h is equivalent to exploring three domains following a straight line whose position depends on l_c/l_v.

For $l_c \gg l_v$, one will observe the successive states (thin-viscous, finite-inviscid, semi-infinite-inviscid) and for $l_c \ll l_v$ (thin-viscous, finite-viscous, semi-infinite-viscous) we observe that the considered instability is related to a Rayleigh–Taylor instability which occurs whenever fluids of different density are subject to acceleration in a direction opposite that of the density gradient (see a review by Sharp in [35]).

Finally, the reader can find in [18] a general formulation for a three-dimensional thin film (subject to van der Waals forces, surface tension, and surfactants) using a curvilinear coordinate system which is defined by the positions of the interfaces (free-free case) of the film; note that a long-wave approximation is used. For practical application, the authors consider a planar, spherical, and cylindrical thin free film, and a bounded film in the form of a catenoid (which is a very straightforward bounded film case – the simple confining geometry giving rise to an uncomplicated set of boundary conditions).

In Chapter 10 the reader can find a discussion and references related to various liquid film problems.

References

1. H. Bénard, Les tourbillons cellulaires dans une nappe liquide. *Revue Générale des Sciences Pures et Appliquées* **11**, 1261–1271 and 1309–1328, 1900. See also: Les tourbillons cellulaires dans une nappe liquide transportant de la chaleur par convection en régime permanent. *Annales de Chimie et de Physique* **23**, 62–144, 1901.

2. S. Chandrasekhar, *Hydrodynamic and Hydromagnetic Stability*. Clarendon Press, Oxford, 1961. See also: Dover Publications, New York, 1981.
3. E.L. Koschmieder, *Bénard Cells and Taylor Vortices*. Cambridge University Press, Cambridge, 1993.
4. C. Marangoni, *Sull'espansione delle gocce di un liquido gallegianti sulla superficie di altro liquido*. Pavia: Tipografia dei fratelli Fusi, 1965 and *Ann. Phys. Chem. (Poggendotff)* **143**, 337–354, 1871.
5. J. Plateau, *Statique Experimentale et Theorique des Liquides Soumis aux Seules Forces Moléculaires*, Vol. 1. Gauthier-Villars, Paris, 1873.
6. R.Kh. Zeytounian, The Bénard–Marangoni thermocapillary-instability problem. *Physics-Uspekhi* **41**(3), 241–267, March 1998 [English edition].
7. D.A. Nield, Surface tension and buoyancy effects in cellular convection. *J. Fluid Mech.* **19**, 341–352, 1964.
8. M.J. Block, Surface tension as the cause of Bénard cells and surface deformation in a liquid film. *Nature* **178**, 650–651, 1956.
9. R.Kh. Zeytounian, The Bénard–Marangoni thermocapillary instability problem: On the role of the buoyancy. *Int. J. Engng. Sci.* **35**(5), 455–466, 1997.
10. M.G. Velarde and R.Kh. Zeytounian (Eds.), *Interfacial Phenomena and the Marangoni Effect*. CISM Courses and Lectures, No. 428, Udine. Springer-Verlag, Wien/New York, 2002.
11. S.H. Davis, Thermocapillary instabilities. *Ann. Rev. Fluid Mech.* **19**, 403–435, 1987.
12. J.R.A. Pearson, On convection cells induced by surface tension. *J. Fluid Mech.* **4**, 489–500, 1958.
13. S.J. Vanhook et al., Long-wavelength instability in surface-tension-driven Bénard convection. *Phys. Rev. Lett.* **75**, 4397, 1995.
14. D.D. Joseph, *Stability of Fluid Motions*, Vol. II. Springer, Heidelberg, 1976.
15. P.M. Parmentier, V.C. Regnier and G. Lebond, Nonlinear analysis of coupled gravitational and capillary thermoconvection in thin fluid layers. *Phys. Rev. E* **54**(1), 411–423, 1996.
16. M. Takashima, *J. Phys. Soc. Japan* **50**(8), 2745–2750 and 2751–2756, 1980.
17. R.Kh. Zeytounian, *Int. J. Engng. Sci.* **27**(11), 1361, 1989.
18. M.P. Ida and M.J. Miksis, The dynamics of thin films. I. General theory; II. Applications. *SIAM J. Appl. Math.* **58**(2), 456–473 (I) and 474–500 (II), 1998.
19. D.J. Needham and J.H. Merkin, *J. Fluid Mech.* **184**, 357–379, 1987.
20. Ch. Bailly, Modélisation asymptotique et numérique de l'écoulement dû à des disques en rotation. Thesis, Université de Lille I, Villeneuve d'Ascq, No. 1512, 160 pp., 1995.
21. D.D. Joseph and Yu.R. Renardy, *Fundamentals of Two-Fluid Dynamics. Part I: Mathematical Theory and Applications*. Springer-Verlag, 1993.
22. C. Ruyer-Quil, B. Scheid, S. Kalliadasis, M.G. Velarde and R.Kh. Zeytounian, Thermocapillary long-waves in a liquid film flow. Part 1: Low-dimensional formulation. *J. Fluid Mech.* **538**, 199–222, 2005.
23. A. Oron, S.H. Davis and S.G. Bankoff, Long-scale evolution of thin liquid film. *Rev. Modern Phys.* **69**(3), 931–980, July 1997.
24. J.K. Platten and J.C. Legros, *Convection in Liquids*, 1st edn. Springer-Verlag, Berlin, 1984.
25. B. Scheid, C. Ruyer-Quil, S. Kalliadasis, M.G. Velarde and R.Kh. Zeytounian, Thermocapillary long-waves in a liquid film flow. Part 2: Linear stability and nonlinear waves. *J. Fluid Mech.* **538**, 223–244, 2005.
26. D. Dowson, *History of Tribology*. Longmans, Green, London/New York, 1979.

27. H. Schlichting, *Boundary-Layer Theory*. McGraw-Hill, New York, 1968.
28. C.-S. Yih, Stability of parallel laminar flow down an inclined plane. *Phys. Fluids* **6**, 321–333, 1963.
29. T.B. Benjamin, Wave formation in laminar flow down an inclined plane. *J. Fluid Mech.* **2**, 554–574, 1957.
30. S. Ndoumbe, F. Lusseyran, B. Izrar, Contribution to the modeling of a liquid film flowing down inside a vertical circular tube. *C.R. Mecanique* **331**, 173–178, 2003.
31. C.M. Gramlich, S. Kalliadasis, G.M. Homsy and C. Messer, Optimal leveling of flow over one-dimensional topography by Marangoni stresses. *Phys. Fluids* **14**(6), 1841–1850, June 2002.
32. L.E. Stillwagon and R.G. Larson, Leveling of thin film over uneven substrates during spin coating. *Phys. Fluids* **A2**, 1937, 1990.
33. S. Kalliadasis, C. Bielarz, and G.M. Homsy, Steady free-surface thin film flows over topography. *Phys. Fluids* **12**, 1889, 2000.
34. L. Limat, Instability of a liquid hanging below a solid ceiling: influence of layer thickness. *C.R. Acad. Sci. Paris, Ser. II* **317**, 563–568, 1993.
35. D.H. Sharp, *Physica D* **12**, 3–18, 1984.

Chapter 5
The Rayleigh–Bénard Shallow Thermal Convection Problem

5.1 Introduction

It is usual in the literature (see, for instance, the book by Drazin and Reid [1]) to denote as Rayleigh–Bénard (RB) shallow thermal convection, the instability problem produced mainly by buoyancy, possibly including the Marangoni and Biot effects in a non-deformable free surface.

In Chapter4, all the material necessary for a rational derivation of the RB equations governing model shallow thermal convection is in fact prepared for obtaining such a result via the Boussinesq limiting process written below in (5.3a). Namely, if we choose to focus on three unknown dimensionless functions

$$\mathbf{u}'(t', x', y', z'), \tag{5.1a}$$

$$\Theta = [T(t', x', y', z') - T_A]/(T_w - T_A), \tag{5.1b}$$

$$\pi(t', x', y', z') = \left(\frac{1}{\mathrm{Fr}_d^2}\right)\left\{\left[\frac{(p - p_A)}{g d \rho_d}\right] + z' - 1'\right\}, \tag{5.1c}$$

and consider our two main small parameters to be

$$\varepsilon' = \alpha(T_A)(T_w - T_A) \equiv -\left[\frac{(T_w - T_A)}{\rho(T_A)}\right]\left(\frac{\mathrm{d}\rho(T)}{\mathrm{d}T}\right)_A, \tag{5.2a}$$

$$\mathrm{Fr}_d^2 = \frac{(\nu(T_A)/d)^2}{gd}, \tag{5.2b}$$

related to a weakly expansible liquid

$$\alpha(T_A) = -\left(\frac{1}{\rho(T_A)}\right)\left(\frac{\mathrm{d}\rho(T)}{\mathrm{d}T}\right)_A \ll 1, \tag{5.2c}$$

and with a not very thin liquid layer

$$\mathrm{Fr}_d^2 \ll 1 \Rightarrow d \gg \left(\frac{\nu_A^2}{g}\right)^{1/3}, \tag{5.2d}$$

then it is necessary to take into account the following Boussinesq limiting process:

$$\varepsilon' \downarrow 0 \quad \text{and} \quad \mathrm{Fr}_d^2 \downarrow 0 \quad \text{such that } \mathrm{Gr} = \frac{\varepsilon'}{\mathrm{Fr}_d^2} = \mathrm{O}(1), \tag{5.3a}$$

the Grashof number Gr being a fixed reference parameter, associated with the asymptotic expansion relative to ε:

$$\begin{aligned}
\mathbf{u}' &= \mathbf{u}_{\mathrm{RB}} + \varepsilon' \mathbf{u}_1 + \cdots, \\
\Theta &= \Theta_{\mathrm{RB}} + \varepsilon' \Theta_1 + \cdots, \\
\pi &= \pi_{\mathrm{RB}} + \varepsilon' \pi_1 + \cdots.
\end{aligned} \tag{5.3b}$$

The dominant dimensionless equations and boundary conditions at $z' = 0$ and $z' = H'(t', x', y')$, for $\mathbf{u}'$, Θ and π, governing (with an error of ε'^2) the exact Bénard problem, heated from below, with an upper, deformable free surface subject to Newton's cooling law, are the following:

$$\nabla' \cdot \mathbf{u}' = \varepsilon' \frac{\mathrm{d}\Theta}{\mathrm{d}t'}; \tag{5.4a}$$

$$\begin{aligned}
&[1 - \varepsilon'\Theta]\frac{\mathrm{d}\mathbf{u}'}{\mathrm{d}t'} + \nabla'\pi - \left(\frac{\varepsilon'}{\mathrm{Fr}_d^2}\right)\left[\Theta - \left(\frac{1}{K_0}\right)\varepsilon'(1 - z')\right]\mathbf{k} \\
&\quad = \nabla'\mathbf{u}' + (1/3)\varepsilon'\nabla'\left(\frac{\mathrm{d}\Theta}{\mathrm{d}t'}\right);
\end{aligned} \tag{5.4b}$$

$$\begin{aligned}
&[1 - \varepsilon' B_d\Theta]\frac{\mathrm{d}\Theta}{\mathrm{d}t'} - \varepsilon'\mathrm{Bo}'\left[\frac{T_d}{(T_w - T_A)} + \Theta\right]\left[\mathrm{Fr}_d^2\frac{\mathrm{d}\pi'}{\mathrm{d}t'} - u_3'\right] \\
&\quad = \left(\frac{1}{\mathrm{Pr}}\right)\nabla'\Theta + (1/2\,\mathrm{Gr})\varepsilon'\mathrm{Bo}'\left[\frac{\partial u_i'}{\partial x_j'} + \frac{\partial u_j'}{\partial x_i'}\right]^2 \\
&\quad - \varepsilon'\left(\frac{1}{\mathrm{Pr}}\right) D_A\frac{\partial}{\partial x_j'}\left(\Theta\frac{\partial\Theta}{\partial x_j'}\right),
\end{aligned} \tag{5.4c}$$

$$\left.\begin{array}{l}\mathbf{u}'=0\\ \Theta=1\end{array}\right\}\quad \text{at } z'=0; \tag{5.5}$$

$$u_3'=\frac{\partial H'}{\partial t'}+u_1'\frac{\partial H'}{\partial x'}+u_2'\frac{\partial H'}{\partial y'},\quad \text{at } z'=H'(t',x',y'); \tag{5.6a}$$

$$\begin{aligned}\pi &= \frac{[H'(t',x',y')-1]}{\mathrm{Fr}_d^2}+\left(\frac{\partial u_i'}{\partial x_j'}+\frac{\partial u_j'}{\partial x_i'}\right)n_i'n_j'-(2/3)\varepsilon'\frac{\mathrm{d}\Theta}{\mathrm{d}t'},\\ &\quad +(\mathrm{We}-\mathrm{Ma}\,\Theta)(\nabla_{\|}'\cdot\mathbf{n}'),\quad \text{at } z'=H'(t',x',y');\end{aligned} \tag{5.6b}$$

$$\left(\frac{\partial u_i'}{\partial x_j'}+\frac{\partial u_j'}{\partial x_i'}\right)t_i'^{(k)}n_j'+\mathrm{Ma}\,t_i'^{(k)}\frac{\partial\Theta}{\partial x_i'}=0,\quad k=1 \text{ and } 2,$$
$$\text{at } z'=H'(t',x',y'); \tag{5.6c}$$

$$(1-\varepsilon' D_A)\frac{\partial\Theta}{\partial n'}+\mathrm{Bi}_{\mathrm{conv}}\Theta=0,\quad \text{at } z'=H'(t',x',y'); \tag{5.6d}$$

and

$$H'(t',x',y')=\frac{H}{d}\equiv 1+\eta h'(t',x',y').$$

During the Boussinesq limiting process (5.3a), the constant parameters $1/K_0$, B_d, D_A, and Prandtl (Pr), Biot ($\mathrm{Bi}_{\mathrm{conv}}$), Marangoni,

$$\mathrm{Ma}=\left[-\frac{\mathrm{d}\sigma(T)}{\mathrm{d}T}\right]_A d\frac{(T_w-T_A)}{\rho_A\nu_A^2} \tag{5.7a}$$

with

$$\sigma(T)=\sigma(T_A)-\left[-\frac{\mathrm{d}\sigma(T)}{\mathrm{d}T}\right]_A(T-T_A), \tag{5.7b}$$

and Boussinesq

$$\mathrm{Bo}'=\frac{gd}{(T_w-T_A)C_{pA}} \tag{5.7c}$$

numbers are fixed, O(1).

The Weber number

$$\mathrm{We}=\frac{\sigma_A d}{\rho_A\nu_A^2}, \tag{5.7d}$$

which is usually large, is assumed such that

$$\eta\,\mathrm{We}=\mathrm{We}^*=\mathrm{O}(1), \tag{5.8}$$

according to the third term in upper condition (5.6c) for π because we want to take into account the effect of the Weber number.

We observe that the free surface amplitude parameter η is necessarily a small parameter because, in the upper, free-surface condition for π, the square of the Froude number $\mathrm{Fr}_d^2 \ll 1$, and as a consequence we assume the following similarity rule between η and Fr_d^2:

$$\frac{\eta}{\mathrm{Fr}_d^2} = \eta^* = O(1), \quad \text{when } \eta \downarrow 0 \text{ and } \mathrm{Fr}_d^2 \downarrow 0. \tag{5.9}$$

After the above *mise en scène*, in this chapter, the next 'tricky step' is to first extract from the above information, (5.1a) to (5.9), via the Boussinesq limiting process (5.3a), associated with the asymptotic expansion (5.3b), a consistent Rayleigh–Bénard model problem for the leading-order functions, $\mathbf{u}_{\mathrm{RB}}$, Θ_{RB} and π_{RB}. Then, a second-order, linear, model problem, a companion to the leading-order RB model, must be derived.

We observe again that the way described above is the only consistent one for a rational derivation of the RB model and, associated with this RB model, a second-order model for $\mathbf{u}_1$, Θ_1 and π_1, according to asymptotic expansion (5.3b).

The rational approach is adopted here to make sure that, terms neglected, in leading and second-order model equations, are really much smaller than those retained. Until this is done, and even now it is possible in part, it will be difficult to convince the detached and possibly skeptical reader of their value as an aid to understanding and in various applications.

A second, important from my point of view, observation, concerns the role of the squared Froude number (5.2b) in the rational process which gives the opportunity to derive successfully the RB or BM model problems in accordance with the value of the thickness d of the liquid layer. Namely, for the RB model problem, as $\mathrm{Fr}_d^2 \ll 1$, then (5.2d) is valid and gives for d a lower bound ($d_{\mathrm{RB}} \gg 1$ mm) which strongly depends on the kinematic viscosity $\nu(T_A)$. However, for this RB model problem, where the viscous dissipation is excluded because $\mathrm{Bo}' \approx 1$, we obtain also an upper bound for the thickness d of the liquid layer; namely from (5.7c),

$$d \approx \frac{(T_w - T_A)}{(g/C_{pA})}, \tag{5.10}$$

and we observe that the value of the thickness d of the liquid layer is also strongly dependent on the temperature difference $(T_w - T_A)$.

In the above mathematical formulation of the classical Bénard thermal-free surface, heated from below, problem, issuing from the unsteady NS–F full problem, there are two mechanisms responsible for driving the convective instability:

- the first one is (in equation (5.4b)) the density variation generated by the thermal (weak) expansion of the liquid (effect of the Grashof number Gr);
- the second results from the free surface tension gradients (in upper conditions (5.6c) and (5.6c), an effect of the Marangoni number Ma) due to temperature dependence of the surface tension at the upper, free surface of the liquid layer.

We also observe that the non-deformable free-surface condition (5.6c), for the dimensionless pressure π, is (asymptotically) only consistent under the condition (5.9), which, in fact, 'smoothes out' the deformations of the upper, free surface!

In derivation of the RB shallow convection model problem it is necessary, in reality, to consider three similarity rules, between the small parameters η, ε', Fr_d^2 and large parameter We, when

$$\eta \to 0, \quad \varepsilon' \to 0, \quad \mathrm{Fr}^2 \to 0, \quad \text{and} \quad \mathrm{We} \to \infty, \tag{5.11a}$$

such that

$$\frac{\eta}{\mathrm{Fr}^2} = \eta^*, \quad \mathrm{Gr} = \frac{\varepsilon'}{\mathrm{Fr}_d^2}, \quad \eta\,\mathrm{We} = \mathrm{We}^*, \tag{5.11b}$$

where η^*, Gr and We^* are all O(1).

With (5.11a, b), we observe also that for the value of the purely static basic dimensionless temperature gradients β_s – given in (4.39) and linked with the Bénard conduction effect – we have the relation

$$\beta_s = \left(\frac{C_{pA}}{g}\right)\left[\frac{\mathrm{Bi}_s}{1+\mathrm{Bi}_s}\right], \tag{5.12}$$

when we replace $(T_w - T_A)/d$ by C_{pA}/g according to (5.10). Relation (5.12) exhibits the dependence of the purely static motionless (in conduction regime) basic temperature gradient β_s from the specific heat of the liquid C_{pA}, at constant pressure and constant temperature T_A.

From the similarity rule $\eta\,\mathrm{We} \approx 1$, with (5.7d) and the similarity rule (5.9), we can also write the following upper bound (instead of (5.10)) for the thickness d of the liquid layer:

$$d \approx \left[\frac{\sigma_A}{g\rho_A}\right]^{1/2} \tag{5.13}$$

and, instead of (5.10), we write in such a case a relation for the temperature difference $(T_w - T_A)$:

$$(T_w - T_A) \approx \frac{gd}{C_{pA}}. \tag{5.14}$$

Thus the above relations (5.2d), (5.12), (5.13) and (5.14), written for the data β_s, d and $(T_w - T_A)$ give some criteria related to various physical parameters characterizing the weakly expansible liquid layer.

5.2 The Rayleigh–Bénard System of Model Equations

Now, the application of the Boussinesq limiting process (5.3a), with asymptotic expansion (5.3b), gives for the leading-order RB three functions: $\mathbf{u}_{\mathrm{RB}}$, Θ_{RB} and π_{RB}, in (5.3b), the following Rayleigh–Bénard system of three model equations (the terms with ε are neglected):

$$\nabla' \cdot \mathbf{u}_{\mathrm{RB}} = 0; \tag{5.15a}$$

$$\frac{\mathrm{d}\mathbf{u}_{\mathrm{RB}}}{\mathrm{d}t'} + \nabla'\pi_{\mathrm{RB}} - \mathrm{Gr}\,\Theta_{\mathrm{RB}}\mathbf{k} = \Delta'\mathbf{u}_{\mathrm{RB}}; \tag{5.15b}$$

$$\frac{\mathrm{d}\Theta_{\mathrm{RB}}}{\mathrm{d}t'} = \left(\frac{1}{\mathrm{Pr}}\right)\Delta'\Theta_{\mathrm{RB}}, \tag{5.15c}$$

where

$$\begin{aligned}\frac{\mathrm{d}}{\mathrm{d}t'} &= \frac{\partial}{\partial t'} + (\mathbf{u}_{\mathrm{RB}} \cdot \nabla') \\ &= \frac{\partial}{\partial t'} + u_{1\mathrm{RB}}\frac{\partial}{\partial x'} + u_{2\mathrm{RB}}\frac{\partial}{\partial y'} + u_{3\mathrm{RB}}\frac{\partial}{\partial z'}.\end{aligned} \tag{5.16}$$

From (5.6a–d) – see also, for instance, our discussion in Section 4.3 – for the above model equations (5.15a–c), in the leading order (again the terms with ε being neglected), we derive the associated upper boundary conditions at a non-deformable free surface $z' = 1$, when we take into account the similarity rule (5.11) and the results of Section 4.3. First, with (4.31a),

$$\mathbf{u}_{\mathrm{RB}} \cdot \mathbf{k} \equiv u_{3\mathrm{RB}} = 0, \quad \text{at } z' = 1, \tag{5.17a}$$

according to (5.6a), and then from (5.6c), with (4.31e–f), we derive the following two conditions:

$$\mathrm{Ma}\frac{\partial\Theta_{\mathrm{RB}}}{\partial x'} = -\left[\frac{\partial u_{1\mathrm{RB}}}{\partial z'} + \frac{\partial u_{3\mathrm{RB}}}{\partial x'}\right], \quad \text{at } z' = 1, \tag{5.17b}$$

and

$$\mathrm{Ma}\frac{\partial \Theta_{\mathrm{RB}}}{\partial y'} = -\left[\frac{\partial u_{2\mathrm{RB}}}{\partial z'} + \frac{\partial u_{3\mathrm{RB}}}{\partial y'}\right], \quad \text{at } z' = 1. \tag{5.17c}$$

But equation (5.15a) gives

$$-\frac{\partial[\partial u_{1\mathrm{RB}}/\partial x' + \partial u_{2\mathrm{RB}}/\partial y']}{\partial z'} = \frac{\partial^2 u_{3\mathrm{RB}}}{\partial z'^2}, \quad \text{at } z' = 1,$$

and, instead of (5.17b, c), with (5.17a), we obtain the following single, upper condition at a non-deformable free surface:

$$\frac{\partial^2 u_{3\mathrm{RB}}}{\partial z'^2} = \mathrm{Ma}\left[\frac{\partial^2 \Theta_{\mathrm{RB}}}{\partial x'^2} + \frac{\partial^2 \Theta_{\mathrm{RB}}}{\partial y'^2}\right], \quad \text{at } z' = 1. \tag{5.17d}$$

Finally, from (5.6d), when we take into account the smallness of the amplitude parameter η, we derive a condition for Θ:

$$\frac{\partial \Theta_{\mathrm{RB}}}{\partial z'} + \mathrm{Bi}_{\mathrm{conv}}\,\Theta_{\mathrm{RB}} = 0, \quad \text{at } z' = 1. \tag{5.17e}$$

The upper conditions (5.17a), (5.17d) and (5.17e), with the boundary conditions at lower plate

$$u_{1\mathrm{RB}} = u_{2\mathrm{RB}} = u_{3\mathrm{RB}} = 0 \quad \text{and} \quad \Theta_{\mathrm{RB}} = 1, \quad \text{at } z' = 0, \tag{5.18}$$

are the associated boundary conditions for the RB, 'rigid-free' model equations (5.15a–c). The upper condition (5.17e) for the dimensionless temperature Θ_{RB}, defined by (5.1b), is used in two recent papers [2, 3]. The RB model problem formulated there, (5.15a–c)–(5.17a, d, e), being relative to a 'rigid-free' case, is a sequel of the Bénard physical starting problem, heated from below, when we take into account the existence of a deformable upper, free surface.

The 'rigid-rigid' case, has been considered, in fact without the Marangoni and Biot effects, in Chapter 3 devoted to Rayleigh's 1916 problem; see, for instance, the model problem (3.25a–d) derived in Section 3.4.

Often a 'free-free', 'unrealistic', case is also considered when the corresponding boundary conditions (again, without the Marangoni and Biot effects) are

$$u_{3\mathrm{RB}} = 0 \text{ and } \frac{\partial^2 u_{3\mathrm{RB}}}{\partial z'^2} = 0, \quad \text{at both free boundaries, } z' = 0, 1, \tag{5.19a}$$

$$\frac{\partial \Theta}{\partial z'} = 0, \quad \text{at } z' = 0, 1, \tag{5.19b}$$

when the boundaries $z' = 0$ and $z' = 1$ are modelled as a perfect insulator (as is the case in [4]) or, if the boundaries are at fixed temperature:

$$\Theta = 1, \quad \text{at } z' = 0, z' = 1. \tag{5.19c}$$

In most cases (in the *ad hoc* approaches), as a 'free-free' RB problem, the following dimensionless equations are considered:

$$\nabla \cdot \mathbf{V} = 0; \tag{5.20a}$$

$$\left(\frac{1}{\mathrm{Pr}}\right)\frac{\mathrm{d}V}{\mathrm{d}t} = -\nabla\Pi + T\mathbf{k} + \nabla^2\mathbf{V}; \tag{5.20b}$$

$$\frac{\mathrm{d}T}{\mathrm{d}t} = \mathrm{Ra}\, w + \nabla^2 T, \tag{5.20c}$$

with Ra (= Pr Gr) the Rayleigh number, for the dimensionless velocity vector $\mathbf{V}$, pressure Π and temperature T, with as boundary conditions

$$T = w = 0, \quad \frac{\partial^2 w}{\partial z^2} = 0 \quad \text{at } z = 0, z = 1. \tag{5.20d}$$

In equation (5.20c) and boundary conditions (5.20d),

$$w = \mathbf{V} \cdot \mathbf{k}. \tag{5.20e}$$

Above, in the derivation of the RB, free-free, model problem (5.15a–c)–(5.17a, d, e), we have not considered the upper condition (5.6c) for π! It is however true that from this upper condition (5.6c) we have concluded the smallness of the amplitude parameter η (see (5.9)).

Indeed, this upper condition (5.6c) gives an equation for the determination of the deformation $h'(x', y')$ of the free surface, when we take into account the similarity rule (5.8). Namely, we have aready noted in Section 1.2 (see, for instance, equation (1.28a)), the emergence of such an equation for $h'(x', y')$ from the upper jump condition, for the dimensionless pressure π.

It seems that such an equation for the deformation of the free surface had not been discovered before! Here, in the framework of our rational analysis and asymptotic modelling approach, such a result is expected and is obtained from (5.6c), when we take into account (4.31g) and the similarity rule (5.8). This equation is written in the following form for the deformation $h'(t', x', y')$ of the upper, free surface:

$$\frac{\partial^2 h'}{\partial x'^2} + \frac{\partial^2 h'}{\partial y'^2} - \left(\frac{\eta^*}{\mathrm{We}^*}\right) h' = -\left(\frac{1}{\mathrm{We}^*}\right) \pi(t', x', y', z' = 1). \tag{5.21}$$

Fig. 5.1 Two 'almost regular' RB convection patterns. Reprinted with kind permission from [10].

Fig. 5.2 Two 'exotic' RB convection patterns. Reprinted with kind permission from [10].

In Figures 5.1–5.3, some 'spectacular' RB convection patterns selected from the cited survey paper [10] are presented.

The linear theory of the RB shallow convection model problem is very well analyzed in the Drazin and Reid book [1], and also in Chandrasekhar's monograph [5]. The RB thermal convection, governed by the above model problem (5.15a–c)–(5.17a, d, e), represents, when Ma and Bi effects are neglected, the simplest (but very 'rich') example of hydrodynamic instability and transition to turbulence in a fluid system. In this case (as $\mathrm{Ma} = 0$ and $\mathrm{Bi} = 0$), the more important effect is linked with buoyancy (Archimedean

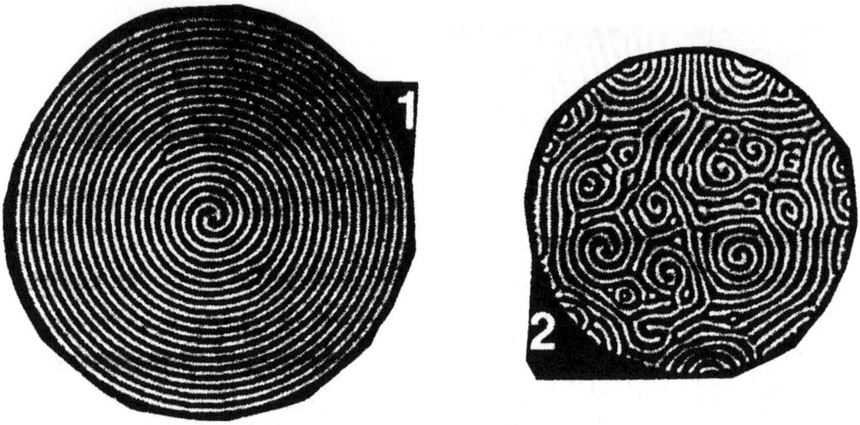

Fig. 5.3 Two 'circular' RB convection patterns. Reprinted with kind permission from [10].

force) via the Grashof Gr number and the reader can find an excellent account of the various features of this buoyancy effect in the book by Turner [6].

A qualitative description of the convection motions is given in a paper by Velarde and Normand [7]. On the other hand, in [8], a physicist's approach to convective instability is presented; this review paper is a pertinent account of the theoretical and experimental results on convective instability up to 1957. In the book by Getling [9] the reader can find a pertinent discussion related to the 'structures and dynamics' of the Rayleigh–Bénard convection. In the survey paper by Bodenschatz et al. [10] (published in 2000), various developments in RB thermal convection are given and, in particular, results for RB convection that have been obtained during the years 1900–2000 are summarized.

An interesting point made in this paper is that it is now well known that thermal convection occurs in a spatially extended system when a sufficiently steep temperature gradient is applied across a fluid layer, and a 'pattern' appears that is generated by the spatial variation of the convection structure; the nature of such convection patterns is at the center of this survey.

Finally, we note that the above derived RB model problem with Marangoni and Biot effects was considered (in 1996) by Dauby and Lebon [11].

5.3 The Second-Order Model Equations, Associated to RB Equations

We return to equations and conditions written in Section 5.1. First, from (5.4a), at once we derive a second-order, divergence non-free, constraint for the velocity vector $\mathbf{u}_1$, in asymptotic expansion (5.3b). Namely,

$$\nabla' \cdot \mathbf{u}_1 = \frac{\mathrm{d}\Theta_{\mathrm{RB}}}{\mathrm{d}t'}. \tag{5.22a}$$

Then, from equation (5.4b) for the velocity vector $\mathbf{u}'$, the terms proportional to ε' give a second-order equation for $\mathbf{u}_1$, of the following form:

$$\begin{aligned}&\frac{\partial u_1}{\partial t'} + (\mathbf{u}_1 \cdot \nabla')\mathbf{u}_{\mathrm{RB}} + (\mathbf{u}_{\mathrm{RB}} \cdot \nabla')\mathbf{u}_1 + \nabla'\pi_1 \\ &\quad - \mathrm{Gr}\left[\Theta_1 - \left(\frac{1}{K_0}\right)(1 - z')\right]\mathbf{k} - \mathbf{\Delta}'\mathbf{u}_1 \\ &\quad = (1/3)\nabla'\left(\frac{\mathrm{d}\Theta_{\mathrm{RB}}}{\mathrm{d}t'}\right) + \Theta_{\mathrm{RB}}\frac{\mathrm{d}u_{\mathrm{RB}}}{\mathrm{d}t'}.\end{aligned} \tag{5.22b}$$

Collecting all the terms proportional to ε', for the dimensionless temperature Θ given by (5.1b) with the second asymptotic expansion of (5.3b), as second-order equation for Θ_1, we obtain next from equation (5.4c):

$$\begin{aligned}&\frac{\partial \Theta_1}{\partial t'} + \mathbf{u}_1 \cdot \nabla'\Theta_{\mathrm{RB}} + \mathbf{u}_{\mathrm{RB}} \cdot \nabla'\Theta_1 - \left(\frac{1}{\mathrm{Pr}}\right)\Delta'\Theta_{\mathrm{RB}} \\ &\quad = (1/2\,\mathrm{Gr})\mathrm{Bo}'\left[\frac{\partial u_{i\mathrm{RB}}}{\partial x'_j} + \frac{\partial u_{j\mathrm{RB}}}{\partial x'_i}\right]^2 + B_d\Theta_{\mathrm{RB}}\frac{\mathrm{d}\Theta_{\mathrm{RB}}}{\mathrm{d}t'} \\ &\quad - \left(\frac{1}{\mathrm{Pr}}\right) D_A \frac{\partial}{\partial x'_j}\left(\Theta_{\mathrm{RB}}\frac{\partial \Theta_{\mathrm{RB}}}{\partial x'_j}\right) \\ &\quad - \mathrm{Bo}'\left[\frac{T_d}{(T_w - T_A)} + \Theta_{\mathrm{RB}}\right] u_{3\mathrm{RB}}.\end{aligned} \tag{5.22c}$$

Now for the above second-order system of equations (5.22a–c) , it is necessary to write consistent boundary conditions for $\mathbf{u}_1$, Θ_1 and π_1; they can be obtained from (5.6a–d) in a straightforward, but tedious manner.

As an example, from (5.6a), because in $H'(t', x', y') \equiv 1 + \eta h'$, with the parameter $\eta \ll 1$ according to (5.9) and with Taylor's formula (up to order η), we write

$$u_3'(t', x', y', z' = 1 + \eta h') \approx u_3'|_{z'=1} + \eta h' \left(\frac{\partial u_3'}{\partial z'}\right)\Bigg|_{z'=1};$$

or, because with (5.9) we have $\eta = (\eta^*/\mathrm{Gr})\varepsilon'$ instead of the above condition (5.6a), when we take into account asymptotic expansion (5.3b) for the velocity at order ε', we obtain

$$\begin{aligned}(\mathbf{u}_1 \cdot \mathbf{k})|_{z'=1} &= \left(\frac{\eta^*}{\mathrm{Gr}}\right)\left\{h'\left[-\frac{\partial(\mathbf{u}_{\mathrm{RB}} \cdot \mathbf{k})}{\partial z'}\right]\Bigg|_{z'=1}\right. \\ &\quad \left. + \frac{\partial h'}{\partial t'} + u_{1\mathrm{RB}}|_{z'=1}\frac{\partial h'}{\partial x'} + u_{2\mathrm{RB}}|_{z'=1}\frac{\partial h'}{\partial y'}\right\}. \qquad (5.23)\end{aligned}$$

As an easy exercise, the reader can find (via an accurate calculation from (5.6b–d) with (5.3b)) the upper free surface conditions at $z' = 1$ for the second-order model equations (5.22a–c).

At $z' = 0$ the conditions for $\mathbf{u}_1$ and Θ_1 are simply

$$\mathbf{u}_1 = \Theta_1 = 0. \qquad (5.24)$$

Hopefully, the above second-order system of equations (5.22a–c) will have an application later on. In any case the presentation here, in Chapter 3 and also further on in this chapter, provides a way for derivation of second-order, consistent, model equations and gives to the reader a methodology which can be used in various fluid mechanics problems where one or several small (or large) parameters are present and govern miscellaneous physical effects.

We observe, finally, that in upper condition (5.23) appears the unknown $h'(t', x', y')$ and, as a consequence, our derived equation (5.21) is a necessary closing equation for obtaining the solution of the second-order problem.

Perhaps, for a particular convection problem, it will be necessary to assume that h' is also (as in (5.3b)) subject to an asymptotic expansion relative to ε':

$$h' = h_{\mathrm{RB}} + \varepsilon' h_1 + \cdots. \qquad (5.25)$$

5.4 An Amplitude Equation for the RB Free-Free Thermal Convection Problem

Below we consider the RB free-free dimensionless model problem (5.20a–e) for the three functions, $\mathbf{V}$, T and Π which depend on dimensionless time-space variables, t, x, y and z. As usual in weakly nonlinear analyses, one is constrained to values of the Ra close to critical Rayleigh Ra_c, and a 'supercriticality' parameter r, of order O(1), is introduced such that

$$\mathrm{Ra} = \mathrm{Ra}_c + \kappa^2 r, \tag{5.26a}$$

with $\kappa \ll 1$ regarded as a measure of the closeness Ra from Ra_c.

In the free-free case the RB system first becomes unstable when

$$\mathrm{Ra}_c = \frac{27\pi^4}{4}, \tag{5.26b}$$

where from a periodic array of convection rolls arises a critical wave number

$$k_c = \frac{\pi}{\sqrt{2}}. \tag{5.26c}$$

First it is necessary to introduce the slow scales

$$\xi = \kappa x, \quad \eta = \kappa^{1/2} y, \quad \tau = \kappa^2 t. \tag{5.27a}$$

The choice (5.27a) of slow scales is motivated by the behavior of the linear growth rate in the vicinity of (k_c, Ra_c), and by the expected form of the leading-order nonlinearity in the final amplitude equation. All dependent variables are expanded according to

$$\begin{aligned}(u, v, w, \Pi, T) \equiv U &= \kappa U_1 + \kappa^{3/2} U_{3/2} + \kappa^2 U_2 \\ &\quad + \kappa^{5/2} U_{5/2} + \kappa^3 U_3 + \cdots .\end{aligned} \tag{5.27b}$$

where

$$U_n = U_n^{(0)}(\xi, \eta, \tau, z) + \mathrm{Real}[\Sigma U_n^{(m)}(\xi, \eta, \tau, z) \exp(imk_c x)], \tag{5.27c}$$

with $m = 1$ to N and $n = 1 + (p/2)$, $p = 0, 1, 2, 4, \ldots$.

In Zeytounian's book [12, pp. 378–392], the reader can find a detailed derivation of an amplitude evolution equation. Namely, for the amplitude $A(\xi, \eta, \tau)$ of the $O(\kappa)$ problem (see (5.28a)). We give below only the main steps of this asymptotic derivation.

The first step is substitution of the above expansion (5.27c) into the governing RB equations (5.20a–c) and boundary conditions (5.20d), after introducing (5.26a) and the slow scales (5.27a).

Because $U_1^{(0)} = 0$, describing the periodic array of convection rolls, the solution of the $O(\kappa)$ problem ($n = 1$, $m = 1$) for $U_1^{(1)}$ is obtained in the following form:

$$u_1^{(1)} = \pi A \cos(\pi z), \tag{5.28a}$$

$$v_1^{(1)} = 0, \tag{5.28b}$$

$$w_1^{(1)} = -ik_c A \sin(\pi z), \tag{5.28c}$$

$$T_1^{(1)} = -\left(\frac{i}{\sqrt{2}}\right)(9/2)\pi^3 A \sin(\pi z), \tag{5.28d}$$

$$\Pi_1^{(1)} = \left(\frac{i}{k_c}\right)\pi(\pi^2 + k_c^2) A \cos(\pi z), \tag{5.28e}$$

where the complex amplitude $A(\xi, \eta, \tau)$ is, at this stage, an unknown function to be determined at higher order by applying a suitable orthogonality condition (elimination of secular terms according to a multiple-scale method – MSM).

For the $O(\kappa^{3/2})$ problem, one finds

$$U_{3/2}^{(0)} = 0$$

in a straightforward manner, and then:

$$u_{3/2}^{(1)} = 0, \tag{5.29a}$$

$$v_{3/2}^{(1)} = -i\left(\frac{\pi}{k_c}\right)\cos(\pi z)\frac{\partial A}{\partial \eta}, \tag{5.29b}$$

$$w_{3/2}^{(1)} = 0, \tag{5.29c}$$

$$T_{3/2}^{(1)} = 0, \tag{5.29d}$$

$$\Pi_{3/2}^{(1)} = 0. \tag{5.29e}$$

At the $O(\kappa^2)$ order, the problem is more complicated since $U_2^{(0)}$ is different from zero, and $w_2^{(1)}$ and $T_2^{(1)}$ are solutions of a non-homogeneous system of two equations. For the components of $U_2^{(0)}$ we derive the following system of equations:

$$\frac{\partial u_2^{(0)}}{\partial z} = 0, \quad \frac{\partial v_2^{(0)}}{\partial z} = 0, \quad \frac{\partial w_2(0)}{\partial z} = 0,$$

$$\frac{\partial \Pi_2^{(0)}}{\partial z} - T_2^{(0)} = -\left(\frac{\pi}{2\,\mathrm{Pr}}\right) k_c^2 |A^2| \sin(2\pi z),$$

$$\frac{\partial^2 T_2^{(0)}}{\partial z^2} = \left(\frac{9}{4\sqrt{2}}\right)\pi^4 k_c |A^2| \sin(2\pi z) - \mathrm{Ra}_c\, w_2^{(0)}. \tag{5.30a}$$

The solution of (5.30a) is simply:

$$u_2^{(0)} = 0, \quad v_2^{(0)} = 0, \quad w_2^{(0)} = 0,$$
$$T_2^{(0)} = -(9/32)\pi^3 |A^2| \sin(2\pi z),$$
$$\Pi_2^{(0)} = (1/8)\left[\left(\frac{1}{\mathrm{Pr}}\right) + (9/8)\right]\pi^2 |A^2| \cos(2\pi z). \qquad (5.30b)$$

For the components of $U_2^{(1)}$, we obtain first

$$v_2^{(1)} = 0, \qquad (5.30c)$$

and then for the two functions $w_2^{(1)}$ and $T_2^{(1)}$ we derive again a non-homogeneous system of two equations, namely:

$$\left(\frac{\partial^2}{\partial z^2} - k_c^2\right) w_2^{(1)} - k_c^2 T_2^{(1)} = F_2, \qquad (5.30d)$$

$$\left(\frac{\partial^2}{\partial z^2} - k_c^2\right) T_2^{(1)} + \mathrm{Ra}_c\, w_2^{(1)} = G_2, \qquad (5.30e)$$

where

$$F_2 = -(3/2)\pi^4 \left[\frac{\partial A}{\partial \xi} - \left(\frac{i}{2k_c}\right)\frac{\partial^2 A}{\partial \eta^2}\right]\sin(\pi z), \qquad (5.30f)$$

$$G_2 = -(9/2)\pi^4 \left[\frac{\partial A}{\partial \xi} - \left(\frac{i}{2k_c}\right)\frac{\partial^2 A}{\partial \eta^2}\right]\sin(\pi z), \qquad (5.30g)$$

with the boundary conditions

$$w_2^{(1)} = \frac{\partial^2 w_2^{(1)}}{\partial z^2} = T_2^{(1)} = 0 \quad \text{at } z = 0 \text{ and } 1. \qquad (5.30h)$$

In order for this above problem (5.30d–h) to admit a non-trivial solution, the forcing terms (5.30f, g) must be orthogonal to the adjoint eigenfunctions of the homogeneous problem, i.e., to the adjoint eigenfunctions of the equations at order $\mathrm{O}(\kappa)$ with boundary conditions similar to the above (but written for $w_1^{(1)}$ and $T_1^{(1)}$). One readily obtains the adjoint eigenfunctions to be $w^* = -3\sin(\pi z)$, and $T^* = \sin(\pi z)$, and the othogonality condition is the following:

$$\int_0^1 [F_2 w^* + G_2 T^*]\,\mathrm{d}z = 0, \qquad (5.30i)$$

which is identically satisfied.

Then, the solution for $w_2^{(1)}$ and $T_2^{(1)}$ is

$$w_2^{(1)} = 0, \tag{5.30j}$$

$$T_2^{(1)} = 3\pi^2 \left[\frac{\partial A}{\partial \xi} - \left(\frac{i}{2k_c} \right) \frac{\partial^2 A}{\partial \eta^2} \right] \sin(\pi z), \tag{5.30k}$$

and for $u_2^{(1)}$ and $\Pi_2^{(1)}$, we have:

$$u_2^{(1)} = -\left(\frac{\pi}{ik_c} \right) \left[\frac{\partial A}{\partial \xi} + \left(\frac{1}{ik_c} \right) \frac{\partial^2 A}{\partial \eta^2} \right] \cos(\pi z), \tag{5.30l}$$

$$\Pi_2^{(1)} = i \left(\frac{\pi}{k_c} \right)^3 \left[2ik_c \frac{\partial A}{\partial \xi} + \frac{\partial^2 A}{\partial \eta^2} \right] \cos(\pi z). \tag{5.30m}$$

At the $O(\kappa^{5/2})$ order, all field variables admit a solution of the form (5.27c) with $p = 3$ and $m = 1$, and in this case we obtain easily the following solution for the components of $U_{5/2}^{(0)}$:

$$u_{5/2}^{(0)} = 0, \tag{5.31a}$$

$$v_{5/2}^{(0)} = -(3/32) \left[\left(\frac{1}{\mathrm{Pr}} \right) + 3/8 \right] \frac{\partial |A_2|}{\partial \eta} \cos(2\pi z), \tag{5.31b}$$

$$w_{5/2}^{(0)} = 0, \tag{5.31c}$$

$$T_{5/2}^{(0)} = 0. \tag{5.31d}$$

Then for the components of $U_{5/2}^{(1)}$ we obtain:

$$u_{5/2}^{(1)} = 0, \tag{5.31e}$$

$$v_{5/2}^{(1)} = \left(\frac{4}{\pi} \right) \frac{\partial}{\partial \eta} \left[\frac{\partial A}{\partial \xi} - \left(\frac{i}{2k_c} \right) \frac{\partial^2 A}{\partial \eta^2} \right] \cos(\pi z), \tag{5.31f}$$

$$w_{5/2}^{(1)} = 0, \tag{5.31g}$$

$$T_{5/2}^{(1)} = 0. \tag{5.31h}$$

At the $O(\kappa^3)$ order, all field variables admit a solution of the form (5.27c) with $p = 4$, $m = 1$ to 3. However, only the components of $U_3^{(0)}$ and $U_3^{(1)}$ are of interest in determining the evolution amplitude equation for the leading-order amplitude $A(\xi, \eta, \tau)$. First, we obtain the following two equations for $u_3^{(0)}$ and $w_3^{(0)}$:

$$\frac{\partial^2 u_3^{(0)}}{\partial z^2} = S_3, \tag{5.32a}$$

$$\frac{\partial^2 w_3^{(0)}}{\partial z^2} = Q_3, \tag{5.32b}$$

where

$$S_3 = \left(\frac{\pi^2}{8}\right)\left[\left(\frac{1}{\mathrm{Pr}}\right) + 9/8\right]\frac{\partial |A|^2}{\partial \eta}\cos(2\pi z), \tag{5.32c}$$

$$Q_3 = (3/32)\left[\left(\frac{1}{\mathrm{Pr}}\right) + 3/8\right]\frac{\partial |A|^2}{\partial \eta}\cos(2\pi z). \tag{5.32d}$$

As a consequence, in order to satisfy the boundary conditions

$$\frac{\partial u_3^{(0)}}{\partial z} = 0, \quad w_3^{(0)} = 0 \quad \text{at } z = 0, 1, \tag{5.32e}$$

we must enforce the following two compatibility conditions on the forcing terms in (5.32a, b):

$$\int_0^1 S_3 \,\mathrm{dz} = 0, \quad \int_0^1 Q_3 \,\mathrm{dz} = 0, \tag{5.32f}$$

which are again identically satisfied.

Next, for $w_3^{(1)}$ and $T_3^{(1)}$ we derive a system of two non-homogeneous equations analogous to the system (5.30d, e) for $w_2^{(1)}$ and $T_2^{(1)}$, but with F_3 and G_3 on the right-hand side, such that

$$\begin{aligned} F_3 \;=\; & 3i\left(\frac{\pi^3}{2\sqrt{2}}\right)\left\{\left(\frac{1}{\mathrm{Pr}}\right)\frac{\partial A}{\partial \tau} + \frac{\partial^2 A}{\partial \xi^2}\right. \\ & \left. - (16/3)\left[\frac{\partial A}{\partial \xi} - \left(\frac{i}{2k_c}\right)\frac{\partial^2 A}{\partial \eta^2}\right]^2\right\}\sin(\pi z), \end{aligned} \tag{5.32g}$$

$$\begin{aligned} G_3 \;=\; & -9i\left(\frac{\pi^3}{2\sqrt{2}}\right)\left\{\frac{\partial A}{\partial \tau} - \frac{\partial^2 A}{\partial \xi^2} + (4/3)\left[\frac{\partial A}{\partial \xi} - \left(\frac{i}{2k_c}\right)\frac{\partial^2 A}{\partial \eta^2}\right]^2\right. \\ & \left. - \left(\frac{2}{9\pi^2}\right) rA - \left(\frac{\pi^2}{8}\right)\cos(2\pi z)A \mid A^2\right\}\sin(\pi z). \end{aligned} \tag{5.32h}$$

The boundary conditions are

$$w_3^{(1)} = \frac{\partial^2 w_3^{(1)}}{\partial z^2} = T_3^{(1)} = 0, \quad \text{at } z = 0 \text{ and } 1. \tag{5.32i}$$

Again, the orthogonality with adjoint eigenfunctions requires that

$$\int_0^1 [F_3 w^* + G_3 \Theta^*]\,\mathrm{d}z = 0, \tag{5.32j}$$

thereby, finally leading to the above evolution equation for the amplitude function $A(\tau, \xi, \eta)$, which appears first the solution (5.28a–e) of the $O(\kappa)$ problem:

$$\left[1+\left(\frac{1}{\mathrm{Pr}}\right)\right]\frac{\partial A}{\partial \tau} = 4\left[\frac{\partial A}{\partial \xi} - \left(\frac{i}{2k_c}\right)\frac{\partial^2 A}{\partial \eta^2}\right]^2 + \left(\frac{2}{9\pi^2}\right) rA - \left(\frac{\pi^2}{16}\right) A|A|^2. \tag{5.33}$$

If

$$x = 2k_c\xi, \quad y = 2k_c\eta, t = 16\left[1+\left(\frac{1}{\mathrm{Pr}}\right)\right]k_c^2\tau, \tag{5.34a}$$

then for the new amplitude function $B(t, x, y)$, such that

$$B = \left(\frac{\pi}{16k_c}\right) A\left(\frac{x}{2k_c}, \frac{y}{2k_c}, \frac{t}{16(1+\frac{1}{\mathrm{Pr}})k_c^2}\right), \tag{5.34b}$$

the evolution equation for $B(t, x, y)$ takes the final form:

$$\frac{\partial B}{\partial t} = \left[\frac{\partial B}{\partial x} - i\frac{\partial^2 B}{\partial y^2}\right]^2 + \mu B - B|B|^2, \tag{5.35}$$

with

$$\mu = \left(\frac{r}{36\pi^4}\right).$$

The reduced amplitude equation (5.35), for B, is the amplitude equation previously derived in 1969 by Newell and Whitehead [13]. Our above derivation of the amplitude equation (5.35) is directly suggested by the paper of Coullet and Huerre [14].

For equation (5.35) we can obtain first a family of stationary periodic (in x) solutions, namely:

$$B_{\mathrm{st}} = Q\exp[iqx), \tag{5.36}$$

where the amplitude Q is given by the relation

$$(\mu - q^2)Q - Q^3 = 0 \Rightarrow Q = (\mu - q^2)^{1/2}. \tag{5.37}$$

In order to study the stability of this pattern, we make a change of variables;

$$B(t, x, y) = [Q + \rho(t, x, y)]\exp[i(qx + \varphi)], \tag{5.38a}$$

with $\varphi = \varphi(t, x)$.

In this case, from (5.35) we obtain two equations

$$\frac{\partial \rho}{\partial t} = -2Q^2\rho - 2\rho Q\frac{\partial \varphi}{\partial x} + \frac{\partial^2 \rho}{\partial x^2} + 2q\frac{\partial^2 \rho}{\partial y^2}; \tag{5.38b}$$

$$\frac{\partial \varphi}{\partial t} = (2q/Q)\frac{\partial \rho}{\partial x} + \frac{\partial^2 \varphi}{\partial x^2} + 2q\frac{\partial^2 \varphi}{\partial y^2}. \tag{5.38c}$$

Thus the spatial pattern may be subject to two possible modes of perturbations. The amplitude mode associated with the variable ρ is governed by equation (5.38b) and in the long-wavelength approximation,

$$\frac{\partial}{\partial x} \ll 1, \quad \frac{\partial}{\partial y} \ll 1,$$

this amplitude mode is highly damped.

By contrast, the remaining variable φ corresponds to the marginal phase mode, its dynamics being governed by equation (5.38c) and, again, in the long-wavelength limit,

$$\frac{\partial \varphi}{\partial t} = 0;$$

this mode is neutrally stable.

To describe the long-wavelength dynamics of the phase mode φ, it is legitimate to assume that the amplitude ρ is adiabatically slaved to the slowly-varying phase. To leading order, the amplitude equation (5.38b) can then be approximated by

$$\rho \sim -(q/Q)\frac{\partial \varphi}{\partial x}, \tag{5.39}$$

and substituting in (5.38c) gives rise to the phase evolution equation

$$\frac{\partial \varphi}{\partial t} = \left[1 - \left(\frac{2q^2}{Q^2}\right)\right]\frac{\partial^2 \varphi}{\partial x^2} + 2q\frac{\partial^2 \varphi}{\partial y^2}, \tag{5.40}$$

where, according to (5.37),

$$1 - \left(\frac{2q^2}{Q^2}\right) = \frac{(\mu - 3q^2)}{(\mu - q^2)} = \beta. \tag{5.41}$$

Finally, phase fluctuations are governed by the single diffusive equation

$$\frac{\partial \varphi}{\partial t} = \beta\frac{\partial^2 \varphi}{\partial x^2} + 2q\frac{\partial^2 \varphi}{\partial x^2}. \tag{5.42}$$

The signs of β and q control, respectively, the so-called Eckhaus and zig-zag instability. We note that q can change sign if the basic pattern is zig-zag unstable, and if $q > 0$, the phase is diffusive in y and no zig-zag instability can take place. If $q < 0$, the medium is zig-zag unstable and additional terms need to be brought into equation (5.42) to describe possible two-dimensional soliton lattices!

A simplified case for the amplitude evolution equation (5.35), is closely linked to the assumption that the amplitude B (assumed real) is a function only of time t. In such a case we derive from (5.35) the so-called Landau–Stuart equation:

$$-\frac{\mathrm{d}B}{\mathrm{d}t} = -\left(\frac{1}{36\pi^4}\right) rB + B|B|^2, \tag{5.43}$$

where $r > 0$.

5.5 Instability and Route to Chaos in RB Thermal Convection

We have already noted that the RB thermal convection in a fluid layer heated from below, with a non-deformable upper, free surface and without surface tension, represents the simplest example of hydrodynamic instability and transition to turbulence (as a temporal chaos) in a fluid system.

The only systematic analytical method for analyzing the manifold of 3D nonlinear steady solutions of the RB equations is the perturbation approach (as in Section 5.4) based on the small parameter κ. This approach is particularly appropriate in the case of convection because the instability occurs in the form of infinitesimal disturbances.

In particular, both evolution amplitude equations (5.35) and (5.43) derived above have played an important role in analytical investigations of hydrodynamic instability in the 60 years from the outset of the so-called 'finite-dimensional dynamical system approach to turbulence'.

In the framework of this approach, the pioneering role (20 years after Landau's theory [15]) is ascribed to the Lorenz dynamical system [16], which is a system of three relatively simple ordinary (but nonlinear) differential equations of the following form, for the three amplitude functions of time t, $A(t)$, $B(t)$ and $C(t)$:

$$\frac{\mathrm{d}A}{\mathrm{d}t} = -10A + 10B, \tag{5.44a}$$

$$\frac{\mathrm{d}B}{\mathrm{d}t} = -AC + 28A - B, \tag{5.44b}$$

$$\frac{dC}{dt} = AB - (8/3)C. \tag{5.44c}$$

Such a Lorenz system (5.44a–c) is derived for the Rayleigh–Bénard two-dimensional problem when, instead of the constraint (5.20a) for the velocity vector V, we consider in the two-dimensional case ($\partial/\partial y \equiv 0$ and $v \equiv 0$) the reduced 2D equation

$$\frac{\partial u}{\partial x} + \frac{\partial w}{\partial z} = 0 \Rightarrow u = \frac{\partial \psi}{\partial z}, \quad w = -\frac{\partial \psi}{\partial x}. \tag{5.45a}$$

In such a case, after the elimination of Π, we derive from (5.20b) for the stream function ψ, the equation

$$\left(\frac{1}{\mathrm{Pr}}\right)\left[\frac{\partial \nabla^2 \psi}{\partial t} + D(\psi; \nabla^2 \psi)\right] + \frac{\partial T}{\partial x} = \nabla^2(\nabla^2 \psi), \tag{5.45b}$$

where

$$D(\psi; f) = \frac{\partial \psi}{\partial z}\frac{\partial f}{\partial x} - \frac{\partial \psi}{\partial x}\frac{\partial f}{\partial z},$$

and for temperature T, from (5.20c), we obtain

$$\frac{\partial T}{\partial t} + D(\psi; T) + \mathrm{Ra}\frac{\partial \psi}{\partial x} = \nabla^2 T. \tag{5.45c}$$

According to Lorenz [16], we write the solution of the system of two equations (5.45b, c), for $\psi(t, x, z)$ and $T(t, x, z)$ as

$$\psi = \mathrm{Pr}\left[A(t)\sin\left(\frac{\pi x}{\lambda}\right)\sin(\pi z)\right] \tag{5.46a}$$

and

$$T = \mathrm{Ra}\left[B(t)\cos\left(\frac{\pi x}{\lambda}\right)\sin(\pi z) + C(t)\sin(2\psi z)\right]. \tag{5.46b}$$

The above approximate form (5.46a, b) for ψ and T is compatible with the following boundary conditions (free-free case):

$$\psi = 0 \quad \text{and} \quad \frac{\partial^2 \psi}{\partial z^2} = 0, \quad T = 0 \quad \text{at } z = 0 \text{ and } z = 1, \tag{5.47a}$$

$$\psi = 0 \quad \text{and} \frac{\partial^2 \psi}{\partial x^2} = 0, \quad \frac{\partial T}{\partial x} = 0 \quad \text{at } x = 0 \text{ and } x = \lambda. \tag{5.47b}$$

Then, by a Galerkin technique, the next step is to substitute the approximate (three amplitudes) solution (5.46a, b), into two 2D equations (5.45b, c), then requiring the residue to be orthogonal to each function of the set (5.46a, b).

More precisely, after this substitution, on the one hand, the equation obtained from (5.45b) is multipied by $\sin(p\pi x/\lambda)\sin(\pi z)$ and, on the other hand, the equation obtained from (5.45c) is multipied by $\cos(p\pi x/\lambda)\sin(\pi z)$. These two new equations are then integrated over x, between $x = 0$ and $x = \lambda$, and over z, between $z = 0$ and $z = 1$. These above two orthogonality conditions with:

$$\left(\frac{2}{\pi}\right)\int_0^{\pi} \sin(iy)\sin(jy)\,\mathrm{d}y = \delta_{ij}, \tag{5.48a}$$

and

$$\left(\frac{2}{\pi}\right)\int_0^{\pi} \cos(iy)\sin(jy)\,\mathrm{d}y = \delta_{ij}, \tag{5.48b}$$

lead to the following system (5.49a) of three equations for the three reduced (see (5.50) time-dependent coefficients $X(t)$, $Y(t)$ and $Z(t)$:

$$\begin{aligned}
\frac{\mathrm{d}X}{\mathrm{d}t} &= \mathrm{Pr}\,(Y - X),\\
\frac{\mathrm{d}Y}{\mathrm{d}t} &= -XZ + r_0 X - Y,\\
\frac{\mathrm{d}Z}{\mathrm{d}t} &= XY - bZ,
\end{aligned} \tag{5.49a}$$

with

$$r_0 = \mathrm{Ra}\,\frac{q^2}{(\pi^2 + q^2)^3}, \tag{5.49b}$$

the ‘bifurcation’ parameter.

The relation between $X(t)$, $Y(t)$ and $Z(t)$, and the amplitudes $A(t)$, $B(t)$, $C(t)$ in (5.46a, b) is

$$\begin{aligned}
X(t) &= \left(\frac{\pi q}{\sqrt{2}}\right)\left[\frac{1}{(\pi^2 + q^2)}\right]A(t),\\
Y(t) &= -\left(\frac{1}{\mathrm{Ra}}\right)\left(\frac{\pi q^2}{\sqrt{2}}\right)\left[\frac{1}{(\pi^2 + q^2)}\right]B(t),\\
Z(t) &= -\left(\frac{1}{\mathrm{Ra}}\right)\left[\frac{\pi q^2}{(\pi^2 + q^2)}\right]C(t).
\end{aligned} \tag{5.50}$$

The system (5.49a), with (5.49b), was first obtained by Lorenz [16]. In a book by Sparrow [17] the reader can find a detailed and careful theory of the above (5.49a) *à la* Lorenz system. We observe that the condition

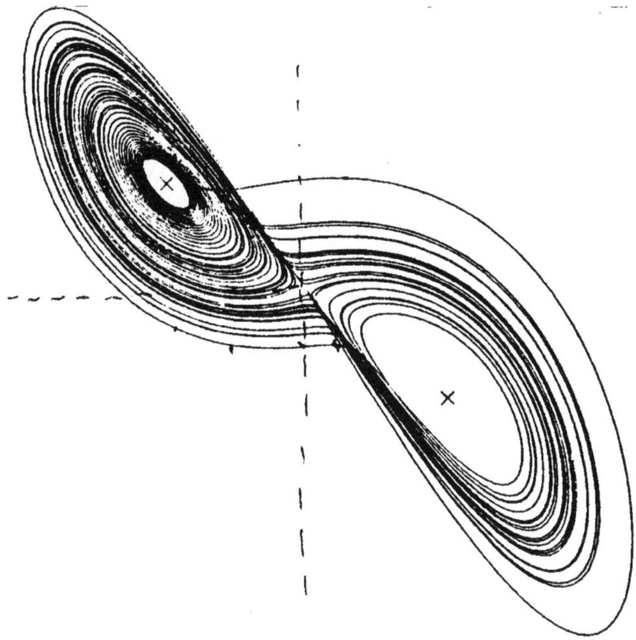

Fig. 5.4a Lorenz strange attractor – cross-section (A–B) in phase space (A, B, C). Reprinted with kind permission from [18].

$$\frac{\partial}{\partial X}\left(\frac{\mathrm{d}X}{\mathrm{d}t}\right)+\frac{\partial}{\partial Y}\left(\frac{\mathrm{d}Y}{\mathrm{d}t}\right)+\frac{\partial}{\partial Z}\left(\frac{\mathrm{d}Z}{\mathrm{d}t}\right)=-[\mathrm{Pr}+b+1] \tag{5.51}$$

shows that the Lorenz system (5.49a) is real-dissipative!

Thanks to system (5.44), Lorenz, in his 1963 paper [16] 'exhibits' for the first time, via a numerical computation of the system (5.44), a 'strange attractor'; see Figures 5.4a–c, reproduced from [18, pp. 480–482].

The Lorenz system (5.49a) in a steady state has as constant solution

$$X_{\mathrm{st}}=\pm[b(r_0-1)],\quad Y_{\mathrm{st}}=\pm[b(r_0-1)],\quad Z_{\mathrm{st}}=r_0-1, \tag{5.52a}$$

$$r_0-1=\frac{(\mathrm{Ra}-\mathrm{Ra}_c)}{\mathrm{Ra}_c}. \tag{5.52b}$$

According to the Routh and Hurwitz criterion, a particular value of r_0 (noted r_0^*) exists, for which the above steady-state solution (5.52) is unstable – from this value for $\mathrm{Pr}=10$ and $b=8/3$ (as in (5.44)) we obtain

$$r_0^*-1=23.74. \tag{5.53}$$

As a consequence, a steady-state solution (5.52) of the Lorenz system is unstable when (5.53) is realized. But, because the Lorenz system has no

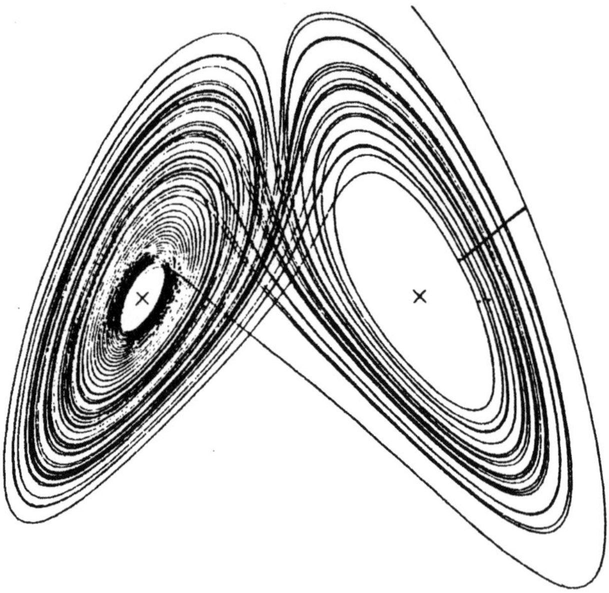

Fig. 5.4b Lorenz strange attractor – cross-section (A–c) in phase space (A, B, C). Reprinted with kind permission from [18].

other steady-state solutions than (5.52), then it is possible to conclude that:

When $r_0 > r_0^*$, the solution of a Lorenz system is necessarily dependent on time t!

In Figures 5.4a–c, the crosses '×' indicate the steady-state values at the given Pr and Ra (related to r_0 by (5.52b)) numbers. In these figures, the system travels along a very irregular and complicated path around the steady-state values, inside a limit region of the phase space, according to (5.51). This chaotic behavior is usually invoked in the transition to tubulence. Indeed, even if the Lorenz system seems well deterministic (i.e. A, B and C, given initial conditions, can be known at any time), due to its sensitivity to initial conditions, the solution is almost unpredictable and obviously this unpredictability of the flow field is also a main feature of turbulence. The Lorenz system is not only the first but also the most famous example of deterministic chaos, and is an explanation of a possible route to turbulence. In Chapter 6, devoted to the 'deep thermal convection problem', the routes (scenarios) to turbulence are discussed.

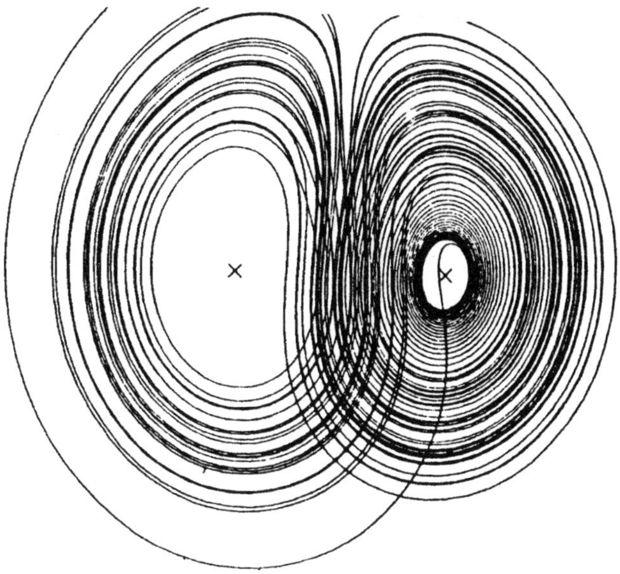

Fig. 5.4c Lorenz strange attractor – cross-section (B–C) in phase space (A, B, C). Reprinted with kidn permission from [18].

Finally, we observe that from the Lorenz system (5.49a) it is possible to derive a Landau single equation, e.g., for $X(t)$ when we assume that the two other amplitudes are independent of time t. Namely, in such a case the first and third equations of (5.49a) give

$$Y = X \quad \text{and} \quad Z = \left(\frac{1}{b}\right) X_2$$

and as a consequencc , from the second equation (since $Y = X$ and $Z = (1/b)X^2$) of the system (5.49a), for $X(t)$ we obtain

$$\frac{\mathrm{d}X}{\mathrm{d}t} = (r_0 - 1)X - \left(\frac{1}{b}\right) X_3. \tag{5.54}$$

Obviously, and unfortunately, this analytic method, a perturbation expansion and a Galerkin technique are of limited usefulness when the Rayleigh number Ra is increased much beyond its critical value. For this case, numerical computations have been performed by various authors, e.g., by Curry et al. [20].

Concerning the fully nonlinear, RB convection problem, direct numerical methods have been used. For this, it is convenient to define five values of the Ra that distinguish various flow regimes; however, in any given system, some or all of these Ra may be non-existent!

First, the linear critical Ra_c is defined so that the heat-conduction motionless basic state of the fluid/liquid is stable under infinitesimal disturbances for $\mathrm{Ra} < \mathrm{Ra}_c$ and is unstable for $\mathrm{Ra} > \mathrm{Ra}_c$. As Ra increases beyond Ra_c, steady-state convection rolls appear and these rolls are 2D in character.

Next, Ra_1 is defined as that Rayleigh number at which these rolls undergo a bifurcation to a periodic, possibly 3D oscillatory state; periodic convection ensues as Ra increases above Ra_1.

At Ra_2 a second (normally incommensurate) frequency appears, so the flow is quasi-periodic, but if this second frequency is commensurable with the first, then a phase locking occurs, so the flow is still periodic but with a new frequency. Then at Ra_t the flow undergoes transition to a chaotic state with broadband frequency response. Of course there may also be transitional Rayleigh numbers, Ra_n, for $n > 3$, in which n distinct incommensurate frequencies are observable.

In fact, the Ruelle et al. scenario, considered in detail [19, section 10.3] and also in Chapter 6 in relation with deep thermal convection, suggests that $\mathrm{Ra}_t = \mathrm{Ra}_3$.

There is another critical Rayleigh number that it is useful to define although its existence is not anticipated by the generic mathematical analysis outlined in [20]. This Ra_1' is defined as that value of Ra at which a reverse transition from quasi- periodic or chaotic flow to periodic flow occurs as Ra increases! Though the flow just below Ra_1' has at least two incommensurate frequencies present, that just above Ra_1' has but one significant frequency. Furthermore, there may even be bands of Rayleigh numbers between Ra_t and Ra_1' in which the flow reverts to quasi-periodic behavior.

Finally, we emphasize once again that some or all of these putative critical Ra values may not exist in any particular realization of a real thermal RB convection flow. Below we present some numerical results obtained by Curry et al. in 1984 [20], with free-slip (no-stress) conditions, and periodic conditions in x and y, through a spectral method (*à la* Orszag [21]). The dependent flow variables are expanded in a Fourier series, and then the nonlinear terms are evaluated by fast-transform methods with aliasing terms usually removed; time-stepping is done by a leapfrog scheme for the nonlinear terms and an implicit scheme for the viscous terms. The pressure term is computed in a Fourier representation by local algebraic manipulation of the constraint $\nabla \cdot \mathbf{u} = 0$.

In Figures 5.5 and 5.6, for the 3D case, some results for the transition are presented for 16^3 and $32^2 \times 16$ runs, respectively. The curves show $u(\mathbf{p}, t)$

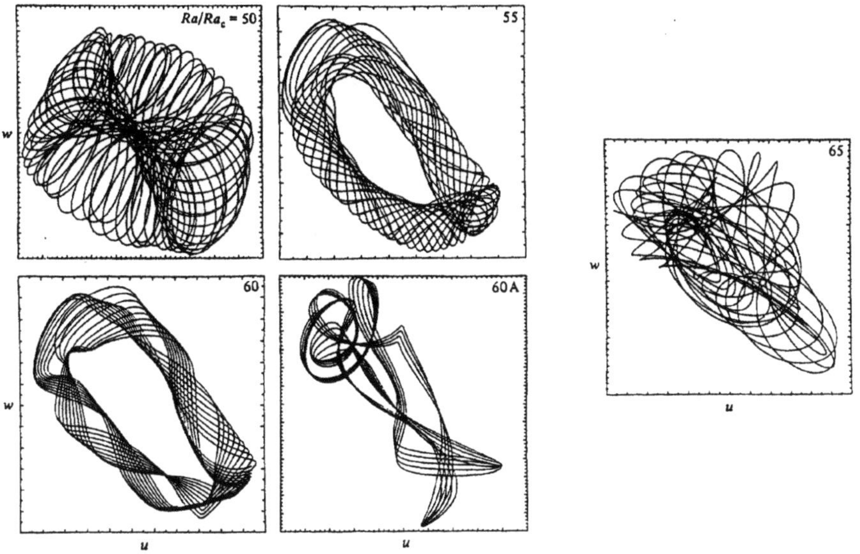

Fig. 5.5 2D phase projections of the (u, w) fields for resolution 16^3 runs. Reprinted with kind permission from [20].

versus $w(\mathbf{p}, t)$, where $\mathbf{p}$ is a coordinate near the midpoint of the box, for $3 < t < 4$.

The two plots in Figure 5.5, in the case of 16^3 runs at $\mathrm{Ra} = 60\mathrm{Ra}_c$, are obtained for different initial conditions that show dependence of the final quasi-periodic state on initial data. This difference may also suggest alternative routes to chaos, in addition to the Ruelle et al. scenario! At $\mathrm{Ra} = 65\mathrm{Ra}_c$ (in the case of 16^3 runs) the phase portrait suggests *chaotic* flow, although the spectrum of the flow is still dominated by phase-locked lines. Only the velocity components have phase plots that project onto a torus; those that involve the temperature appear much more random.

In contrast with the 16^3 results plotted in Figure 5.5, the plot of (u, w) in Figure 5.6, for the $32^2 \times 16$ runs at $50\mathrm{Ra}_c$, is now a simple circle, corresponding to the presence of only a single frequency. The phase plot at $\mathrm{Ra} = 60\mathrm{Ra}_c$ has much the same appearance as with 16^3 resolution. At $\mathrm{Ra} = 70\mathrm{Ra}_c$, we again observe a chaotic regime.

The transition scenario reported here for 3D closely parallels route I described by Gollub and Benson in [22]; the qualitative differences are related to the existence or non-existence of phase-locked regimes. Although such regimes may be present for some range of parameters, they have not been

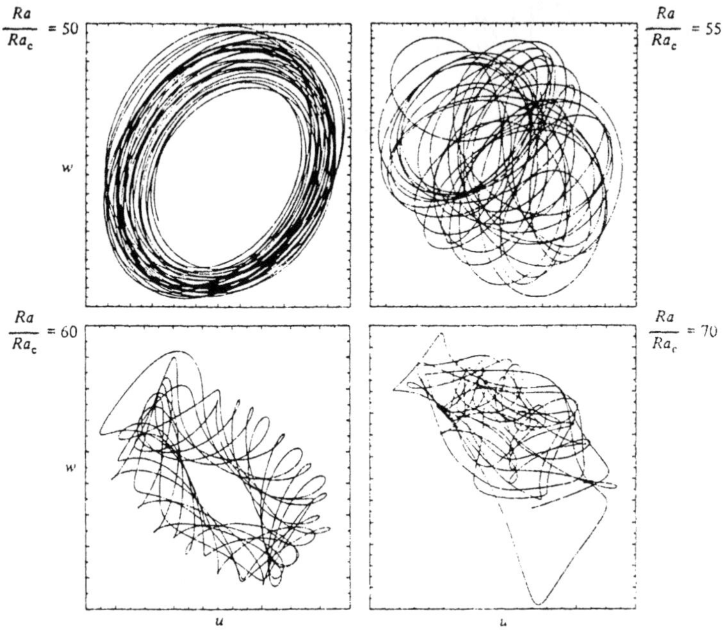

Fig. 5.6 2D phase projections of the (u, w) fields for resolution $32^2 \times 16$ runs. Reprinted with kind permission from [20].

observed by the above authors because of the coarseness of the partition through parameter space.

In Howard and Krishnamurti's paper [23], large-scale flow in (turbulent) convection is considered, and to this end the three Fourier components that lead to Lorenz's famous three equations (5.44c), were augmented with three additional components, leading to a sixth-order system. Namely:

$$\frac{\mathrm{d}A}{\mathrm{d}t} = -\mathrm{Pr}\,aA + \mathrm{Pr}\,bD + cBC, \tag{5.55a}$$

$$\frac{\mathrm{d}B}{\mathrm{d}t} = -\mathrm{Pr}\,B - dAC, \tag{5.55b}$$

$$\frac{\mathrm{d}C}{\mathrm{d}t} = -\mathrm{Pr}eC - \mathrm{Pr}\,fF - gAB, \tag{5.55c}$$

$$\frac{\mathrm{d}D}{\mathrm{d}t} = -hD + \mathrm{Ra}\,\alpha A - \alpha AE - \left(\frac{\alpha}{2}\right) BF, \tag{5.55d}$$

$$\frac{\mathrm{d}E}{\mathrm{d}t} = -4E + \left(\frac{\alpha}{2}\right) AD, \tag{5.55e}$$

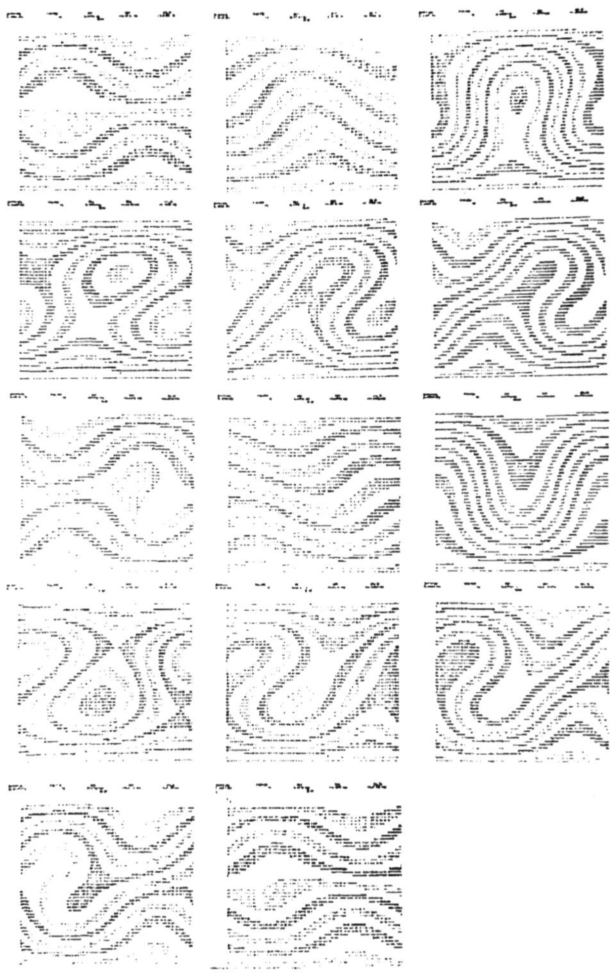

Fig. 5.7 Temperature field at successive time intervals within one oscillation period; Pr = 1.0, $\alpha = 1.2$ and Ra = 55. Reprinted with kind permission from [23].

$$\frac{\mathrm{d}F}{\mathrm{d}t} = -mF - \mathrm{Ra}\,\alpha C + \left(\frac{\alpha}{2}\right) BD. \tag{5.55f}$$

In this sixth-order dynamical system (5.55a–f), the scalars a, b, c, d, e, f, g, h and m are scalars depending on wave number α. One example of the temperature field at times equally spaced within one period is shown in Figure 5.7 for Pr = 1.0, Ra = 55, and similar orbits were found for Pr = 0.1 and Pr = 10, for Ra slightly in excess of the critical Ra for the onset of oscillatory convection

The main results of a study of the bifurcations of this six-amplitudes system (5.55) are that: after the second bifurcation, steady tilted cells are the stable flow, and after the third bifurcation, stable limit cycles are found for a range of Ra. But within this range of Ra, where stable limit cycles are found, there are narrow sub-ranges of aperiodic flows, and the occurrence of this chaotic behaviour is shown to be related to the existence of heterocline orbit pairs. A hot plume or bubble is seen to form in the lower part of the region, then rise and tilt from lower left to upper right. Later, a cold plume forms in the upper part, sinks and tilts from upper right to lower left. It also shows a leftward-propagating wave in the isotherms near the bottom of the layer and a rightward-propagating wave near the top of the layer.

Concerning the scenarios/routes to chaos, we mention here that usually the investigations are linked with three prominent routes which have been theoretically and experimentally successful:

- first, the *Ruelle–Takens–Newhouse scenario* in which, after a few bifurcations, an invariant point set in phase space appears; this set is not a torus but a strange attractor, the motion being aperiodic;
- second, the *Feigenbaum scenario* in which case the route to chaos involves successive periodic doubling (subharmonic) bifurcations of a (simple) periodic flow, the chaotic attractor being not (strictly) a strange attractor (*à la* Ruelle–Takens); and
- third, the *Pomeau–Manneville* scenario, when transition to 'turbulence' is realized through intermittency.

In Chapter 7 we will return to these three routes to chaos in the case of the Bénard deep thermal convection problem. In [19, chapter 10, pp. 387–448] the reader can find an overview relative to a 'finite-dimensional dynamical system approach to turbulence' which contains mostly arguments about current research but is mainly discursive from a fluid dynamicist's point of view.

5.6 Some Complements

In the vicinity of the threshold of RB thermal convection, it is commonly known that the dynamics can be described by means of an *amplitude equation*; see, for example, the evolution equations (5.35) and (5.43), derived above in Section 5.4.

In the 2D case when the starting RB equations are (5.45b) and (5.45c), with boundary conditions (5.47a), and we investigate the nonlinear stability

2D, (x, z), problem of an ideal pattern of straight rolls parallel to the y direction, we can derive a so-called 'Landau–Ginzburg equation' which has the following reduced form (see e.g., [24]):

$$\frac{\partial A}{\partial t} = \frac{\partial^2 A}{\partial x^2} + rA - |A|^2 A. \tag{5.56}$$

Usually, a multiple-scales perturbation method is used to compute the coefficients in the non-reduced amplitude equation obtained at κ^3 order via the Fredholm alternative. In derivation of (5.56) the same arguments as in [25] have been used. There is a lot of literature devoted to the analysis of (5.56); for instance, stability analysis of this equation was accomplished in [26] and the question of existence of a maximal attractor for this equation and its characterization was dealt with in [27]. The Landau–Ginzburg equation was also analysed numerically in [28]. All these known results for the Landau–Ginzburg equation apply as such to equation (5.56), sometimes with only slight modifications, such as a change of variables, are necessary.

The *RB problem in rarefied gases* has in recent years attracted considerable interest as a model problem for studying such fundamental issues as the mechanisms of instability and self-organization at the molecular level and their relation to macrocospic phenomena (according to [29], where the reader can find various recent references).

In [29] the transition to convection in the RB problem at small Knudsen (Kn) number is studied via a linear temporal stability analysis of the compressible 'slip-flow' problem. Indeed, significant convection only occurs at small $O(10^{-2})$ Knudsen numbers. In a Cartesian system of coordinates (x_1, x_2, x_3) whose origin lies on the lower wall, $x_2 = 0$, and whose x_2 axis is pointing upwards – opposite to the direction of g, the acceleration of gravity – the following dimensionless equations are used as starting equations:

$$\frac{\partial \rho}{\partial t} + \frac{\partial}{\partial x_i}(\rho u_i) = 0, \tag{5.57a}$$

$$\rho\frac{Du_i}{Dt} = -(1/2)\frac{\partial p}{\partial x_i} + \mathrm{Kn}\frac{\partial}{\partial x_j}\left[2\mu\left(e_{ij} - (1/3)\frac{\partial u_i}{\partial x_i}\right)\right] - \left(\frac{1}{\mathrm{Fr}}\right)\rho\delta_{i_2}, \tag{5.57b}$$

$$\rho\frac{DT}{Dt} = \left(\frac{\gamma}{\mathrm{Pr}}\right)\mathrm{Kn}\frac{\partial}{\partial x_j}\left[\kappa\frac{\partial T}{\partial x_j}\right] - (\gamma-1)p\left(\frac{\partial u_i}{\partial x_i}\right) + 2(\gamma-1)\mathrm{Kn}\,\Phi, \tag{5.57c}$$

$$p = \rho T. \tag{5.57d}$$

Appearing in (5.57b) and (5.57c) are the rate-of-strain tensor,

$$e_{ij} = (1/2)\left[\frac{\partial u_i}{\partial x_j} + \frac{\partial u_j}{\partial x_i}\right]$$

and the rate of dissipation,

$$\Phi = 2\mu\left[e_{ij}e_{ij} - (1/3)\left(\frac{\partial u_i}{\partial x_i}\right)^2\right].$$

The Knudsen number is

$$\mathrm{Kn} = \frac{l}{d},$$

which is the ratio of the mean free path l to the macroscopic scale d, the distance between the walls $x_2 = 0$ and $x_2 = 1$.

The Froude number (describing the relative magnitudes of gas inertia and gravity) is

$$\mathrm{Fr} = \frac{U_{\mathrm{th}}^2}{gd},$$

where $U_{\mathrm{th}} = (2RT_h)^{1/2}$ is the mean thermal speed, T_h the absolute temperature of the lower (hot) wall and R is the gas constant.

The Prandtl number is

$$\mathrm{Pr} = \frac{\mu_h C_p}{\kappa_h}$$

and

$$\gamma = \frac{C_p}{C_v}.$$

The pressure is normalized by $\rho_h RT_h$ and as model of molecular interaction in [29] the authors chose:

$$\gamma = 5/3, \quad \mathrm{Pr} = 2/3 \quad \text{and} \quad \mu(T) = \kappa(T) = \left(\frac{5\pi^{1/2}}{16}\right)T^{1/2}. \tag{5.58a}$$

Finally, the above equations are supplemented by the normalization condition

$$\int_0^1 \rho \mathrm{d}x_1\, \mathrm{d}x_2 = 1, \tag{5.58b}$$

specifying the total amount of gas, between the walls, and by the boundary conditions

$$u_2 = 0, \quad u_{1,3} = \zeta\frac{\partial u_{1,3}}{\partial x_2}, \quad T = 1 + \tau\frac{\partial T}{\partial x_2} \quad \text{at } x_2 = 0, \tag{5.59a}$$

and

$$u_2 = 0, \quad u_{1,3} = -\zeta \frac{\partial u_{1,3}}{\partial x_2}, \quad T = R_T - \tau \frac{\partial T}{\partial x_2} \quad \text{at } x_2 = 1, \tag{5.59b}$$

respectively.

In (5.59a, b), $R_T = T_c/T_h$ denotes the ratio of cold-and hot-wall temperatures, $\zeta = 1.1466\,\mathrm{Kn}$ and $\tau = 2.1904\,\mathrm{Kn}$, according to Cercignani's classical book [30], where the reader can find for the above problem, the steady 'pure convection'

$$u_s = 0$$

the following solution:

$$T_s = (Ax_2 + B)^{2/3}, \quad \rho_s = \left(\frac{C}{T_s}\right) \exp\left[-\left(\frac{6}{A\,\mathrm{Fr}}\right) T_s^{1/2}\right], \tag{5.60}$$

in which the constants A, B and C are determined by use of (5.58)–(5.59b).

In [29] each of the above-mentioned fields is generically represented by the sum: steady reference state (5.60) plus perturbed part, and neglecting nonlinear terms a perturbation (linear) problem is derived. From my point of view, before any numerical simulation, it would be interesting to derive, instead of the above problem, a rational approximate model problem? As a first (simple) example it seems possible to consider the limiting case when, $\mathrm{Kn} \to 0$ and investigate the 'passage' to a continuum regime!

If we observe that $\mathrm{Kn} = \mathrm{M}/\mathrm{Re}$, then $\mathrm{Kn} \ll 1$ is realized if

(1) the Mach number $\mathrm{M} \ll 1$ (hyposonic regime [32], but see also [33]) and Re fixed to not very low (Stokes and Oseen case); or,
(2) The Reynolds number $\mathrm{Re} \gg 1$ (Prandtl boundary layer approximation) with M fixed to not very large (hypersonic case).

However, it seems that this 'passage' to a continuum regime is not uniformly valid and may locally fail at certain parts of the flow field (see [34, 35]). In [30], the reader can find a very pertinent modern presentation of *Rarefied Gas Dynamics* by Cercignani, and in [31] the 2D problem relative to RB flow of a rarefied gas has been studied by means of a direct numerical simulation method. Finally, we note that in Chapter 9, devoted to 'Atmospheric Thermal Convection Problems', the reader can find a detailed asymptotic (when $\mathrm{M} \ll 1$) derivation of the Boussinesq approximate equations for a thermally perfect gas – this derivation for a gas being rather different from the derivation, given in Section 3.3, for a weakly expansible liquid.

Below for the standard RB model problem written in the following form (see, for instance, [36, pp. 51–55]):

$$\frac{\partial \mathbf{u}}{\partial t} + (\mathbf{u} \cdot \nabla)\mathbf{u} = -\nabla p + \Delta \mathbf{u} + \mathrm{Ra}\,\theta \mathbf{k},$$
$$\nabla \cdot \mathbf{u} = 0,$$
$$\mathrm{Pr}\left[\frac{\partial \theta}{\mathrm{d}\partial t} + \mathbf{u} \cdot \nabla \theta\right] = \mathrm{Ra}\, w + \Delta \theta, \tag{5.61a}$$

with

$$\mathbf{u} = \theta = 0, \quad \text{at } z = 0 \text{ and } z = 1, \tag{5.61b}$$

we shall now show by energy stability theory that *sub-critical instability is not possible* (as has been already noted in Section 3.4).

For this, as in [36], we consider the simplest, natural 'energy', for the RB system of three equations (5.61a), formed by adding the kinetic and thermal energies of the perturbations, and so we define:

$$E(t) = (1/2)\|\mathbf{u}\|^2 + (1/2)\,\mathrm{Pr}\,\|\theta\|^2, \tag{5.62}$$

where

$$\|f\|^2 = \int_V f^2\,\mathrm{d}V.$$

We differentiate $E(t)$, substitute for $\partial \mathbf{u}/\partial t$ and $\partial\theta/\partial t$ from (5.61a), and use the boundary conditions, (5.61b), to find

$$\frac{\mathrm{d}E}{\mathrm{d}t} = 2\,\mathrm{Ra}\langle w\theta\rangle - [D(\mathbf{u}) + D(\theta)], \tag{5.63a}$$

where $D(f)$ denotes the Dirichlet integral, e.g.,

$$D(f) = \|\nabla f\|^2 = \int_V |\nabla f|^2\,\mathrm{d}V,$$

and

$$\langle fg\rangle = \int_V fg\,\mathrm{d}V.$$

Then we introduce

$$I = 2\langle w\rangle \quad \text{and} \quad D = D(\mathbf{u}) + D(\theta),$$

such that, from (5.63a), we can write:

$$\frac{\mathrm{d}E}{\mathrm{d}t} = \mathrm{Ra}\, I - D \le \mathrm{Ra}\, D\left[\left(\frac{1}{\mathrm{Ra}}\right) - \left(\frac{1}{R_E}\right)\right], \tag{5.63b}$$

where R_E is defined by

$$\frac{1}{R_E} = \max_H \left(\frac{I}{D}\right), \quad \text{with } H \text{ the space of admissible solutions.}$$

If now $\mathrm{Ra} < R_E$, then

$$\mathrm{Ra}\left[\left(\frac{1}{\mathrm{Ra}}\right) - \left(\frac{1}{R_E}\right)\right] = \left[\frac{(R_E - \mathrm{Ra})}{R_E}\right] = a > 0,$$

and from (5.63b),

$$\frac{\mathrm{d}E}{\mathrm{d}t} \leq -aD \leq -2a\lambda_1 E, \tag{5.63c}$$

where we have also used the classical Poincaré's inequality; from this inequality

$$\lambda_1 \|\mathbf{u}\|^2 \leq \|\nabla \mathbf{u}\|^2, \quad \lambda_1 > 0.$$

Finally, by integration,

$$E(t) \leq E(0) \exp[-2a\lambda_1 t], \tag{5.63d}$$

from which we see that

$$E \to 0 \quad \text{at least exponentially fast as } t \to \infty. \tag{5.64}$$

This demonstrates that, provided $\mathrm{Ra} < R_E$ is satisfied, the conduction motionless solution $\mathbf{u}_s = 0$ and $T = T_s(z) = T_{z=0} - \beta_s z$, is nonlinearly stable for all initial disturbances.

One the other hand, the Euler–Lagrange equations for the maximum, in the above definition, of R_E are found by using the calculus of variations and gives

$$\Delta \mathbf{u} + R_E \theta \mathbf{k} = -\nabla p; \quad \nabla \cdot \mathbf{u} = 0; \quad \Delta\theta + R_E w = 0,$$

together with the same boundary conditions as (5.61b) and the 'periodicity' conditions.

In such a case R_E satisfies the eigenvalue problem for the classical linear problem, but with $\sigma = 0$ (see (3.30a) with (3.30c–e)). Thus, for the standard RB problem

$$R_E \equiv \mathrm{Ra}_L \quad \text{– the lowest eigenvalue of the linearized theory.}$$

Furthermore, we have the confirmation that [36]:

the linear instability boundary ≡ the nonlinear stability boundary, and so no sub-critical instabilities are possible.

In [37] the reader can find various aspects of theoretical analysis of the stability of convective motions.

It is now very well established, theoretically and experimentally, that in the classical Bénard simple problem for a weakly expansible liquid layer, heated from below, separated from ambient air by an upper, deformable free surface, buoyancy (a volume-temperature-dependent density effect) is more important for a relatively thick layer, while the thermocapillarity (an interfacial/thermocapillary-temperature-dependent free surface tension-Marangoni effect) plays the dominant role in the case of significantly thin layers (or under microgravity conditions).

However, the case where both effects should be taken into account is the most typical in various (other than the classical Bénard convection problem) convection problems. Here, concerning this question, we mention the papers by Braunsfurth and Homsy [38], and Boeck et al. [39] and the reader can find various other references in both these papers. For instance, the convective phenomena in the presence of an interface in a two-layer system have attracted great attention specifically due to numerous technological applications (see, for example [39, 40]).

Usually, when the temperature grows, the interfacial tension can decrease (the normal Marangoni effect), but in some cases – for special liquids, see, e.g., [41] and references therein – the interfacial tension increase (the anomalous Marangoni effect). As this is well explained in [38], in a one-layer system (heated from below, as in Bénard experiments), the buoyancy volume forces and thermocapillary interfacial stresses act in the same direction and produce together a stationary instability, provided the Marangoni effect (for a thin layer) is normal.

For two-layer systems with an interface, the situation is more intricate and requires obviously a carefully rational analysis to obtain an approximate theoretical model with possibly both, Rayleigh and Marangoni, effects. Indeed, if the flow in the lower layer is dominant, the actions of buoyancy and thermocapillary effect are similar to those in the one-layer system (in [38] the liquids are situated between rigid horizontal plates that are kept at different temperatures). If the flow in the upper layer is dominant, the buoyancy forces and thermocapillary stresses act in oposite directions – their competitions lead to a stabilization of the stationary instability, as well as to the generation of a specific kind of linear oscillatory instability, which has been predicted theoretically [42] and observed in experiments [43]. In the case of the anomalous thermocapillary effect, one can obtain an oscillatory instability when the flow in the lower layer is dominant, and only a stationary instability in the opposite case [44].

$$A_x = d/h$$
$$A_y = w/h$$
$$Ma = \gamma\, \Delta T d/\mu\kappa$$
$$Ra = g\, \alpha\, \Delta T d^3/\nu\kappa$$
$$Pr = \nu/\kappa$$
$$G = Ra\, A_x/Ma$$
$$Ca = \gamma\, \Delta T/\sigma$$

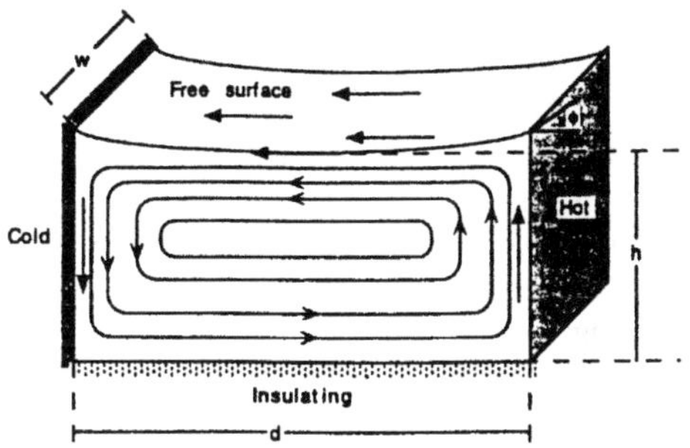

Fig. 5.8 Schematic view of the geometry of the problem considered in [38]. Reprinted with kind permission from [38].

In fact, combined thermocapillary-buoyancy convection occurs in a variety of different applications. In cavity (see Figure 5.8) this problem has been investigated in [38], where the reader can find various references related mainly to various experimental studies. A dimensionless analysis for a rectangular container is described by two aspects ratios: $A_x = d/h$ and $A_y = w/h$.

The strength of the buoyancy forces is represented by Ra, and the strength of the surface tension driving forces is given by Ma.

The dynamic Bond number G is often used as a measure of the relative strength of buoyancy to thermocapillarity driving forces; the capillary number, Ca, gives an indication of the possible surface deformation due to the surface forces. Obviously, a rational modelling of the above problem is an interesting, but certainly difficult, task!

References

1. P.G. Drazin and W.H. Reid, *Hydrodynamic Stability*. Cambridge University Press, Cambridge, 1981.
2. C. Ruyer-Quil, B. Scheid, S. Kalliadasis, M.G. Velarde and R.Kh. Zeytounian, Thermocapillary long-waves in a liquid film flow; Part 1: Low-dimensional formulation. *J. Fluid Mech.* **538**, 199–222, 2005.
3. B. Scheid, C. Ruyer-Quil, S. Kalliadasis, M.G. Velarde and R.Kh. Zeytounian, Thermocapillary long-waves in a liquid film flow; Part 2: Linear stability and nonlinear waves. *J. Fluid Mech.* **538**, 223–244, 2005.
4. E.M. Sparrow, R.J. Goldstein and V.K. Jonsson, Thermal instability in a horizontal fluid layer: Effect of boundary conditions and non-linear temperature profile. *J. Fluid Mech.* **18**, 513–528, 1964.

5. S. Chandrasekhar, *Hydrodynamic and Hydromagnetic Stability*. Clarendon Press, Oxford, 1961. See also Dover Publications, New York, 1981.
6. J.S. Turner, *Buoyancy Effects in Fluids*. Cambridge, Cambridge University Press, 1973.
7. M.G. Velarde and C. Normand, *Sci. Amer.* **243**(1), 92, 1980.
8. C. Normand, Y. Pomeau and M.G. Velarde, Convective instability: A physicist's approach. *Rev. Mod. Phys.* **49**(3), 581–624, 1977.
9. A.V. Getling, *Rayleigh Bénard Convection: Structure and Dynamics*. World Scientific, Singapore, 1998.
10. E. Bodenschatz, W. Pesch and G. Ahlers, Recent developments in Rayleigh–Bénard convection. *Annu. Rev. Fluid Mech.* **32**, 709–778, 2000.
11. P.C. Dauby and G. Lebon, Bénard–Marangoni instability in rigid rectangular containers. *J. Fluid Mech.* **329**, 25–64, 1996.
12. R.Kh. Zeytounian, *Mécanique des Fluides Fondamentale*. Springer-Verlag, Heidelberg 1991.
13. A.C. Newell and J. Whitehead, Finite bandwidth, finite amplitude convection. *J. Fluid Mech.* **38**(2), 279–303, 1966.
14. P. Coullet and P. Huerre, Resonance and phase solitons in spatially-forced thermal convection. *Physica D* **23**, 27–44, 1986.
15. L.D. Landau, On the problem of turbulence. *C.R. Acad. Sci. URSS* **44**, 311–314, 1944. See also *Collected Papers*, 387–391, Oxford, 1965.
16. E.N. Lorenz, Deterministic nonperiodic flow. *J. Atmospheric Sci.* **20**, 130–141, 1963.
17. C. Sparrow, *The Lorenz Equations: Bifurcations, Chaos and Strange Attractors*. Springer, 1982.
18. J.K. Platten and J.C. Legros, *Convection in Liqids*. Springer-Verlag, New York, 1984.
19. R.Kh. Zeytounian, *Theory and Applications of Viscous Fluid Flows*. Springer-Verlag, Heidelberg, 2004.
20. J.H. Curry et al., Order and disorder in two- and three-dimensional Bénard convection. *J. Fluid Mech.* **147**, 1–38, 1984.
21. S. Orszag, *Studies Appl. Math.* **L4**, 293–327, 1971.
22. J.P. Gollub and S.V. Benson, Many routes to turbulent convection. *J. Fluid Mech.* **100**, 449–470, 1980.
23. L.N. Howard and R. Krishnamurti, Large-scale flow in turbulent convection: A mathematical model. *J. Fluid Mech.* **170**, 385–410, 1986.
24. Z. Charki and R.Kh. Zeytounian, The Bénard problem for deep convection: Derivation of the Landau–Ginzburg equation. *Int. J. Engng. Sci.* **33**(12), 1839–1847, 1995.
25. D. Siggia and A. Zippelius, Stability of finite-amplitude convection. *Phys. Fluids* **26**, 2905, 1983.
26. P. Coullet and S. Fauve, Propagative phase dynamics for systems with Galilean invariance. *Phys. Rev. Lett.* **55**, 2857–2859, 1985.
27. C. Doering, J. Gibbon, D. Holm and B. Nicolaenko, Low-dimensional behaviour in the complex Ginzburg–Landau equation. *Nonlinearity* **1**, 279–309, 1988.
28. L.R. Keefe, Dynamics of perturbed wavetrain solutions to the Ginzberg–Landau equation. *Stud. Appl. Math.* **73**, 91, 1985.
29. A. Manela and I. Frankel, On the Rayleigh–Bénard problem in the continuum limit: Effects of temperature differences and model of interaction. *Phys. Fluids* **17**, O36101-1–O36107, 2005.
30. C. Cercignani, *Rarefied Gas Dynamics*. Cambridge, Cambridge University Press, 2000.
31. S. Stefanov, V. Roussinov and C. Cercignani, Rayleigh–Bénard flow of a rarefied gas and its attractors. I. Convection regime. *Phys. Fluids* **14**, 2255, 2002.

32. R.Kh. Zeytounian, *Topics in Hyposonic Flow Theory*. Lecture Notes in Physics, Vol. 672. Springer-Verlag, Berlin/Heidelberg, 2006.
33. J. Frölich, P. Laure and R. Peyret, *Phys. Fluids A* **4**, 1355, 1992.
34. G. Bird, *Molecular Gas Dynamics and the Direct Simulation of Gas Flows*. Clarendon, Oxford, 1994.
35. I. Boyd, In: *Rarefied Gas Dynamics*, A.D. Ketsdever and E.P. Muntz (Eds.), American Institute of Physics, New York, p. 899, 2003.
36. B. Straughan, *The Energy Method, Stability, and Nonlinear Convection*. Springer-Verlag, New York, 1992.
37. D.D. Joseph. *Stability of Fluid Motions II*. Springer-Verlag, Berlin, 1976.
38. M.G. Braunsfurth and G.M. Homsy, Combined thermocapillary-buoyancy convection in a cavity. Part II. An experimental study. *Phys. Fluids* **9**(5), 1277–1287, 1997.
39. T. Boeck, A. Nepomnyashchy, I. Simanovskii, A. Golovin, L. Braverman and A. Thess, *Phys. Fluids* **14**(11), 3899–3911, 2002.
40. L. Ratke, H. Walter and B. Feuerbacher (Eds.), *Materials and Fluids under Low Gravity*. Springer-Verlag, Berlin, 1996.
41. H.C. Kuhlmann, *Thermocapillary Convection in Models of Crystal Growth*. Springer-Verlag, Berlin, 1999.
42. J.C. Legros, *Acta Astron.* **13**, 697, 1986.
43. I.B. Simanovskii and A.A. Nepomnyaschy. *Convective Instabilities in Systems with Interface*. Gordon and Breach, 1993.
44. A. Juel et al., Surface tension-driven convection patterns in two liquid layers. *Physica D* **143**, 169–186, 2000.
45. L.M. Bravermam et al., Convection in two-layer systems with an anomalous thermocapillary effect. *Phys. Rev. E* **62**, 3619–3631, 2000.

Chapter 6
The Deep Thermal Convection Problem

6.1 Introduction

As written in Straughan's book [1] (first published in 1993):

> an interesting model of thermal convection for a deep layer of fluid is developed by Zeytounian in 1989 [2]. A linear and weakly nonlinear theory for this model is presented by Errafiy and Zeytounian [3], and transition to chaos results by Errafiy and Zeytounian [4]; sharp nonlinear energy stability bounds are derived by Franchi and Straughan [5].

However, in this same direction, at the University of Lille I (Laboratoire de Mécanique de Lille, Bâtiment 'Boussinesq'), Charki published during the years 1993–1965 three papers relative to: stability [6], existence and uniqueness [7] and the well-posedness of the initial value problem [8]. Finally, we mention the two papers by Charki and Zeytounian [9, 10], where the Lorenz system of three equations and the Landau–Ginzburg amplitude equation, both associated to deep convection equations, were derived.

The major interest of 'deep (Bénard) convection' (DC) equations (as opposed to 'shallow (RB) convection' equations) is, on the one hand, the presence of the viscous dissipation (non-Boussinesq) term in these equations and, on the other hand, the fact that these DC, *à la* Zeytounian, convection equations contain a new parameter related to the depth of the layer and called the 'depth parameter'.

Concerning the Hills and Roberts [11] approach, the reader is once more invited to re-read Sections 2.7 and 3.6. For the Hills and Roberts model, linear and nonlinear stability results were obtained by Richardson [12].

In this chapter, we assume from the start that the deep fluid layer is limited by two horizontal rigid plates, $z = 0$ and $z = d$, in a Cartesian system of coordinates (x, y, z); this seems a limitation, but this limitation is justified, because the main driving, buoyancy force is again governed by the Grashof–Rayleigh number and the thickness of the layer is large.

Indeed, we now see that such an assumption is well founded, because the smallness of the Froude number ensures the significant driving role to buoyancy, via the Grashof number, which is the ratio of the expansibility small parameter ε to the small squared Froude number.

Below, in Section 6.2, as dominant dimensionless equations we choose, for the dimensionless (delete the prime) functions $\mathbf{u}$, π and θ, according to (3.6b, c, d), equations (3.11), (3.14) and (3.15). The boundary conditions at $z = 0$ and $z = 1$ are (3.20a, b).

6.2 The Deep Bénard Thermal Convection Problem

The deep dissipative convective layer case is strongly related to (see (4.27c)) the conditions

$$\mathrm{Fr}^2 \ll 1, \quad \varepsilon \ll 1 \quad \text{and} \quad \mathrm{Bo} \gg 1, \tag{6.1a}$$

with

$$\varepsilon\,\mathrm{Bo} = \mathrm{O}(1), \tag{6.1b}$$

and also

$$\frac{\varepsilon}{\mathrm{Fr}^2} = \mathrm{Gr} = O(1). \tag{6.1c}$$

With relations (6.1a–c), when ε and Fr^2 both tend to zero and Bo tends to infinity, we derive (ε being the main small parameter, when we take into account (6.1b) and (6.1c)) from dominant dimensionless equations (3.11), (3.14) and (3.15), for the leading functions

$$\lim_{\varepsilon \to 0} (\mathbf{u}, \pi, \theta) = (\mathbf{u}_D, \pi_D, \theta_D), \tag{6.2}$$

the following system of deep convection (DC) equations:

$$\nabla \mathbf{u}_D = 0, \tag{6.3a}$$

$$\frac{\mathrm{d}\mathbf{u}_D}{\mathrm{d}t} + \nabla \pi_D - \mathrm{Gr}\,\theta_D \mathbf{k} = \Delta \mathbf{u}_D, \tag{6.3b}$$

$$\{1 - Di[(p_d)' + 1 - z]\}\frac{\mathrm{d}\theta_D}{\mathrm{d}t} = \left(\frac{1}{\mathrm{Pr}}\right) \Delta\theta_D + (1/2\,\mathrm{Gr})Di \left[\frac{\partial u_{\mathrm{Di}}}{\partial x_j} + \frac{\partial u_{Dj}}{\partial x_i}\right]^2. \tag{6.3c}$$

where

$$Di = \frac{g\alpha_d d}{C_d} \tag{6.4}$$

is our 'depth' parameter defined by δ in [2]. For the DC equations (6.3a–c) we write as boundary conditions:

$$\mathbf{u}_D = 0 \quad \text{and} \quad \Theta_D = 1 \quad \text{on } z = 0; \mathbf{u}_D = 0 \text{ and } \Theta_D = 0 \text{ on } z = 1. \tag{6.5}$$

In [2], in equation (6.3c) the term $(p_d)'$ is absent because we have neglected p_d in (3.6c) which defines π. As in [2], it is judicious to introduce, instead of θ_D, a temperature perturbation

$$\Xi = \left(\frac{1}{\text{Pr}}\right) \text{Gr}\,[\Theta_D + z - 1]$$

and instead of π_D, a pressure perturbation

$$\Pi = \left(\frac{1}{\text{Pr}}\right) \left\{\pi_D + \text{Gr}\, z \left[\left(\frac{z}{2}\right) - 1\right]\right\}.$$

If we change also $\mathbf{u}_D$ to $(1/\text{Pr})\mathbf{v}$ and t to τ and omit the term $(p_d)'$ in (6.3c), then we find again our DC system of equations (4.6) from [2]. Namely:

$$\nabla \cdot \mathbf{v} = 0, \tag{6.6a}$$

$$\left(\frac{1}{\text{Pr}}\right) \frac{\mathrm{d}\mathbf{v}}{\mathrm{d}\tau} + \nabla\Pi - \Xi\mathbf{k} = \Delta\mathbf{v}, \tag{6.6b}$$

$$[1 + Di(1 - z)]\left[\frac{\mathrm{d}\Xi}{\mathrm{d}\tau} - \text{Ra}\,(\mathbf{v} \cdot \mathbf{k})\right] = \Delta\Xi + 2Di[\mathbf{D}(\mathbf{v}) : \mathbf{D}(\mathbf{v})], \tag{6.6c}$$

$$\mathbf{v} = 0 \quad \text{and} \quad \Xi = 0 \quad \text{at } z = 0 \text{ and } z = 1, \tag{6.6d}$$

where

$$\mathbf{D}(\mathbf{v}) = (1/2)[\nabla\mathbf{v} + (\nabla\mathbf{v})^T] \tag{6.6e}$$

is the rate of deformation tensor.

If we now consider the 2D case (which is judicious at the onset of deep convection) where parallel convective rolls originate, the velocity vector field $\mathbf{v}$, becomes perpendicular to the rolls axis and the 3D equations (6.6a–c) are invariant under the action of translation along the rolls axis.

In such a case, in time-space $(\tau, x_1 = x, x_3 = z)$, for u and w components of a 2D velocity vector, we write

$$u = \frac{\partial\psi}{\partial z} \quad \text{and} \quad w = -\frac{\partial\psi}{\partial x}, \tag{6.7}$$

since, instead of (6.6a), we have in the 2D case

$$\frac{\partial u}{\partial x} + \frac{\partial w}{\partial z} = 0.$$

In (6.7), $\psi(t, x, z)$ is the stream function and with $\Xi(t, x, z)$, both are solutions of the following system of two equations (after the elimination of Π):

$$\left(\frac{1}{\mathrm{Pr}}\right)\left[\frac{\partial}{\partial t} + \left(\frac{\partial \psi}{\partial z}\right)\frac{\partial}{\partial x} - \left(\frac{\partial \psi}{\partial x}\right)\frac{\partial}{\partial z}\right]\Delta_2\psi + \frac{\partial \Xi}{\partial x} = \Delta_2(\Delta_2\psi); \quad (6.8a)$$

$$\chi(z)\left[\frac{\partial}{\partial t} + \left(\frac{\partial \psi}{\partial z}\right)\frac{\partial}{\partial x} - \left(\frac{\partial \psi}{\partial x}\right)\frac{\partial}{\partial z}\right]\Xi + \mathrm{Ra}\left(\frac{\partial \psi}{\partial x}\right)$$
$$= \Delta_2\Xi + \delta\left[4\left(\frac{\partial^2 \psi}{\partial x \partial z}\right)^2 + \left(\frac{\partial^2 \psi}{\partial z^2} - \frac{\partial^2 \psi}{\partial x^2}\right)^2\right], \quad (6.8b)$$

with $\Delta_2 = \partial^2/\partial z^2 + \partial^2/\partial x^2$ and $\chi(z) = 1 + \delta(1 - z)$, $Di \equiv \delta \in [0, 1]$.

For these DC equations (6.8a, b) we write as boundary conditions:

$$\psi = 0, \quad \frac{\partial \psi}{\partial z} = 0, \quad \text{or} \quad \frac{\partial^2 \psi}{\partial z^2} = 0, \quad \text{and} \quad \Xi = 0 \quad \text{at } z = 0 \text{ and } z = 1 \quad (6.8c)$$

according to the nature of the boundary.

We observe that the Lorenz system in [9], and the amplitude Landau–Ginzburg equation in [10], have been derived for the 2D system (6.8a, b) with the conditions $\psi = 0$, $\partial^2\psi/\partial z^2 = 0$, and $\Xi = 0$ at $z = 0$ and $z = 1$.

Obviously when $Di \equiv \delta \to 0$, one obtains, instead of (6.6a–c) and (6.8a, b), the corresponding 3D and 2D classical RB system of shallow convection equations.

6.3 Linear – Deep – Thermal Convection Theory

The linear theory, when the deep convection equations (6.6a–c) are linearized, relative to the zero solution ($\mathbf{v} = 0$, $\Pi = 0$, $\Xi = 0$), for small perturbations ($\mathbf{v}'$, Π', Ξ') – ignoring the nonlinear terms – gives the following system of three equations:

$$\nabla \cdot \mathbf{v}' = 0, \quad (6.9a)$$

$$\left(\frac{1}{\mathrm{Pr}}\right)\frac{\partial \mathbf{v}'}{\partial \tau} + \nabla \Pi' - \Xi'\mathbf{k} = \Delta \mathbf{v}', \tag{6.9b}$$

$$[1 + Di(1 - z)]\left[\frac{\partial \Xi'}{\partial \tau} - \mathrm{Ra}\,(\mathbf{v}' \cdot \mathbf{k})\right] = \Delta \Xi'. \tag{6.9c}$$

The linear system (6.9a–c) was investigated carefully by Errafiy [13] in the 2D case when $\mathbf{v}' = (u', w')$ in time space (τ, x, z). In this case, with the stream function $\psi'(\tau, x, z)$ such that

$$u' = \left(\frac{1}{\mathrm{Pr}}\right)\frac{\partial \psi'}{\partial z}, \quad w' = -\left(\frac{1}{\mathrm{Pr}}\right)\frac{\partial'}{\partial x},$$

after the elimination of the pressure Π', we get for $\psi'(\tau, x, z)$ and $\Xi'(\tau, x, z)$ the following system of two linear equations ($\mu = \delta/[1 + (\delta/2)]$):

$$\mathrm{Pr}\frac{\partial \psi'}{\partial \tau} + \frac{\partial \Xi'}{\partial x} = \mathrm{Pr}\,\Delta_2(\Delta_2\psi'); \tag{6.10a}$$

$$\mathrm{Pr}\,\lambda(z)\left[\frac{\partial \Xi'}{\partial \tau} + \mathrm{Ra}\frac{\partial \psi'}{\partial z}\right] = \left[1 - \left(\frac{\mu}{2}\right)\right]\Delta_2\Xi', \tag{6.10b}$$

where

$$\lambda(z) = 1 + \mu[(1/2) - z] \quad \text{and} \quad \frac{\mu}{\delta} = \left[1 - \left(\frac{\mu}{2}\right)\right].$$

In the free-free case, the boundary conditions are

$$\psi' = \frac{\partial^2 \psi'}{\partial z^2} = \Xi' = 0, \quad \text{at } z = 0, 1, \tag{6.11a}$$

$$\frac{\partial \Xi'}{\partial x} = \psi' = \frac{\partial^2 \psi'}{\partial x^2} = 0, \quad \text{at } x = 0 \text{ and } x = l_0, \tag{6.11b}$$

where l_0 is the horizontal dimensionless length of a DC cell.

A Galerkin formulation, with as representation

$$\psi' = \Sigma A_n(t)\sin(n\pi z)\sin(q_0 x), \tag{6.12a}$$

$$\Xi' = \Sigma B_n(t)\sin(n\pi z)\sin(q_0 x), \tag{6.12b}$$

where $n = 1$ to N, and $q_0 = \pi/l_0$ – after a quite lengthy but straightforward calculation (analogous to derivation of the Lorenz system of three equations in Section 5.5 – gives a system of $2N$ ordinary differential equations which can be represented via the matrix $\mathbf{D}_f^f$ as

$$\frac{\mathrm{d}\mathbf{F}}{\mathrm{d}\tau} = \mathbf{D}_f^f\mathbf{F}, \quad \text{where } \mathbf{F} = (X_1, X_2, \ldots, X_N; Y_1, Y_2, \ldots, Y_N), \tag{6.13}$$

$\mathbf{D}^f_f = -$ 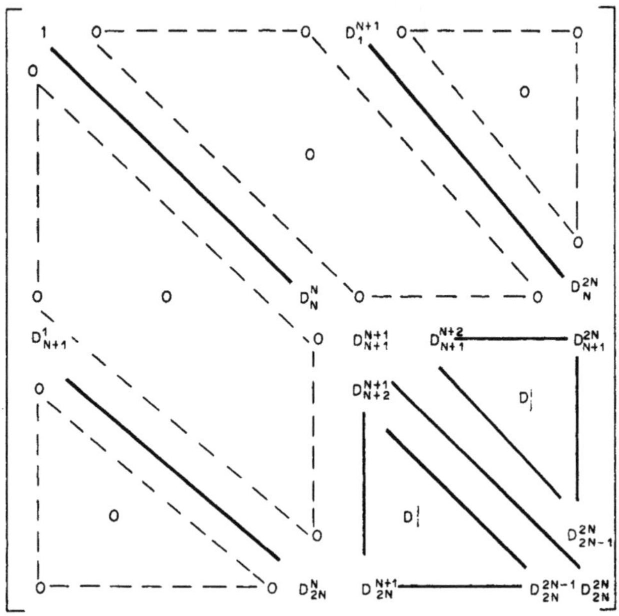

Fig. 6.1 Matrix of $\mathbf{D}^f_f$ for the system of $2N$ ODE. Reprinted with kind permission from [13].

and the general structure of this matrix $\mathbf{D}^f_f$ in (6.12), is represented in Figure 6.1. In this figure, the coefficients $D^i_j = D^i_j$ of the matrix D^f_f for $N = 3$ are given in [3, p. 628].

For further details, see the paper by Errafiy and Zeytounian [3] and Errafiy's doctoral thesis [13]. We see that, if the matrix $\mathbf{D}^f_f$ is real and symmetric then can have only real eigenvalues. In this case oscillatory instability is impossible, and this is a classical proof of the principle of exchange of stabilities for our linear DC problem (6.10a, b), with (6.11a, b) when $\delta \neq 0$. As a consequence, the principle of exchange of stabilities being proved for the above linear DC free-free problem (6.10a, b) and (6.11a, b), it is possible to consider only the stationary linear problem to seek the marginal states and the critical value of the Rayleigh number.

In the rigid-free case (the case of 'oceanic circulation') the boundary conditions are

$$\psi' = \frac{\partial \psi'}{\partial z} = \Xi' = 0, \quad \text{at } z = 0, \tag{6.14a}$$

$$\psi' = \frac{\partial^2 \psi'}{\partial z^2} = \Xi' = 0, \quad \text{at } z = 1. \tag{6.14b}$$

In this case, using again the Galerkin technique, but with a modified representation for ψ', instead of (6.12a) we write

$$\psi' = \Sigma A_n(t)\psi_n(z)\sin(q_0 x), \quad n \geq 1, \tag{6.15a}$$

where for $\psi_n(z)$ we have an explicit formula (see [3, (2.13), with (2.14)]. We can write, again, the resulting system of ordinary differential equations in the form (6.13). In the particular case when $n = 1$ (one component representation for ψ' and Ξ'), we easily show that the corresponding matrix is real and symmetric and can have only two real eigenvalues. In this case, at the steady linear state, we have that the neutral stability curve is given by the following equation:

$$\mathrm{Ra} - \alpha\mu\,\mathrm{Ra} - \beta\left(\frac{\mu}{\delta}\right) = 0, \tag{6.16}$$

where the coefficients α and β are functions of four scalars which appear in the formula for the function $\psi_n(z)$, in representation (6.15a) of ψ'.

When $n = 2$ (two-components solution) the matrix is symmetric but not real. In this case oscillatory instability is possible and it is necessary to compare the critical Rayleigh number for stationary instability $\mathrm{Ra}_c^{\mathrm{st}}$ with that for oscillatory instability $\mathrm{Ra}_c^{\mathrm{osc}}$. For this we can take into account the classical Routh–Hurwitz criterion and Orlando's formula (see [14, pp. 231–234]). Numerical calculation shows that, for all δ,

$$\mathrm{Ra}_c^{\mathrm{st}} < \mathrm{Ra}_c^{\mathrm{osc}} \tag{6.17}$$

and the validity of the principle of exchange of stabilities is clearly evident, independently of the value of the depth parameter. According to the above result we can consider a stationary DC linear problem, which is written below for ψ' and the function

$$T = \left(\frac{1}{\mathrm{Pr}}\right)\frac{\partial\,\Xi'}{\partial x}.$$

Namely

$$\Delta_2(\Delta_2\psi') = T, \tag{6.18a}$$

$$\Delta_2 T = \mathrm{Ra}[1 + \delta(1 - z)]\frac{\partial^2\psi'}{\partial z'^2}. \tag{6.18b}$$

By analyzing the disturbance ψ' and T into normal modes, we seek solutions of (6.18a, b) which are of the form

$$\psi' = W(z)f(x) \quad \text{and} \quad T = \Theta(z)f(x), \tag{6.19a}$$

with

$$\frac{\mathrm{d}^2 f}{\mathrm{d}x^2} + q_0^2 f = 0, \quad q_0^2 = \text{const.} \tag{6.19b}$$

In particular, for the rigid-free case (which is the more difficult case) we obtain the following problem for $W(z)$ and $\Theta(z)$:

$$\left(\frac{\mathrm{d}^2}{\mathrm{d}z^2} - q_0^2\right) W = \Theta, \tag{6.20a}$$

$$\left(\frac{\mathrm{d}^2}{\mathrm{d}z^2} - q_0^2\right) \Theta = -\mathrm{Ra}\, q_0^2[1 + \delta(1 - z)]W, \tag{6.20b}$$

with

$$0 = W = \frac{\mathrm{d}W}{\mathrm{d}z} = 0 \quad \text{at } z = 0,$$

$$\Theta = W = \frac{\mathrm{d}^2 W}{\mathrm{d}z^2} = 0 \quad \text{at } z = 1.$$

As in the classic, *à la* Chandrasekhar [15], approach, we suppose that for Θ the solution is

$$\Theta = \Sigma\Theta_n \alpha_n^2 \sin(n\pi z), \quad n > 1.$$

Substituting this solution for Θ in equation (6.20a), we can write the following solution for W:

$$W = \Sigma\Theta_n \psi_n(z), \quad n > 1,$$

where $\psi_n(z)$, for the rigid-rigid case, is an explicit function of z which appears also in the representation (6.15a) of ψ' using the Galerkin technique.

Now substituting for Θ and W the above expressions and taking into account the explicit form of $\psi_n(z)$, we obtain the following condition:

$$\Sigma\Theta_n \left\{\frac{\alpha_n^3}{\mathrm{Ra}\, q_0^2[1 + (\delta/2)]} \sin(n\pi z) - \psi_n(z)\right\} = 0, \quad n > 1. \tag{6.21a}$$

Multiplying (6.21a) by $\sin(m\pi z)$ and integrating over the range of z, we obtain a system of linear homogeneous equations for the constants Θ_n – the requirement that these constants are not all zero leads to the secular equation

$$\det\left\{\frac{\alpha_n^3}{\mathrm{Ra}\, q_0^2[1 + (\delta/2)]} \delta_{nm} + 2K_n^m\right\} = 0, \tag{6.21b}$$

where $2K_n^m$ is determined in an explicit form (see [3, (3.11)]). With the aid of (6.21b) and the expression for $2K_n^m$, Errafiy obtained the critical Rayleigh numbers for different values of δ for the three cases: free-free, rigid-rigid and rigid-free. We note that the critical Rayleigh numbers, for different values of δ, decrease when δ increases from the value $\delta = 0$ to $\delta = 1$. As a result: *when the depth parameter δ increases, the layer of (weakly) expansible liquid becomes very much more stable.*

In [3], the reader can find also an approximate solution of the steady-state problem (6.20a, b) for

$$\mu = \frac{2\delta}{(2+\delta)} \ll 1,$$

by a perturbation method.

In particular for the rigid-rigid case, for the critical Ra, we obtain the following approximate formula (with $q_0 = 3.117$):

$$\mathrm{Ra} = \left[\frac{1707.9}{(1+\delta/2)}\right]\left\{1 - 7.61 \times 10^{-3}\left[\frac{\delta}{(2+\delta)}\right]\right\}, \tag{6.21c}$$

and this formula (6.21c) gives good values for Ra_c even when $\delta = 1$!

Finally, in [3], the reader can find also a direct proof of the Principle Exchange of Stabilities for the free-free case, deduced from an explicit relation obtained from the linear system of two equations:

$$\Pr\left[\lambda - \left(\frac{\mathrm{d}^2}{\mathrm{d}z^2} - q_0^2\right)\right]\left(\frac{\mathrm{d}^2 W}{\mathrm{d}z^2} - q_0^2 W\right) = -\Theta, \tag{6.22a}$$

$$\left\{\Pr\lambda - \frac{[(\mathrm{d}^2/\mathrm{d}z^2 - q_0^2)]}{[1+\delta(1-z)]}\right\}\Theta = \Pr\mathrm{Ra}\, q_0^2 W. \tag{6.22b}$$

Namely, as a direct consequence of (6.22a, b), we derive the following integral relation:

$$\mathrm{Imag}\,(\lambda)\left\{\int_0^1\left[\left|\frac{\mathrm{d}\Theta}{\mathrm{d}z}\right|^2 + q_0^2\,\mathrm{Ra}\Pr\left|\frac{\mathrm{d}^2 W}{\mathrm{d}z^2} - q_0^2 W\right|^2 \mathrm{d}z\right]\right\} = 0; \tag{6.23}$$

the quantity inside the curly brackets being positive definite for $\mathrm{Ra} > 0$, we have obviously

$$\mathrm{Imag}\,(\lambda) = 0, \tag{6.24}$$

and this establishes that λ is real for $\mathrm{Ra} > 0$ and for all $\delta > 0$, and that the principle of the exchange of stabilities is valid for the thermal DC problem in the free-free case. An interesting observation is linked with the system of two equations (6.20a, b), for $W(z)$ and $\Theta(z)$, which appears as very similar to the adjoint system for the classical Couette flow (see [15, sections 71, 130]) – this remark possibly leads to a complementary method for investigation of the problem (6.20a, b).

6.4 Routes to Chaos

In [4], the partial differential equations governing two-dimensional thermal DC, with free-free conditions has been reduced, again according to the Galerkin method, to a set of two ordinary nonlinear differential evolution (in time) equations for two amplitudes $A_{pq}(t)$ and $B_{pq}(t)$. To slightly change the definition of $A_{pq}(t)$ and $B_{pq}(t)$ by the introduction of two new amplitudes, $X_{pq}(t)$ and $Y_{pq}(t)$, and then replacing the double subscript (pq) by a single subscript (n), the coupled equations for $(X_{pq}(t), Y_{pq}(t))$ can be written as

$$\frac{\mathrm{d}Z_n}{\mathrm{d}t} = \Sigma_m a_{mn} Z_m + \Sigma_m \Sigma_l Q_{nml} Z_m Z_l, \tag{6.25}$$

$m = 1$ to N and $1 = m$ to N, where the amplitude variables $(Z_1, Z_2, \ldots, Z_N)$ are respectively the Fourier amplitudes $(X_{pq}(t), Y_{pq}(t))$. In [4], for $N = 15$, all coefficients a_{nm} and Q_{nml} have been calculated (as functions of Pr, Ra and δ). Errafiy [13] adopts as a truncation scheme:

$$p + q < K \quad \text{with} \quad K = 4.$$

It is interesting to observe that, if in the classical RB thermal shallow convection case (when $\delta = 0$) the amplitudes with $(p+q)$ odd do not contribute – they tend to zero when time tends to infinity even if initially they were different from zero – on the contrary this is not true in the thermal DC case. Indeed, the new (five) odd components which appear in the case when $\delta \neq 0$ are responsible ('open the door') for the appearance 'in a new space' of a variety of strange attractors via the three main routes to chaos: Ruelle–Takens [16], Feigenbaum [17] and Pomeau–Manneville [18] scenarii.

With $N = 15$, we have ten equations corresponding to RB convection (only even amplitudes) and five equations connecting with $\delta \neq 0$ (odd amplitudes, as a direct consequence of the non-equivariance of the exact DC 2D system).

We wish to point out two interesting features of our thermal DC model. The first relates to the interactions between the even and odd amplitudes in system (6.25), even though the depth parameter δ is very small. As a result we obtain numerically all three of the routes (scenarios) to chaos, for various values of Pr, δ and

$$\kappa = \frac{(\mathrm{Ra} - \mathrm{Ra}_c)}{\mathrm{Ra}_c}, \tag{6.26}$$

The second point relates to the 'chaotic configuration' of strange attractors when δ is not small, e.g., for $\delta = 0.6$ and $\delta = 1$. When δ is not small the

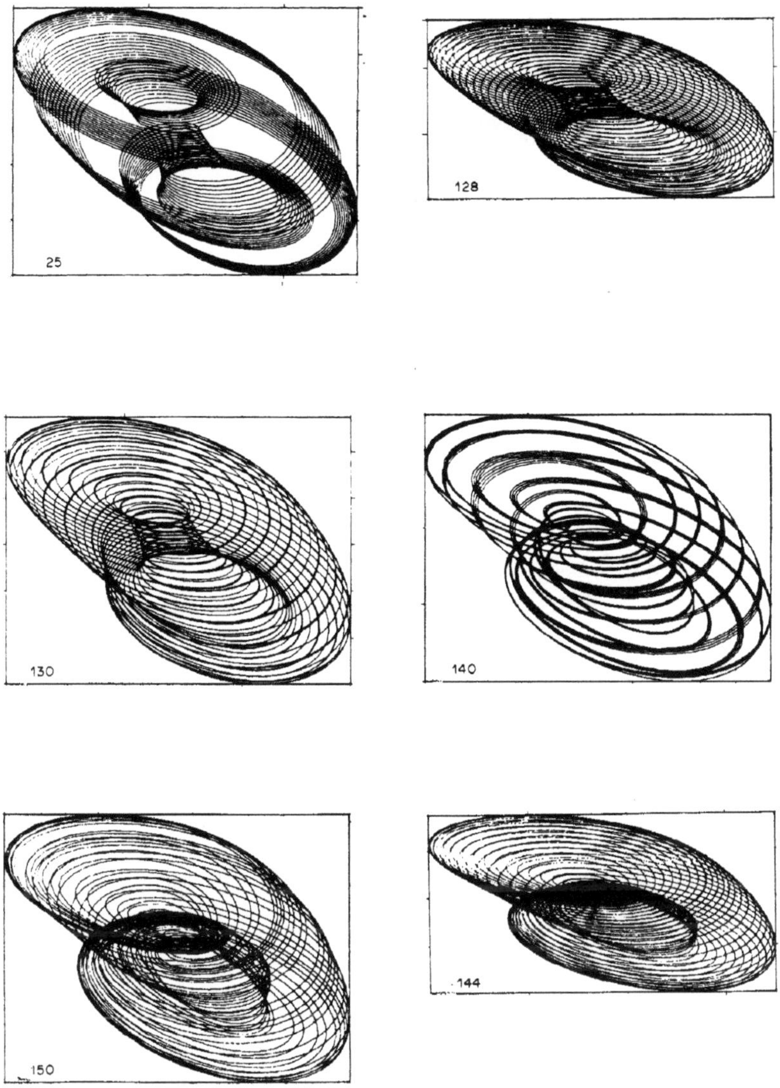

Fig. 6.2 Projection onto the (Y, X) plane of successive torii for Pr $= 20$ and $\delta = 0.2$. Reprinted with kind permission from [13].

chaos appears more rapidly and the corresponding strange attractor is more complex (see for example Figures 6.8 and 6.9 below) and it seems that, via the depth parameter δ a 'space effect on the temporal chaos is being taken into account'. In Figure 6.2 above, the successive attractors (for various val-

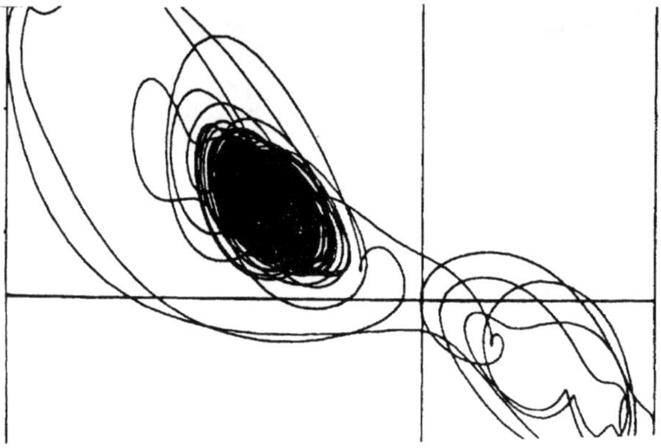

Fig. 6.3 Strange attractor *à la* Ruelle–Takens for $\kappa = 154$, Pr $= 20$ and $\delta = 0.2$. Reprinted with kind permission from [13].

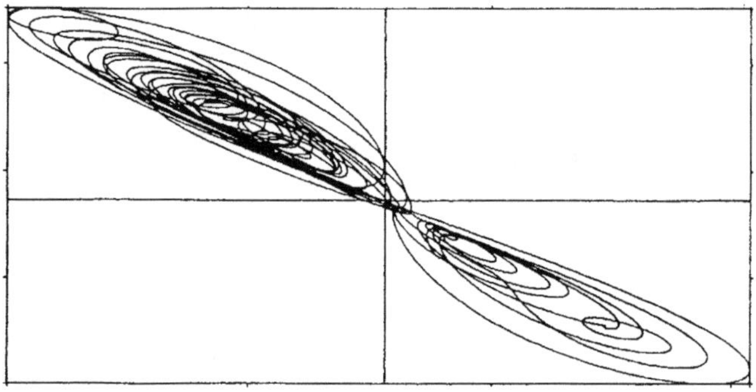

Fig. 6.4 'Chaotic Feigenbaum' attractor for $\kappa = 290$ and Pr $= 100$ and $\delta = 0.1$. Reprinted with kind permission from [13].

ues of κ between 125 and 150) are the 2D torii $\mathbf{T}^2$ and the strange attractor, which is represented in Figure 6.3, results from the 'destroying' of the last torus for the value of κ, quite near to $\kappa = 154$, for which the strange attractor, *à la* Ruelle–Takens, appears. We see that this strange attractor is rather similar to the well-known Lorenz attractor (see Figures 5.4a–c, in Section 5.5) and seems to have many of the gross features observed in the Lorenz model. Therefore it is an excellent candidate for a higher dimensional analogue. For a pertinent discussion concerning bifurcations of periodic solutions onto invariant torii, see [19].

In Figure 6.4, we have represented the Feigenbaumm 'chaotic' attractor for $k = 290$, Pr $= 100$ and $\delta = 0.1$ corresponding to successive period doubling of Figure 6.5. In Figure 6.5, we have represented the same projections as those in Figure 6.2, for various values of κ between 200 and 250, but for

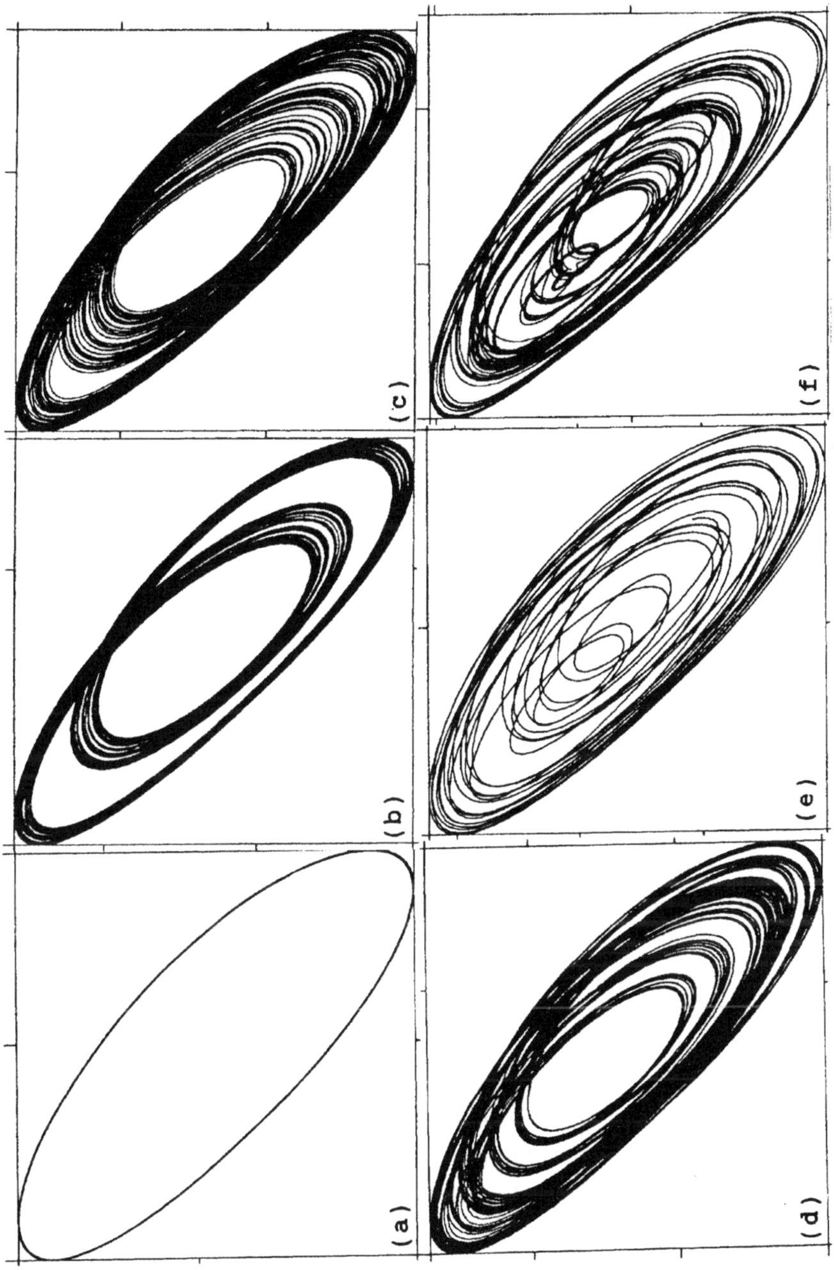

Fig. 6.5 Successive period doubling for Pr = 100 and $\delta = 0.1$. Reprinted with kind permission from [13].

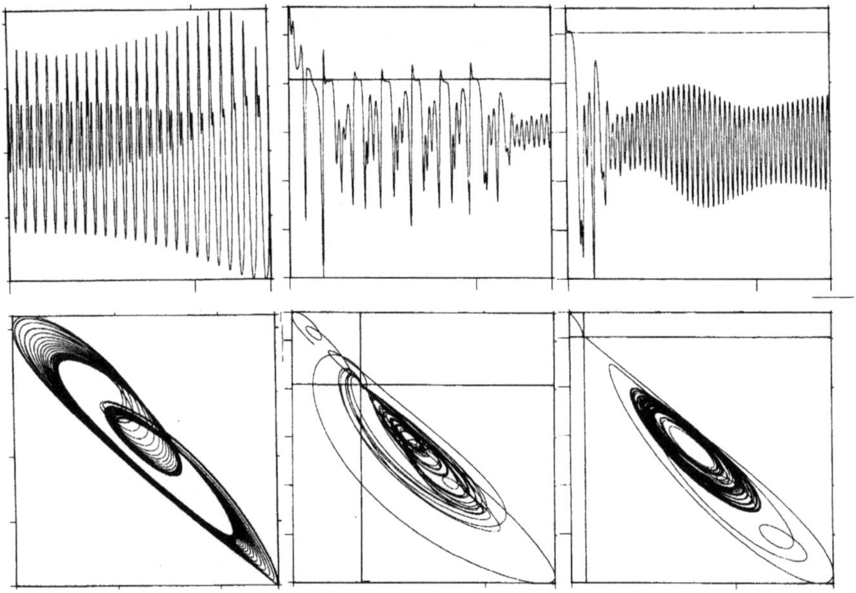

Fig. 6.6 Temporal evolutions of $X(t)$ and associated attractor in three time intervals ($\kappa = 100$, $\mathrm{Pr} = 100$, $\delta = 1$). Reprinted with kind permission from [13].

$\mathrm{Pr} = 100$ and $\delta = 0.1$. In this case, curiously, instead of a series of 2D torii $\mathbf{T}^2$, we have a series of periodic regimes with period doublings and the chaos appears for $\kappa = 270$.

Strictly speaking, the Feigenbaum chaotic attractor is not a strange attractor (see Schuster's book [20] for the precise definition of this 'object') but it is very representative of the chaos when the power spectrum is continuous. In this case the route to chaos involves successive period doubling (subharmonic) bifurcations of a periodic deep thermal convection. It is also interesting to observe that for $\delta = 1$ and $\mathrm{Pr} = 100$ (see Figure 6.6) we have for $\kappa = 100$, a phenomenon of intermittency between two periodic regimes when the time increases; in such a case we are confronted with a strong influence of the depth parameter $\delta = 1$.

In Figure 6.7, according to the Pomeau–Manneville scenario, we have numerical evidence of the intermittency for $\mathrm{Pr} = 10$ and $\delta = 0.1$. For $\kappa = 130$ the bursts are relatively large but for $\kappa = 120$ we have a pure periodic regime. The instability occurs through the intermittent regime and for $\kappa = 130$ the attractor is chaotic.

The numerical route to obtaining chaos via intermittency is 'fascinating' and confirms very well the Pomeau–Manneville scenario. For $\kappa = 130$ we

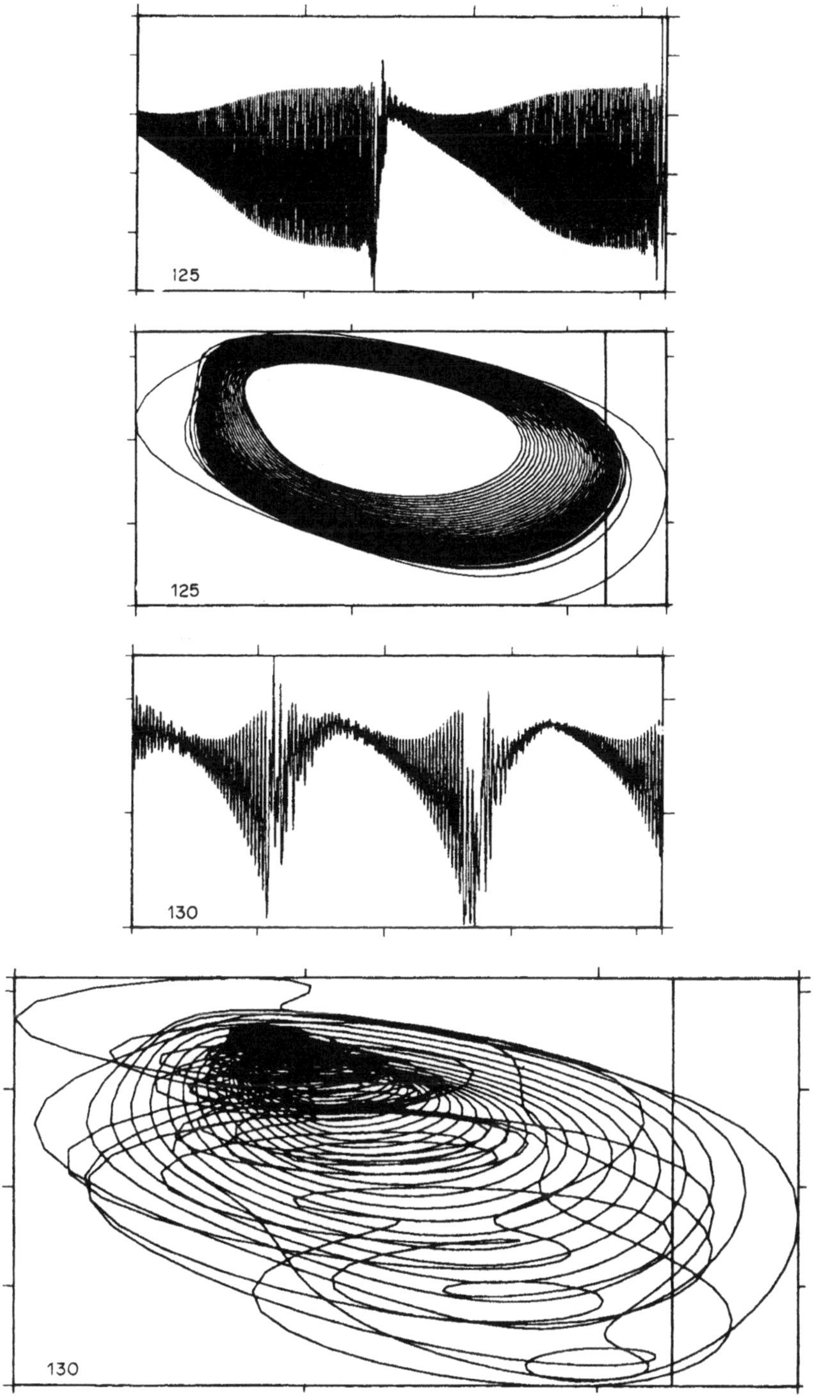

Fig. 6.7 Numerical evidence of the Pomeau–Manneville scenario. Reprinted with kind permission from [13].

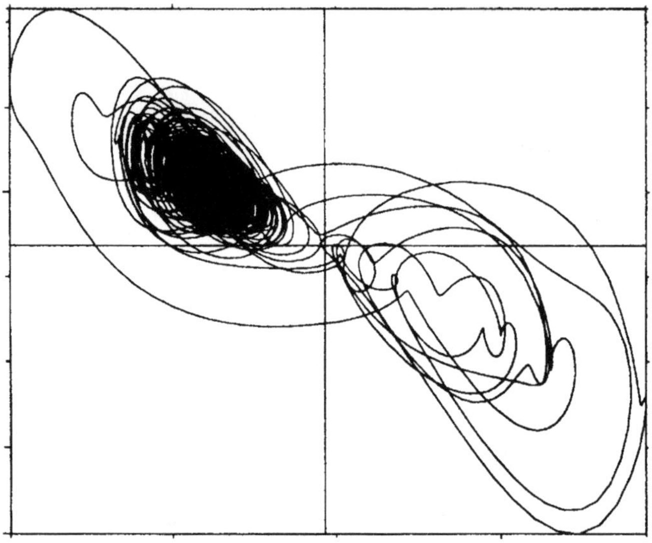

Fig. 6.8 Strange attractor with $\delta = 1$ ($\kappa = 115$ and $\mathrm{Pr} = 10$). Reprinted with kind permission from [13].

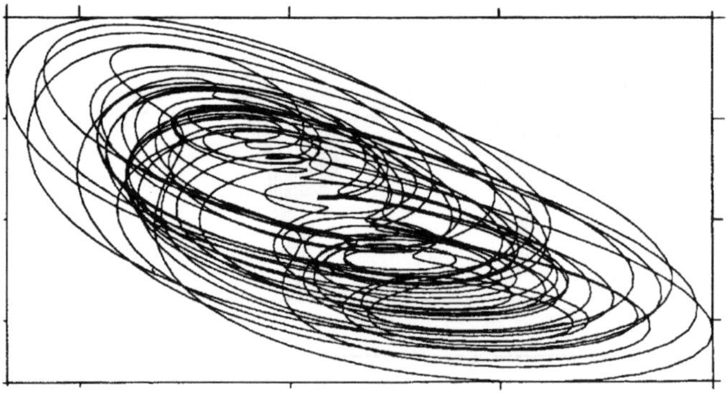

Fig. 6.9 Strange attractor with $\delta = 0.6$ ($\kappa = 139$ and $\mathrm{Pr} = 20$). Reprinted with kind permission from [13].

have the occurrence of a temporal evolution (linked with $X(\tau)$) which alternates randomly between 'long' regular (laminar) phases (so-called 'intermissions') and relatively short irregular bursts. We also observe that the number of chaotic bursts increases with an 'external' parameter, which means that intermittency offers a continuous route from regular to chaotic convection; for our case the increase is from 122 to 130.

As a first example of a strong influence of the depth parameter δ on the route to chaos, we have represented in Figure 6.8 a strange attractor (when

Pr $= 10$ and $\delta = 1$), for $\kappa = 115$; but for a very close value, $\kappa = 114.66$, instead of this strange attractor we have a simple limit cycle (periodic regime) and more surprisingly for $\kappa = 114.80$ the regime is already chaotic.

A second example of the influence/dependence of the depth parameter δ on the appearance of a strange attractor is the strange very chaotic attractor in Figure 6.9.

Indeed, when Pr $= 20$ but $\delta = 0.1$, only for a high value of $\kappa = 160$ do we obtain a strange attractor similar to the one represented in Figure 6.3, which is much less chaotic than the one in Figure 6.8! Maybe it is interesting to obtain a numerically strange attractor for larger (than $\delta = 1$?) values of δ! But in such a case, the probability for appearance of attractors being in an explosive manner, the value seems very high?

6.5 Rigorous Mathematical Results

After the publication of [2], where the thermal DC equations were first derived, and the two papers by Errafyi and Zeytounian [3, 4] concerning the linear theory and routes to chaos for these DC equations had been published, some authors considered the existence, uniqueness and stability of solutions for these DC convection equations for various thermal convection problems. In particular, Franchi and Straughan [5] applied a nonlinear energy stability analysis for these deep convection equations. On the other hand, Charki during the years 1994–1996, at the University of Lille I published three papers relative to: stability [6], existence and uniqueness of solutions for the steady problem [7], and the initial value problem [8].

Before, in 1992, Richardson [12] used a nonlinear stability analysis of convection in a generalized (*à la* Hills and Roberts [11]) incompressible fluid, the equations governing such a fluid being a particular *ad hoc* case of our DC equations. In [6] by Charki, existence and uniqueness of a local in time strong solution for the unsteady DC problem is proved through use of a semigroup theory. The bifurcation problem (for linear and nonlinear cases) is also dealt with and the possibility of existence of periodic and quasiperiodic solutions to the DC problem is analyzed.

We observe that under the same assumptions as in Iooss' paper [21], all the results there concerning the existence and stability of periodic solutions are valid for the DC problem; in particular, 'for the DC problem, subcritical periodic motions are unstable, while supercritical periodic motions are stable in the linearized theory'. Charki [7] deals mainly with the steady DC problem in a bounded domain. Existence and uniqueness of solutions is established there for both the linear and the nonlinear problems, subject either to homogeneous or non-homogeneous boundary conditions. The proof is based on estimates for the linear problem, followed by a fixed point argument. A fixed point argument is also used by Charki in [8] to prove the existence and uniqueness of solutions for the unsteady DB convection equations in a bounded domain. Using some methods of Solonnikov [22], Charki first proves a global existence theorem for the linear deep convection equations in L_q spaces. Then, using classical estimates for the nonlinear terms, he also proves a local existence theorem for the nonlinear DB convection equations. As this is mentioned by Padula (University of Ferrara) it is worthy of notice that the summability exponent, p, must be greater than 5/2, unlike the case of thin layers where p is only required to be greater than 5/3; but, obviously, the existence of a 'thin layer' is excluded in the DC case!

Concerning the paper by Franchi and Straughan [5], the analysis of these authors is completely rigorous and, due to the nonlinearities of the system of equations governing deep convection, requires a generalized energy theory. For this, the authors derive from the equations of Zeytounian [2] the following equation for energy:

$$\frac{\mathrm{d}E}{\mathrm{d}t} = R\mathbf{I} - D + 2\left(\frac{\delta}{R}\right)\langle\mu(z)\theta\,\mathrm{d}_{ij}\,\mathrm{d}_{ji}\rangle, \tag{6.27}$$

where $R = \sqrt{\mathrm{Ra}}$ and, in [5], as temperature $\theta = R\Xi$ is used, with δ and Ξ, defined in [2].

In (6.27), E, $\mathbf{I}$ and D are

$$E = (1/2\,\mathrm{Pr})\|\mathbf{u}\|^2 + (1/2)\|\theta\|^2; \quad \mathbf{I} = 2\langle\theta w\rangle, \tag{6.28a}$$

$$D = \|\mathbf{u}\|^2 + \langle\mu|\nabla\theta|^2\rangle - \delta^2\langle\mu^3\theta^2\rangle >, \tag{6.28b}$$

D being positive-definite and $[1/(1+\delta)] \leq \mu(z) \leq 1$.

Employing Poincaré's inequality (where π is the pressure)

$$\pi^2\|\theta\|^2 \leq \|\nabla\theta\|^2 \Rightarrow D \geq \|\nabla\mathbf{u}\|^2 + k\|\nabla\theta\|^2, \tag{6.29}$$

where

$$k = \left[\frac{1}{(1+\delta)}\right] - \left(\frac{\delta^2}{\pi^2}\right) > 0.$$

The difficulty in proceeding from (6.27) is the nonlinear term involving

$$\langle\mu(z)\theta\,\mathrm{d}_{ij}\,\mathrm{d}_{ji}\rangle?$$

To handle the nonlinear term in (6.27), it is necessary to consider an identity for $\|\nabla \mathbf{u}\|^2$ (see [23, 24]). After a quite 'long manipulation' using various inequalities (in particular, from Adams' book [25], and also [24], 'Cauchy–Schwarz' and also again Poincaré) and employing the arithmetic mean, the authors prove that, provided

$$E_G(0) < \frac{1}{A}, \quad \text{then } E_G(t) \to 0 \text{ as } t \to \infty, \tag{6.30a}$$

see [23, chapter 2].

In [5] Charki introduces a generalized energy

$$E_G(t) = E + \left(\frac{\lambda}{2\,\mathrm{Pr}}\right) \|\nabla \mathbf{u}\|^2 \tag{6.30b}$$

R_E is defined by

$$R_E = \max\left(\frac{1}{D}\right) \quad \text{on the space of 'admissible solutions'.}$$

Finally, provided that

(a)
$$R < R_E \tag{6.31a}$$

and

(b)
$$E_G(0) < \frac{1}{A}, \tag{6.31b}$$

with

$$A = \left(2^{3/2}\frac{c}{R}\right)\left\{\left[\frac{R}{(\lambda\,\mathrm{Pr})^{1/2}}\right] + \left[\left(\frac{2}{k}\right)\left(\frac{\delta^2}{\lambda}\right)\right]^{1/2} + \delta + \frac{2\delta}{\lambda[ak]^{1/2}}\right\}, \tag{6.31c}$$

the authors rigorously established nonlinear stability.

In (6.31c), a value for 'c' in the current context is contained in [24] and

$$a = \frac{(R_E - R)}{R_E} > 0,$$

because it is assumed that $R < R_E$.

Of course, the stability so obtained is conditional (on the size of the initial amplitudes), but, due to the nonlinear nature of the DC equation for θ, this is not unexpected. The number $R_E(\delta)$ is the nonlinear stability threshold and, in [5], $R_E(\delta)$ is found from the following system for $W(z)$ and $\Theta(z)$:

$$(d^2 - a^2)^2 W - R_E a^2 \Theta = 0, \tag{6.32a}$$

$$(d^2 - a^2)^2 \Theta + \delta\, \mathrm{d}\Theta + \delta^2 a^2 [\mu(z)]^2 \Theta - R_E \left[\frac{a^2}{\mu(z)}\right] W = 0, \tag{6.32b}$$

with

$$W(z) = \Theta(z) = d^2 W = 0, \tag{6.32c}$$

where

$$\theta = \Theta(z) G(x, y), \quad w = W(z) G(x, y) \quad \text{and} \quad \frac{\partial^2 G}{\partial x^2} + \frac{\partial^2 G}{\partial y^2} = -a^2 G,$$

and $\mathrm{d} = \mathrm{d}/\mathrm{d}z$, a being the horizontal wave number.

Due to the dependence of μ on z this system, (6.32a, b) with (6.32c), would have to be solved numerically and then we find

$$\mathrm{Ra}_E = \min[R_E^2(a^2, \delta^2)], \tag{6.33}$$

a minimum, relative to the square of the horizontal wave number, a^2 – Ra_E being the critical Rayleigh number of energy stability theory.

Finally, in [5] the reader can find an asymptotic analysis for small δ (strongly inspired by Errafyi and Zeytounian [3]). By analogy with this approach in [5], the following result was derived:

$$\mathrm{Ra}_E = (27/4)\pi^4[1 - (1/2)\delta + \mathrm{O}(\delta^2)], \tag{6.34}$$

and the authors write that

> (6.34) agrees exactly with the linear relation given in [3] (see (4.11) in [3]) to $\mathrm{O}(\delta)$.

As a consequence (according to Franchi and Straughan [5]):

> to order δ the linear instability and nonlinear energy stability critical Rayleigh numbers are the same.

However, here we observe that, in a paper by Errafyi and Zeytounian [3, eq. (4.11)] for the free case, we have for the critical Rayleigh number

$$\mathrm{Ra}^* = \left[\frac{657.5}{(1 + (\delta/2))}\right]\left[1 - 4.99 \times 10^{-3}\left(\frac{\delta}{(2+\delta)}\right)\right]. \tag{6.35}$$

References

1. B. Straughan, *Mathematical Aspects of Penetrative Convection*. Longman, 1993.
2. R.Kh. Zeytounian, *Int. J. Engng. Sci.* **27**(11), 1361–1366, 1989.
3. M. Errafyi and R.Kh. Zeytounian, *Int. J. Engng. Sci.* **29**(5), 625, 1991.
4. M. Errafyi and R.Kh. Zeytounian, *Int. J. Engng. Sci.* **29**(11), 1363, 1991.
5. F. Franchi and B. Straughan, *Int. J. Engng. Sci.* **30**, 739–745, 1992.
6. Z. Charki, Stability for the deep Benard problem. *J. Math. Sci. Univ. Tokyo* **1**, 435–459, 1994.
7. Z. Charki, *ZAMM* **75**(12), 909–915, 1995.
8. Z. Charki, The initial value problem for the deep Benard convection equations with data in Lq. *Math. Models Meth. Appl. Sci.* **6**(2), 269–277, 1996.
9. Z. Charki and R.Kh. Zeytounian, *Int. J. Engng. Sci.* **32**(10), 1561–1566, 1994.
10. Z. Charki and R.Kh. Zeytounian, *Int. J. Engng. Sci.* **33**(12), 1839–1847, 1995.
11. R. Hills and P. Roberts, *Stab. Appl. Anal. Continuous Media* **1**, 205–212, 1991.
12. L. Richardson, *Geophys. Astrophys. Fuid Dynamics* **66**, 169–182, 1992.
13. M. Errafyi, Transition vers le chaos dans le problème de Bénard profond. Thèse de Doctorat en Mécanique des Fluides, No. 540, Université des Sciences et Technologies de Lille, LML, Villeneuve d'Ascq, 125 pp., 1990.
14. F.R. Gantmacher, *Applications of the Theory of Matrices*. Interscience, New York, 1959.
15. S. Chandrasekhar, *Hydrodynamic and Hydromagnetic Stability*. Oxford University Press, 1961.
16. D. Ruelle and F. Takens, *Comm. Math. Phys.* **20**, 167–192, and **23**, 343–344, 1971.
17. M.J. Feigenbaum, *J. Statist Phys.* **19**, 25–52, 1978; and *Physica D* **7**, 16–39, 1983.
18. Y. Pomeau and P. Manneville, *Comm. Math. Phys.* **77**, 189–197, 1980. See also, P. Manneville and Y. Pomeau, *Phys. Lett. A* **75**, 1–2, 1979.
19. O.E. Lanford III, Lecture Notes in Mathematics, Vol. 322, Springer-Verlag, Heidelberg, 1973.
20. H.G. Schuster, *Deterministic Chaos, An Introduction*. Physik-Verlag, Weinheim, 1984.
21. G. Iooss, *Arch. Rat. Mech. Anal.* **47**, 301–329, 1972.
22. V. Solonnikov, *J. Soviet Math.* **8**, 467–529, 1977.
23. B. Straughan, *The Energy Method, Stability, and Nonlinear Convection*, Applied Mathematical Sciences, Vol. 91. Springer, Berlin, 1992.
24. G.P. Galdi and B. Straughan, *Proc. Roy. Soc. London A* **402**, 257–283, 1995.
25. R.A. Adams, *Sobolev Spaces*. Academic Press, New York, 1975.

References

Chapter 7
The Thermocapillary, Marangoni, Convection Problem

7.1 Introduction

It seems very judicious (at least from my point of view) to quote again some remarks about 'the dynamics of thin liquid films' from the preface of the recent special issue of *Journal of Engineering Mathematics* [1]:

> A detailed understanding of flows in thin liquid films is important for a wide range of modern engineering processes. This is particularly so in chemical and process engineering, where the liquid films are encountered in heat-and-mass-tranfer devices (e.g. distillation columns and spinning-disk reactors), and in coating processes (e.g. spin coating, blade coating, spray painting and rotational moulding). In order to design these processes for safe and efficient operation it is important to build mathematical models that can predict their performance, to have confidence in the predictions of the models, and to be able to use the models to optimize the design and operation of the devices involved. Thin liquid films also occur in a variety of biological contexts. On the other hand, when liquids flow in thin films, the interface (free surface) between the liquid and surrounding (passive!) gas can adopt a rich variety of interesting waveforms – these shapes are determined by a balance of the principle driving forces, usually including gravity, (temperature-dependent Marangoni phenomena) surface tension and viscous effects.

See, the paper by Trevelyan and Kalliadasis in [1, pp. 177–208]. The Marangoni effects due to the presence of surfactants are also the subject of many investigations, see, for instance, the papers by Edmonstone et al. in [1, pp. 141–156] and by Schwartz et al. in [1, pp. 157–175].

Over the last 40 years the nonlinear dynamics of a thin liquid film flowing down an inclined plane have been extensively studied using the famous Benney equation (see [2] and Section 7.5). In the paper by Oron and Gottlieb in [1, pp. 121–140], the problem of the stability threshold predicted by this Benney equation is revisited. On the other hand:

> the lubrication theory and its various extensions is an interesting and challenging one. Beyond the incorporation of different and more varied physical effects, there remain many mathematical challenges in the field of thin-film flows. The overriding mathematical advantage of thin-film theories is that they take account of a wide separation of scales in the geometrical configuration under consideration – this affords valuable simplification, obviating the need for computationally-expensive fully numerical simulations while preserving essential elements of the physics of the starting system.

Undoubtedly, the 12 papers in [1] are very valuable and very well illustrate both the wide variety of mathematical methods that have been employed and the broad range of their application; from this point of view, this special double issue of the *Journal of Engineering Mathematics* [1] is an 'recommended' complement to the present chapter. But, once again, various authors use *ad hoc* (non-rational) methods to derive various approximate models and this, unfortunately, strongly reduces their validity for practical applications!

During recent years, many books have been published in which thermocapillary, Marangoni convection is discussed. Of particular interest are, first, the book by Colinet et al. [3] which appeared in 2001 and then the book by Nepomnyashchy et al. [4] that appeared in 2002. In CISM Courses and Lectures, No. 428 [5], edited by Velarde and Zeytounian, the readers can find also various contributions relative to 'Interfacial Phenomena and the Marangoni effect' (presented at a Summer Course held at CISM, Udine, in July 2000).

Concerning the thermocapillary effect – driving the BM convection – we observe that, if the (free) surface tension σ changes with temperature T: $\sigma = \sigma(T)$, then

$$\nabla_{\|}\sigma(T) = \left[\frac{\mathrm{d}\sigma(T)}{\mathrm{d}T}\right]\nabla_{\|}T,$$

where $\nabla_{\|}$ indicates a gradient operator; but the subscript '$\|$' restricts the corresponding gradient vector to its surface components. The liquid tends to move in the direction from lower to higher surface tension (Marangoni effect, see Figure 7.1). The above quantity $\mathrm{d}\sigma(T)/\mathrm{d}T$ is negative for practically all

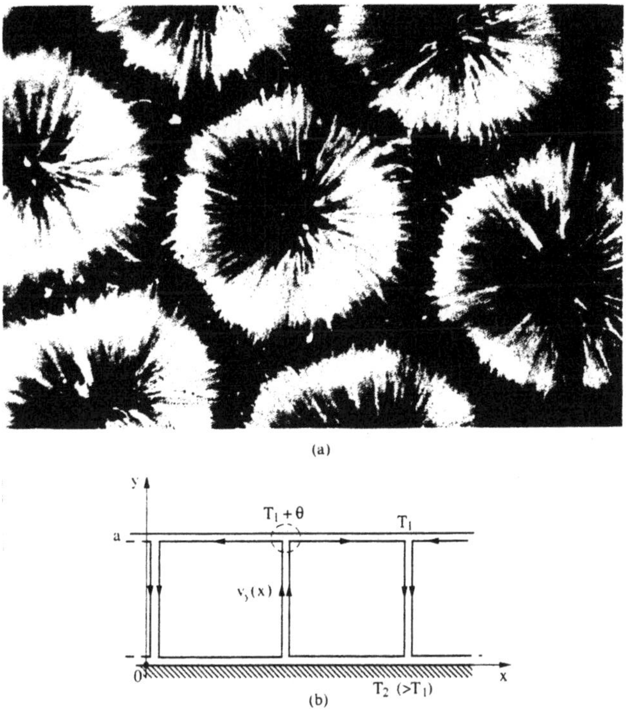

Fig. 7.1 BM instability. Reprinted with kind permission from [6].

substances which are relatively easily obtained in an almost pure state (the order of magnitude is 10^{-1} to 10^{-2}).

In Figure 7.1, the reader can find a visualization relative to Bénard–Marangoni instability; (a) shows convection cells visible from above in a thin liquid layer and (b) gives a scheme of the convection in BM cells. Thanks to the Marangoni effect, the free surface (or interface) becomes active in driving flow or instability in thin liquid layer, films or drops and also bubbles.

We also observe that the ratio between buoyancy ('Archimedean' effect) and surface-tension-gradient (Marangoni effect) forces is the dynamic Bond number, and when density $\rho = \rho(T)$, this number is given by

$$\mathrm{Bd} = gd^2\left[\frac{(-\mathrm{d}\rho(T)/\mathrm{d}T)_A}{(-\mathrm{d}\sigma(T)/\mathrm{d}T)_A}\right] \equiv \frac{\mathrm{Gr}}{\mathrm{Ma}}. \tag{7.1a}$$

where, with (7.1b) and (7.2a),

$$\mathrm{Gr} = \frac{\varepsilon'}{\mathrm{Fr}_A^2 d}, \tag{7.1b}$$

and

$$\mathrm{Ma} = \left[-\frac{\mathrm{d}\sigma(T)}{\mathrm{d}T}\right]_A d\frac{(T_w - T_A)}{\rho(T_A)\nu(T_A)^2}, \tag{7.1c}$$

$$\varepsilon' = \left(\frac{1}{\rho(T_A)}\right)\left[-\frac{\mathrm{d}\rho(T)}{\mathrm{d}T}\right]_A (T_w - T_A), \tag{7.1d}$$

$$\mathrm{Fr}^2_{Ad} = \frac{(\nu_A/d)^2}{gd}, \tag{7.1e}$$

when as dimensionless temperature we have (see (1.17c)),

$$\Theta = \frac{(T - T_A)}{(T_w - T_A)}.$$

For usual values we obtain

$$\mathrm{Bd} \approx 1, \quad \text{for } d \approx (\gamma_{\sigma A}/g\rho_A\alpha_A)^{1/2} \approx (1/10)\ \mathrm{cm}$$

which is the same value for d obtained (see (1.11)) when

$$\mathrm{Fr}^2_{Ad} \approx 1 \Rightarrow d \approx \left(\frac{\nu_A^2}{g}\right)^{1/3} \approx 1.00\ \mathrm{mm}.$$

From a physicist's point of view (see, for instance, [7]) the ratio (7.1a) gives a measure of the relative effectiveness of buoyancy and of surface tension effects, each of which results from variation in temperature.

For a given temperature difference, this ratio varies with d^2, and as a result *surface tension effects dominate for small thickness of the fluid layer, and gravitational ones for very thick layers*. Ten years ago, when I first read the above sentence in [6], I understood that these two effects should be related (certainly) to two particular values of a single dimensionless reference parameter. A little later on I discovered that, in fact, the Grashof number (7.1b) is a ratio of two dimensionless parameters, $\varepsilon' \ll 1$ and Fr^2_{Ad}, and this has been for me an indication that the squared Froude number (where the thickness d of the liquid layer is present) plays this role. This remark allowed me to formulate the following 'alternative' [8], published in 1998:

> Either the buoyancy is taken into account, and in this case the free-surface deformation effect is negligible and we rediscover the classical Rayleigh–Bénard (RB) shallow convection rigid-free problem or, the free-surface deformation effect is taken into account and, in a such case, at the leading-order approximation for a weakly expansible fluid, the buoyancy does not play a significant rôle in the Bénard–Marangoni (BM) thermocapillary instability problem.

In this chapter, our main objective is to take into account accurately – in the case when the squared Froude number Fr^2_{Ad} is fixed and of order 1 (the liquid film layer being thin and weakly expansible) – the various significant results obtained mainly in Chapter 5, for a presentation of a rational theory of the BM thermocapillary convection. First, because the fact that

$$\mathrm{Fr}^2_{Ad} = 1, \quad \text{and as a consequence, Gr} \approx \varepsilon', \text{ with } \varepsilon' = \alpha(T_A)(T_w - T_A), \tag{7.2a}$$

the Boussinesq limiting process is not necessary and, instead we use simply the fact that

$$\text{the expansibility parameter } \varepsilon' \text{ tends to zero!} \tag{7.2b}$$

In Section 7.2 we thus consider the case when the dimensionless temperature is given by Θ $[= (T - T_A)/(T_w - T_A)]$, T_A being the passive air constant temperature above the free surface and T_w the constant temperature of the lower heated plate on $z = 0$. The dimensionless upper free-surface condition associated with Θ is, in such a case (see (5.6d))

$$\frac{\partial \Theta}{\partial n'} + \mathrm{Bi}_{\mathrm{conv}}\Theta = 0 \quad \text{at } z' = H'(t', x', y'), \tag{7.3a}$$

when we assume that $D_A \equiv 0$.

We observe that in a motionless steady conduction state, the dimensionless 'conduction' temperature $\Theta_S(z')$ satisfies the upper condition (instead of (7.3a))

$$\frac{\mathrm{d}\Theta_S}{\mathrm{d}z'} + \mathrm{Bi}_s\Theta_S = 0 \quad \text{at } z' = 1. \tag{7.3b}$$

As a consequence, we obtain

$$\Theta_S = 1 - \left[\frac{\mathrm{Bi}_s}{(1 + \mathrm{Bi}_s)}\right] z'. \tag{7.3c}$$

In Section 7.3 we return to full formulation of the BM dimensionless thermocapillary convection model, given in Section 7.2, keeping in mind the goal of obtaining a simplified 'BM long-wave reduced model problem'.

In Section 7.4, thanks to the results of the preceding section (Section 7.3), we derive accurately a 'new' lubrication equation for the thickness of the thin liquid film. In particular, taking into account our 'two Biot (for conduction and convection regimes) numbers' approach, we show that the consideration of a variable (for instance, function of the thickness $H'(t', x', y')$, of the liquid film) convective Biot number, gives the possibility to take into account, in a 'newly derived' lubrication equation, the thermocapillary/Marangoni effect, even if the convective Biot number is vanishing (zero)!

Section 7.5 is devoted to an asymptotic detailed derivation of a generalized *à la* Benney equation, to the Kuramoto–Sivashinsky (KS) equation, and KS–KdV equation with a dispersive term and also our 1998 IBL non-isothermal system of three averaged equations, for thickness (H), flow rate (q) and (Σ) related to the temperature across a layer.

In Section 7.6, various aspects of the linear and weakly nonlinear stability analysis of the thermocapillary convection are discussed.

In Section 7.7, devoted to 'some complementary remarks', various results derived in Sections 7.4 and 7.5 are re-considered and compared with the results obtained when the dimensionless temperature is given by θ [$= (T - T_d)/(T_w - T_d)$], where T_d [$\equiv T_s(z = d)$] is the temperature linked with the steady-state motionless conduction [$T_s(z) = T_w - \beta_s d$]. The upper free-surface condition associated with θ (see (2.48)) being

$$\frac{\partial\theta}{\partial n'} + \left\{\frac{\mathrm{Bi}_{\mathrm{conv}}}{\mathrm{Bi}_s(T_d)}\right\}[1 + \mathrm{Bi}_s(T_d)\theta] = 0 \quad \text{at } z' = H'(t', x', y') \tag{7.4a}$$

and the associated $\theta_S(z')$ satisfy the upper condition (instead of (7.4a), because in a conduction state $\mathrm{Bi}_{\mathrm{conv}} \Rightarrow \mathrm{Bi}_s(T)d)$],

$$\frac{\mathrm{d}\theta_S}{\mathrm{d}z'} + 1 + \mathrm{Bi}_s(T_d)\theta_S = 0 \quad \text{at } z' = 1, \tag{7.4b}$$

which leads to

$$\theta_S = 1 - z'. \tag{7.4c}$$

The above observation shows that it is necessary for each case to be precise and to use the associated steady motionless conduction solution (with a subscript 'S') with as Biot number Bi_s instead of $\mathrm{Bi}_{\mathrm{conv}}$.

We again stress that the 'usual, *à la* Davis' [41] upper free-surface condition for the dimensional temperature θ is obtained from (7.4a) when we identify (or perhaps confuse) the convection Biot number (variable, $\mathrm{Bi}_{\mathrm{conv}}$) with the conduction Biot number (constant, $\mathrm{Bi}_s(T_d)$).

7.2 The Formulation of the Full Bénard–Marangoni Thermocapillary Problem

Now, our starting dominant, approximate, dimensionless system of three equations (where the terms proportional to ε' are taken into account) is given by (5.4a–c) and boundary conditions by (5.5a, b), (5.6a–d). With the above limiting process (7.2b), we associate three asymptotic expansions for velocity vector $\mathbf{u}'$, dimensionless pressure

$$\pi = \left(\frac{1}{\mathrm{Fr}_d^2}\right)\left\{\left[\frac{(p - p_A)}{gd\rho_A}\right] + z' - 1\right\},$$

and dimensionless temperature Θ. Namely:

$$\mathbf{u}' = \mathbf{u}_{\mathrm{BM}} + \varepsilon'\mathbf{u}_1 + \cdots, \quad \pi = \pi_{\mathrm{BM}} + \varepsilon'\pi_1 + \cdots, \quad \Theta = \Theta_{\mathrm{BM}} + \varepsilon'\Theta_1 + \cdots, \tag{7.5}$$

The reader may observe that, only in Chapter 3 (see (3.6c)), for the simple Rayleigh thermal convection problem, we have defined π with $(p - p_d)$, and this is justified for a constant thickness liquid layer, d. With (7.2a, b) and (7.5) we obtain as leading-order BM equations (all primes $'$ have been dropped, see (5.4a–c)) the following three equations:

$$\nabla \cdot \mathbf{u}_{\mathrm{BM}} = 0, \tag{7.6a}$$

$$\frac{\mathrm{d}\mathbf{u}_{\mathrm{BM}}}{\mathrm{d}t} + \nabla\pi_{\mathrm{BM}} = \Delta\mathbf{u}_{\mathrm{BM}}, \tag{7.6b}$$

$$\mathrm{Pr}\,\frac{\mathrm{d}\Theta_{\mathrm{BM}}}{\mathrm{d}t} = \Delta\Theta_{\mathrm{BM}}, \tag{7.6c}$$

With $\mathrm{d}/\mathrm{d}t = \partial/\partial t + \mathbf{u}_{\mathrm{BM}} \cdot \nabla$, which are the usual Navier equations (7.6a, b) relative to $\mathbf{u}_{\mathrm{BM}}$ and π_{BM}, for an incompressible fluid supplemented *à la* Fourier by temperature equation (7.6c) for Θ_{BM} where Pr is the usual Prandtl number at constant temperature T_A,

$$\mathrm{Pr} = \frac{\nu(T_A)}{\kappa(T_A)} \quad \text{with} \quad \kappa(T_A) = \frac{k(T_A)}{\rho(T_A)C_v(T_A)},$$

T_A being the temperature of the ambiant passive air above the upper free surface.

On the other hand, if the boundary conditions for $\mathbf{u}_{\mathrm{BM}}$ and Θ_{BM}, at lower horizontal plane $z = 0$, according to (5.5a, b), are simply

$$\mathbf{u}_{\mathrm{BM}}|_{z=0} \equiv (u_{1\mathrm{BM}}|_{z=0}, u_{2\mathrm{BM}}|_{z=0}, u_{3\mathrm{BM}}|_{z=0}) = 0, \tag{7.7a}$$

and

$$\Theta_{\mathrm{BM}}|_{z=0} = 1, \tag{7.7b}$$

on the contrary, at the deformable free surface $z = H(t, x, y)$, the upper conditions are very complicated. These upper free-surface conditions are derived from (5.6a–d), written with Θ (see (5.6c–d)).

First, for the dimensionless pressure $\pi_{\mathrm{BM}}|_{z=H}$ we obtain

$$\pi_{\mathrm{BM}} = \left(\frac{1}{\mathrm{Fr}^2_{Ad}}\right)(H-1) + \left(\frac{2}{N}\right)\left\{\left(\frac{\partial u_{1\mathrm{BM}}}{\partial x}\right)\left(\frac{\partial H}{\partial x}\right)^2\right.$$
$$+\left(\frac{\partial u_{2\mathrm{BM}}}{\partial y}\right)\left(\frac{\partial H}{\partial y}\right)^2 + \frac{\partial u_{3\mathrm{BM}}}{\partial z}$$
$$+\left[\frac{\partial u_{1\mathrm{BM}}}{\partial y} + \frac{\partial u_{2\mathrm{BM}}}{\partial x}\right]\left(\frac{\partial H}{\partial x}\right)\left(\frac{\partial H}{\partial y}\right) - \left[\frac{\partial u_{1\mathrm{BM}}}{\partial z} + \frac{\partial u_{3\mathrm{BM}}}{\partial x}\right]\left(\frac{\partial H}{\partial x}\right)$$
$$\left.-\left[\frac{\partial u_{2\mathrm{BM}}}{\partial z} + \frac{\partial u_{3\mathrm{BM}}}{\partial y}\right]\left(\frac{\partial H}{\partial y}\right)\right\} - \left(\frac{1}{N}\right)^{3/2}[\mathrm{We} - \mathrm{Ma}\,\Theta_{\mathrm{BM}}]\left\{N_2\left(\frac{\partial^2 H}{\partial x^2}\right)\right.$$
$$\left.-2\left(\frac{\partial H}{\partial x}\right)\left(\frac{\partial H}{\partial y}\right)\left(\frac{\partial^2 H}{\partial x \partial y}\right) + N_1\left(\frac{\partial^2 H}{\partial y^2}\right)\right\}$$
$$\text{at } z = H(t, x, y), \tag{7.8a}$$

with

$$N = 1 + \left(\frac{\partial H}{\partial x}\right)^2 + \left(\frac{\partial H}{\partial y}\right)^2;$$
$$N_1 = 1 + \left(\frac{\partial H}{\partial x}\right)^2;$$
$$N_2 = 1 + \left(\frac{\partial H}{\partial y}\right)^2.$$

In the above upper free-surface condition (7.8a) for π_{BM}, the first term takes into account a gravity effect (via the squared Froude number Fr^2_{Ad}, which is of order 1) and we have also a Weber (We) effect and a Marangoni (Ma) effect with (see (5.7d) and (5.7a)/(7.1c)),

$$\mathrm{We} = \frac{\sigma(T_A)d}{\rho(T_A)\nu(T_A)^2},$$
$$\mathrm{Ma} = \left[-\frac{\mathrm{d}\sigma(T)}{\mathrm{d}T}\right]_A d\frac{(T_w - T_A)}{\rho(T_A)\nu(T_A)^2}$$

and

$$\sigma(T) = \sigma(T_A)\left[1 - \left(\frac{\mathrm{Ma}}{\mathrm{We}}\right)\Theta_{\mathrm{BM}}\right].$$

Then, as tangential upper free-surface we obtain two conditions with Marangoni effect:

$$
\begin{aligned}
&\left[\frac{\partial u_{1\mathrm{BM}}}{\partial x}-\frac{\partial u_{3\mathrm{BM}}}{\partial z}\right]\left(\frac{\partial H}{\partial x}\right)+(1/2)\left[\frac{\partial u_{1\mathrm{BM}}}{\partial y}+\frac{\partial u_{2\mathrm{BM}}}{\partial x}\right]\left(\frac{\partial H}{\partial y}\right)\\
&\quad+(1/2)\left[\frac{\partial u_{2\mathrm{BM}}}{\partial z}+\frac{\partial u_{3\mathrm{BM}}}{\partial y}\right]\left(\frac{\partial H}{\partial x}\right)\left(\frac{\partial H}{\partial y}\right)\\
&\quad-(1/2)\left[1-\left(\frac{\partial H}{\partial x}\right)^2\right]\left[\frac{\partial u_{3\mathrm{BM}}}{\partial x}+\frac{\partial u_{1\mathrm{BM}}}{\partial z}\right]\\
&\quad=\left(\frac{N^{1/2}}{2}\right)\mathrm{Ma}\left[\frac{\partial\Theta_{\mathrm{BM}}}{\partial x}+\left(\frac{\partial H}{\partial x}\right)\frac{\partial\Theta_{\mathrm{BM}}}{\partial z}\right]\\
&\qquad\text{at } z=H(t,x,y),
\end{aligned}
\tag{7.8b}
$$

and

$$
\begin{aligned}
&\left[\frac{\partial u_{1\mathrm{BM}}}{\partial x}-\frac{\partial u_{2\mathrm{BM}}}{\partial y}\right]\left(\frac{\partial H}{\partial y}\right)\left(\frac{\partial H}{\partial x}\right)^2+\left[\frac{\partial u_{2\mathrm{BM}}}{\partial y}-\frac{\partial u_{3\mathrm{BM}}}{\partial z}\right]\left(\frac{\partial H}{\partial y}\right)\\
&\quad+\left[\frac{\partial u_{1\mathrm{BM}}}{\partial z}+\frac{\partial u_{3\mathrm{BM}}}{\partial x}\right]\left(\frac{\partial H}{\partial x}\right)\left(\frac{\partial H}{\partial y}\right)\\
&\quad+(1/2)\left[1+\left(\frac{\partial H}{\partial x}\right)^2-\left(\frac{\partial H}{\partial y}\right)^2\right]\left(\frac{\partial u_{1\mathrm{BM}}}{\partial y}+\frac{\partial u_{2\mathrm{BM}}}{\partial x}\right)\left(\frac{\partial H}{\partial x}\right)\\
&\quad-(1/2)\left[1-\left(\frac{\partial H}{\partial x}\right)^2-\left(\frac{\partial H}{\partial y}\right)^2\right]\left[\frac{\partial u_{2\mathrm{BM}}}{\partial z}+\frac{\partial u_{3\mathrm{BM}}}{\partial y}\right]\\
&\quad=\left(\frac{N^{1/2}}{2}\right)\mathrm{Ma}\left\{-\left(\frac{\partial H}{\partial x}\right)\left(\frac{\partial H}{\partial y}\right)\frac{\partial\Theta_{\mathrm{BM}}}{\partial x}\right.\\
&\quad\left.+\left[1+\left(\frac{\partial H}{\partial x}\right)^2\right]\frac{\partial\Theta_{\mathrm{BM}}}{\partial y}+\left(\frac{\partial H}{\partial y}\right)\frac{\partial\Theta_{\mathrm{BM}}}{\partial z}\right\}\\
&\qquad\text{at } z=H(t,x,y).
\end{aligned}
\tag{7.8c}
$$

Finally, from (7.3b), when we take into account (4.22a), we derive for Θ_{BM} the following upper free-surface boundary condition:

$$
\begin{aligned}
&\frac{\partial\Theta_{\mathrm{BM}}}{\partial z}+N^{1/2}\mathrm{Bi}_{\mathrm{conv}}\Theta_{\mathrm{BM}}\\
&\quad=\frac{\partial\Theta_{\mathrm{BM}}}{\partial x}\left(\frac{\partial H}{\partial x}\right)+\frac{\partial\Theta_{\mathrm{BM}}}{\partial y}\left(\frac{\partial H}{\partial y}\right)\quad\text{at } z=H(t,x,y),
\end{aligned}
\tag{7.8d}
$$

the convective (or perhaps variable) Biot number, $\mathrm{Bi}_{\mathrm{conv}}$, being different from the conduction, (constant) Biot number, Bi_s, while the kinematic condition

(5.6a) is unchanged (but written without the primes ′).

$$u_{3\mathrm{BM}} = \frac{\partial H}{\partial t} + u_{1\mathrm{BM}}\left(\frac{\partial H}{\partial x}\right) + u_{2\mathrm{BM}}\left(\frac{\partial H}{\partial y}\right) \quad \text{at } z = H(t, x, y). \quad (7.8\mathrm{e})$$

Concerning the conduction (constant) Biot number Bi_s, it appears in function $\Theta_s(z)$:

$$\Theta_s(z) = 1 - \left[\frac{\mathrm{Bi}_s}{(1+\mathrm{Bi}_s)}\right] z, \quad (7.9)$$

according to relation (4.39) for β_s, because

$$\Theta_s(z) \equiv \frac{[T_s(z) - T_A]}{(T_w - T_A)} = \frac{[T_w - T_A - \beta_s dz]}{(T_w - T_A)},$$

for the steady motionless conduction regime.

In a recent paper by Ruyer-Quil et al. [9], if the above upper free-surface condition (7.8d) is well used, unfortunately again a confusion is present between Bi_s and $\mathrm{Bi}_{\mathrm{conv}}$; when the formula for $\Theta_s(z)$ is written, only a single Biot number (a Bi) appears.

For instance if, in particular, $\mathrm{Bi}_{\mathrm{conv}} \equiv B(H)$ and if we take into account that in a steady-state motionless conduction regime we have $H = 1$, then, and only for this case, we have the relation

$$B(H = 1) \equiv \mathrm{Bi}_s.$$

Unfortunately, this confusion has various, certainly undesirable, consequences in derivation of the so-called ‘boundary-layer’ (BL) equations (for instance, equations (4.18a–c) and (4.19a–h) in [9]). Because the relation (2.9) in [9], where the same Bi appears as in the upper/interface condition (2.8) in [9], seems to be used for the derivation of the above-mentioned BL equations in [9].

Obviously, if this is really the case, then the results given in the paper by Scheid et al. [10] will be ‘unreliable’, especially for a very small (and the more for zero) Biot number. This, ‘unreliability’ being related to the fact that the conduction, constant, Biot number Bi_s (always different from zero and often confused with the Biot number Bi, in upper free-surface conditions for the dimensionless temperature) explicitly does not appear in the derived BL equations.

The above model problem (7.6a–c), (7.7a, b) and (7.8a–e) formulated for the BM thermocapillary convection, even in the framework of a numerical simulation, is a very difficult, awkward and tedious problem! It is clear that simplifications in a rational approach are necessary, obviating the need for computationally-expensive (in time and also in money) fully numerical simulations, while preserving essential elements of the physics of the above formulated BM thermocapillary convection model problem.

Among various approaches linked with the BM thermocapillary convection, the formation of long waves (with respect to a very thin film layer) at the surface of a falling film – for instance, free-falling down a uniformly heated vertical plane – is a challenging problem. The waves resulting at the free surface, as a consequence of the interfacial stress generated by (the temperature-dependent) surface tension gradient (Marangoni effect), induce thermocapillary instability modes and various stability results can be obtained. In a very thin film, obviously, a typical length λ of the (long) waves is large in comparison with the thickness, $d \ll \lambda$, of the thin film, so that the slope of the free surface is always small.

In such a case (see Section 7.3) it is necessary to introduce a 'long-wave dimensionless parameter' (see (4.49))

$$\delta = \frac{d}{\lambda} \ll 1. \tag{7.10}$$

The essential advantage of this 'long-wave approximation' is a drastic simplification of the full dimensionless BM model problem formulated in Section 7.2 of this chapter. The two recent papers [9, 10] are precisely devoted to 'low-dimensional formulation' and 'linear stability and nonlinear waves' for this long- wave case. We observe here, also, that in our survey paper [8], an 'integral-boundary-layer' (IBL) model was first suggested and derived for the non-isothermal case in which we have considered three averaged evolution equations for local film thickness, flow rate and mean temperature across the layer (see Section 7.5).

7.3 Some 'BM Long-Wave' Reduced Convection Model Problems

Here we return to Section 7.2, and consider the full derived BM dimensionless thermocapillary convection model problem (7.6a–c), (7.7a, b) and (7.8a–e), keeping in mind that we want to obtain a simplified 'BM long-wave' reduced model. With (7.10) we introduce the following new coordinates and functions:

$$X = \delta x, \quad Y = \delta y, \quad Z \equiv z, \quad T = \delta\, \mathrm{Re}_d t;$$

$$U = \frac{u_{1\mathrm{BM}}}{\mathrm{Re}_d}, \quad V = \frac{u_{2\mathrm{BM}}}{\mathrm{Re}_d} \quad \text{and} \quad W = \frac{u_{3\mathrm{BM}}}{\delta\,\mathrm{Re}_d};$$

$$\Pi = \frac{\pi_{\mathrm{BM}}}{\mathrm{Re}_d^2}, \quad \Xi \equiv \Theta_{\mathrm{BM}} \quad \text{and} \quad \chi(T, X, Y) \equiv H\left(\frac{T}{\delta\,\mathrm{Re}_d}, \frac{X}{\delta}, \frac{Y}{\delta}\right). \tag{7.11}$$

In (7.11) we have introduced a Reynolds number (based on the thickness d) such that

$$\mathrm{Re}_d = \frac{U_c d}{\nu_A}, \tag{7.12}$$

where the characteristic velocity U_c is determined below (see (7.18)) by a significant similarity rule. The following two operators are also introduced:

$$\frac{D}{DT} \equiv \frac{\partial}{\partial T} + U\frac{\partial}{\partial X} + V\frac{\partial}{\partial Y} + W\frac{\partial}{\partial Z}, \tag{7.13a}$$

$$\mathbf{D} = \left(\frac{\partial}{\partial X}, \frac{\partial}{\partial Y}\right) \quad \text{and} \quad \mathbf{D}^2 = \frac{\partial^2}{\partial X^2} + \frac{\partial^2}{\partial Y^2}, \tag{7.13b}$$

and also the horizontal velocity vector

$$\mathbf{V} = (U, V). \tag{7.13c}$$

With (7.13a–c) we write

$$\frac{\mathrm{d}}{\mathrm{d}t} = \delta\,\mathrm{Re}_d\left(\frac{D}{DT}\right), \quad \nabla = \delta\mathbf{D} + \left(\frac{\partial}{\partial Z}\right)\mathbf{k}, \quad \Delta = \delta^2\mathbf{D}^2 + \frac{\partial^2}{\partial Z^2}. \tag{7.14}$$

To begin, instead of the BM model equations (7.6a–c) we can write the following *first* set of equations for $\mathbf{V}$, W, Π, and Ξ:

$$\mathbf{D}\cdot\mathbf{V} + \frac{\partial W}{\partial Z} = 0; \tag{7.15a}$$

$$\frac{D\mathbf{V}}{DT} + \mathbf{D}\Pi = \left(\frac{1}{\delta\,\mathrm{Re}_d}\right)\left[\delta^2\mathbf{D}^2\mathbf{V} + \frac{\partial^2\mathbf{V}}{\partial Z^2}\right]; \tag{7.15b}$$

$$\delta^2\frac{DW}{DT} + \frac{\partial\Pi}{\partial Z} = \left(\frac{\delta^2}{\delta\,\mathrm{Re}_d}\right)\left[\delta^2\mathbf{D}^2\mathbf{V} + \frac{\partial^2 W}{\partial Z^2}\right]; \tag{7.15c}$$

$$\mathrm{Pr}\frac{D\Xi}{DT} = \left(\frac{1}{\delta\,\mathrm{Re}_d}\right)\left[\delta^2\mathbf{D}^2\Xi + \frac{\partial^2\Xi}{\partial Z^2}\right]. \tag{7.15d}$$

However, we can also define a Reynolds number based on the length λ such that (see (7.10)), $\delta = d/\lambda \ll 1$,

$$\frac{U_c \lambda}{\nu_A} \equiv \frac{\mathrm{Re}_d}{\delta} = \mathrm{Re}_\lambda. \tag{7.16}$$

With (7.16) instead of the above first set of equations (7.15a–d), we obtain the following *second* set of equations:

$$\mathbf{D} \cdot \mathbf{V} + \frac{\partial W}{\partial Z} = 0; \tag{7.17a}$$

$$\frac{D\mathbf{V}}{DT} + \mathbf{D}\Pi = \left(\frac{1}{\delta^2 \mathrm{Re}_\lambda}\right)\left[\delta^2 \mathbf{D}^2 \mathbf{V} + \frac{\partial^2 \mathbf{V}}{\partial Z^2}\right]; \tag{7.17b}$$

$$\delta^2 \frac{DW}{DT} + \frac{\partial \Pi}{\partial Z} = \left(\frac{\delta^2}{\delta^2 \mathrm{Re}_\lambda}\right)\left[\delta^2 \mathbf{D}^2 \mathbf{V} + \frac{\partial^2 W}{\partial Z^2}\right]; \tag{7.17c}$$

$$\mathrm{Pr}\frac{D\Xi}{DT} = \left(\frac{1}{\delta^2 \mathrm{Re}_\lambda}\right)\left[\delta^2 \mathbf{D}^2 \Xi + \frac{\partial^2 \Xi}{\partial Z^2}\right]. \tag{7.17d}$$

An obvious case, to simplify the first set (with Re_d) of equations (7.15a–d), is linked with the following limiting process:

$$\delta\, \mathrm{Re}_d = R^*, \text{ fixed, when } \delta \to 0 \text{ and } \mathrm{Re}_d \to \infty \text{ simultaneously,} \tag{7.18a}$$

and as a result, we obtain (with ‘0’ subscript) at the leading order:

$$\begin{aligned}
&\mathbf{D} \cdot \mathbf{V}_0 + \frac{\partial W_0}{\partial Z} = 0;\\
&\frac{D\mathbf{V}_0}{DT} + \mathbf{D}\Pi_0 - \left(\frac{1}{R^*}\right)\frac{\partial^2 \mathbf{V}_0}{\partial Z^2} = 0;\\
&\frac{\partial \Pi_0}{\partial Z} = 0;\\
&\mathrm{Pr}\frac{D\Xi_0}{DT} - \left(\frac{1}{R^*}\right)\frac{\partial^2 \Xi_0}{\partial Z^2} = 0.
\end{aligned} \tag{7.18b}$$

Concerning the above second set (with Re_λ) of equations (7.17a–d) we can simplify if we assume that

$$\delta^2\, \mathrm{Re}_\lambda = R^{**}, \text{ fixed, when } \delta \to 0 \text{ and } \mathrm{Re}_\lambda \to \infty \text{ simultaneously,} \tag{7.19}$$

and, as a result, we obtain at the leading order (with ‘0’ subscript), again, the reduced set of equations (7.18b); but, in front of the viscous and heat conducting terms, instead of $(1/R^*)$ we have $(1/R^{**})$. We observe that the third equation, $\partial \Pi_0/\partial Z = 0$, in the set of equations (7.18b), which replace the full equation (7.15c) or (7.17c) for W, is typically a 'boundary layer equation' and we see (from the classical theory of high Reynolds number, vanishing viscosity, fluid flow) that the system (7.18b) is certainly (at least) singular in the vicinity to initial time, $T = 0$, where it is necessary to write initial data for $\mathbf{V}$, W and Ξ, which are solutions of the three evolution (in time) starting equations (7.15b–d) or (7.17b–d).

Indeed, the simplified equations (7.18b) are (only) outer in time equations and it is easy to verify that the limit process, $T \to 0$, and limiting process (7.18a), do not permute; near $T = 0$ it is necessary to derive a new, significant simplified set of equations, local in time. For the derivation of these significant local equations, near $T = 0$, usually some changes in (7.11) are introduced, relative to time and vertical variable and also to vertical velocity component and pressure. Namely, if we introduce

$$\tau = \frac{T}{\delta^2}, \quad \zeta = \frac{Z}{\delta}, \quad \omega = \delta W, \quad P = \delta \Pi, \tag{7.20}$$

in such a case, instead of starting equations (7.17a–d) we obtain a set of dimensionless equations adapted for the vicinity of $T = 0$. Here we do not consider in detail these local (inner) equations, valid near $T = 0$, and their relations with the above (outer), because this local asymptotic model and its 'matching' with the outer model according to the relation $\tau \to \infty$ and $\zeta \to \infty \Leftrightarrow T = 0$ and $Z = 0$, deserves a careful approach. We note only that, near $T = 0$, with (7.20) we obtain for the horizontal velocity vector $V_0^l(\tau, X, Y, \zeta)$ and temperature $\Xi_0^l(\tau, X, Y, \zeta)$ the following two local/inner equations (where the two horizontal coordinates X and Y play the role of two parameters):

$$\frac{\partial V_0^l}{\partial \tau} - \left(\frac{1}{R^*}\right)\frac{\partial^2 V_0^l}{\partial \zeta^2} = 0 \quad \text{and} \quad \frac{\partial \Xi_0^l}{\partial \tau} - \left(\frac{1}{R^*}\right)\frac{\partial^2 \Xi_0^l}{\partial \zeta^2} = 0, \tag{7.21a}$$

with the matching conditions

$$\lim_{\tau\to\infty,\zeta\to\infty} [V_0^l(\tau, X, Y, \zeta)] = \mathbf{V}_0(T = 0, X, Y, Z = 0), \tag{7.21b}$$

$$\lim_{\tau\to\infty,\zeta\to\infty} [\Xi_0^l(\tau, X, Y, \zeta)] = \Xi_0(T = 0, X, Y, Z = 0). \tag{7.21c}$$

See Section 10.8 for a similar example relative to formation of a thin liquid film on a rotating disk.

Below we consider the set of reduced equations (7.18b) and, first, from (7.7a, b) for $\mathbf{V}_0$ and Ξ_0 we have the two boundary conditions:

$$\mathbf{V}_0 = 0 \tag{7.22a}$$

and

$$\Xi_0 = 1 \quad \text{at } Z = 0. \tag{7.22b}$$

Now, from the kinematic upper condition (7.8e), we obtain

$$W_0 = \frac{\partial \chi}{\partial T} + \mathbf{V}_0 \cdot \mathbf{D}\chi \quad \text{at } Z = \chi(T, X, Y),$$

and with

$$W_0 = 0 \quad \text{at } Z = 0,$$

we derive the following averaged evolution equation for the function $\chi(T, X, Y)$, which characterizes the deformation of the free surface,

$$\frac{\partial \chi}{\partial T} + \mathbf{D} \cdot \left(\int_0^{Z=\chi} \mathbf{V}_0 \, \mathrm{d}Z \right) = 0, \tag{7.23}$$

which plays a central role in lubrication theory (see Section 7.4). However, it is then necessary for this that the horizontal velocity $\mathbf{V}_0$, under the integral in (7.23), was expressed in terms of the thickness $\chi(T, X, Y)$? As a general rule, the average over the film thickness of the horizontal velocity vector $\mathbf{V}_0$ cannot be expressed in terms of χ or any of its spatial derivatives and for this, equation (7.23) is not a closed form evolution equation for thickness $\chi(T, X, Y)$. Our main goal in Section 7.4 is to examine rationally some situations that arise mainly from distinguished limiting processes, from which a closed equation may be obtained for $\chi(T, X, Y)$!

In the present section we only want to derive some simplified model problems which can be computed numerically more easily (but do not lead to an explicit form for $\mathbf{V}_0$). Therefore it is necessary for the reduced system of equations (7.18b), with two boundary conditions on $Z = 0$ (7.22a, b) that we derive from (7.8a–e) the simplified upper free-surface conditions associated with (7.18b).

First, from the upper condition for π_{BM} (7.8a), we obtain for the new dimensionless pressure Π the following full condition at upper free-surface $Z = \chi(T, X, Y)$:

$$\Pi \approx \left[\frac{1}{\mathrm{Re}_d^2 \, \mathrm{Fr}_{Ad}^2}\right](\chi - 1) + 2\frac{\delta^2}{R^*}\left\{1 - \delta^2\left[\left(\frac{\partial \chi}{\partial X}\right)^2 + \left(\frac{\partial \chi}{\partial Y}\right)^2\right]\right\}\left[\frac{\partial W}{\partial Z}\right.$$

$$\left. - \frac{\partial U}{\partial Z}\left(\frac{\partial \chi}{\partial X}\right) - \frac{\partial}{\partial Z}\left(\frac{\partial \chi}{\partial Y}\right) + \mathrm{O}(\delta^2)\right] - \left\{1 - \left(\frac{3}{2}\right)\delta^2\left[\left(\frac{\partial \chi}{\partial X}\right)^2\right.\right.$$

$$+\left(\frac{\partial \chi}{\partial Y}\right)^2\Bigg]\Bigg\}\left\{\frac{\delta^2}{\mathrm{Re}_d^2}[\mathrm{We}-\mathrm{Ma}\,\Xi][\mathbf{D}^2\chi+\mathrm{O}(\delta^2)]\right\}.$$

However,

$$(\mathrm{Re}_d\,\mathrm{Fr}_{Ad}^2)\equiv F^2=\frac{U_c^2}{gd},$$

which is the squared Froude number defined with the characteristic velocity U_c and constant thickness d. We assume that we have the following similarity rule:

$$\frac{R^*}{F^2}=G=\mathrm{O}(1), \tag{7.24a}$$

When $G\approx 1$, from (7.24a) this gives the following relation for the characteristic velocity:

$$U_c=\frac{gd^3}{\lambda\nu_A}. \tag{7.24b}$$

By analogy with the case of the squared Froude number (F^2), defined with characteristic velocity U_c, we introduce the corresponding *modified* Weber (W) and Marangoni (M) numbers, defined also with this velocity U_c,

$$W=\frac{\sigma(T_A)}{\rho T_A)dU_c^2}\equiv\frac{\mathrm{We}}{\mathrm{Re}_d^2},\ M=\left[-\frac{\mathrm{d}\sigma(T)}{\mathrm{d}T}\right]_A\frac{(T_w-T_A)}{d\rho(T_A)U_c^2}\equiv\frac{\mathrm{Ma}}{\mathrm{Re}_d^2}. \tag{7.25a}$$

In such a case, under the limiting process (7.18a), we obtain at $Z=\chi(T,X,Y)$, from the above full condition for Π the following leading-order reduced upper condition, for Π_0 at $Z=\chi(T,X,Y)$:

$$\Pi_0\approx\left[\frac{G}{R^*}\right](\chi-1)-W^*\mathbf{D}^2\chi+\delta^2M\,\Xi_0\mathbf{D}^2\chi+\mathrm{O}(\delta^2), \tag{7.26a}$$

if the following similarity rule is satisfied (for a large Weber number W)

$$\delta^2W=W^*, \tag{7.26b}$$

with W^* as the new Weber number characterizing the reference constant surface tension effect.

For the moment, we do not know whether it is judicious or not to consider also the case of a large Marangoni number such that $\delta^2M=\mathrm{O}(1)$. In order to give an answer to this question, it is necessary to consider the two tangential upper free-surface conditions (7.8b, c). Namely, instead of these two conditions we derive in the long-wave approximation, at the upper free-surface $Z=\chi(T,X,Y)$, respectively:

$$-(1/2)\frac{\partial U_0}{\partial Z} = (1/2)R^*M\left[\frac{\partial \Xi_0}{\partial X} + \left(\frac{\partial \chi}{\partial X}\right)\frac{\partial \Xi_0}{\partial Z}\right], \tag{7.27a}$$

and

$$-(1/2)\frac{\partial V_0}{\partial Z} = (1/2)R^*M\left[\frac{\partial \Xi_0}{\partial Y} + \left(\frac{\partial \chi}{\partial Y}\right)\frac{\partial \Xi_0}{\partial Z}\right], \tag{7.27b}$$

when we have assumed that $M = \mathrm{Ma}/\mathrm{Re}_d^2$ as in (7.25a). From (7.27a, b), it seems more judicious to assume (because R^* is fixed) that

$$M = \mathrm{O}(1), \tag{7.27c}$$

and in such a case, in (7.26a) as leading-order terms on the right-hand side we have only the two first terms proportional to (G/R^*) and W^*.

However, it is also necessary to take into account the upper free-surface condition (7.8d) for Ξ. Again under the limiting process (7.18a), the following reduced upper condition for the temperature Ξ is derived:

$$\frac{\partial \Xi_0}{\partial Z} + \mathrm{Bi}_{\mathrm{conv}}\Xi_0 = 0 \quad \text{at } Z = \chi(T, X, Y). \tag{7.28}$$

Finally, instead of the 'unwieldy' full BM model problem (7.6a–c) with (7.7a, b) and (7.8a–e) formulated in Section 7.2, we derive at the leading order in a long-wave approximation under the limiting process (7.18a) the following strongly 'relieved' BM long-wave reduced consistent model problem with an error of $\mathrm{O}(\delta^2)$:

$$\begin{aligned}
&\mathbf{D}\cdot\mathbf{V}_0 + \frac{\partial W_0}{\partial Z} = 0;\\
&\frac{D\mathbf{V}_0}{DT} + \mathbf{D}\Pi_0 = \left(\frac{1}{R^*}\right)\frac{\partial^2 \mathbf{V}_0}{\partial Z^2};\\
&\mathrm{Pr}\frac{D\Xi_0}{DT} = \left(\frac{1}{R^*}\right)\frac{\partial^2 \Xi_0}{\partial Z^2};\\
&\frac{\partial \Pi_0}{\partial Z} = 0,
\end{aligned} \tag{7.29a}$$

with as boundary conditions, at $Z = 0$,

$$\mathbf{V}_0 = 0 \quad \text{and} \quad \Xi_0 = 1, \tag{7.29b}$$

and, at $Z = \chi(T, X, Y)$,

$$\Pi_0 \approx \left[\frac{G}{R^*}\right](\chi - 1) - W^*\mathbf{D}^2\chi \equiv \Pi_0(\chi);$$

$$\frac{\partial \mathbf{V}_0}{\partial Z} = -R^*M\left[\mathbf{D}\Xi_0 + (\mathbf{D}\chi)\frac{\partial \Xi_0}{\partial Z}\right];$$

$$\frac{\partial \Xi_0}{\partial Z} + \mathrm{Bi}_{\mathrm{conv}}\,\Xi_0 = 0. \tag{7.29c}$$

We have also the evolution equation (7.23),

$$\frac{\partial \chi}{\partial T} + \mathbf{D}\left(\int_0^{Z=\chi} \mathbf{V}_0\,\mathrm{d}Z\right) = 0, \tag{7.29d}$$

for the thickness $\chi(T, X, Y)$, but with an indeterminate velocity $\mathbf{V}_0$?

Obviously, problem (7.29a–d) can be significantly simplified, but even in the given form it is not at all bad for a start as a reduced BM model problem, subject to a numerical computation.

On the other hand, for the derivation from (7.29a–d) of a more simplified reduced long-wave model, it is necessary to pose some complementary simplifying assumptions. First, in the second equation for $\mathbf{V}_0$ in system (7.29a) while taking into account the fourth relation in (7.29a), with the first upper condition for Π_0, at $Z = \chi$ in (7.29c), we can make the term with pressure $\mathbf{D}\Pi_0$ explicit, in the second equation of (7.29a) and obtain the following dominant dimensionless non-homogeneous equation for the horizontal velocity vector $\mathbf{V}_0$:

$$\frac{D\mathbf{V}_0}{DT} - \left(\frac{1}{R^*}\right)\frac{\partial^2 \mathbf{V}_0}{\partial Z^2} = -\left(\frac{G}{R^*}\right)\mathbf{D}\chi + W^*\mathbf{D}(\mathbf{D}^2\chi), \tag{7.30a}$$

where

$$\frac{D\mathbf{V}_0}{DT} = \frac{\partial \mathbf{V}_0}{\partial T} + (\mathbf{V}_0 \cdot \mathbf{D})\mathbf{V} + W_0\frac{\partial \mathbf{V}_0}{\partial Z}$$

with

$$W_0 = -\int_0^Z (\mathbf{D} \cdot \mathbf{V}_0)\,\mathrm{d}Z.$$

As boundary conditions for $\mathbf{V}_0$ we have

$$\mathbf{V}_0 = 0 \quad \text{at } Z = 0, \tag{7.30b}$$

and

$$\frac{\partial \mathbf{V}_0}{\partial Z} = -R^*M\left[\mathbf{D}\Xi_0 + (\mathbf{D}\chi)\frac{\partial \Xi_0}{\partial Z}\right] \quad \text{at } Z = \chi(T, X, Y), \tag{7.30c}$$

where the function Ξ_0 satisfies the following reduced problem:

$$\mathrm{Pr}\frac{D\Xi_0}{DT} = \left(\frac{1}{R^*}\right)\frac{\partial^2 \Xi_0}{\partial Z^2}, \tag{7.31a}$$

with as conditions

$$\Xi_0 = 1 \quad \text{at } Z = 0, \tag{7.31b}$$

and

$$\frac{\partial \Xi_0}{\partial Z} + \mathrm{Bi}_{\mathrm{conv}}\Xi_0 = 0 \quad \text{at } Z = \chi(T, X, Y). \tag{7.31c}$$

The two problems (7.30a–c) and (7.31a–c) govern a second BM long-wave, reduced model problem for $\mathbf{V}_0$ and Ξ_0 which is only a slightly modified alternate version of the above reduced, first, BM model long-wave, reduced problem (7.29a–d). In fact, the above two nonlinear problems are strongly coupled, because in the material derivative $D/DT = \partial/\partial T + \mathbf{V}_0 \cdot \mathbf{D} + W_0\partial/\partial Z$, the horizontal velocity $\mathbf{V}_0$ is present in equation (7.31a) and in the upper free-surface condition (7.30c) we have the function Ξ_0!

The case of a low Prandtl number, when $\mathrm{Pr} \to 0$, greatly simplifies (*linearizes*) the problem (7.31a–c) for Ξ_0 and, in particular, 'decouples' the problem (7.30a–c) for $\mathbf{V}_0$ from the problem (7.31a–c) for Ξ_0. Indeed, in this low Prandtl number case, we have the possibility to determine in explicit form the function Ξ_0 which appears in upper free-surface condition (7.30c). Namely, in a such case the function Ξ_0 is determined from a very simple linear problem:

$$\begin{aligned}
&\frac{\partial^2 \Xi_0}{\partial Z^2} = 0,\\
&\Xi_0 = 1 \quad \text{at } Z = 0,\\
&\frac{\partial \Xi_0}{\partial Z} + \mathrm{Bi}_{\mathrm{conv}}\Xi_0 = 0 \quad \text{at } Z = \chi(T, X, Y),
\end{aligned} \tag{7.32a}$$

which has the solution

$$\Xi_0(\chi, Z) = 1 - \left[\frac{B(\chi)}{(1 + \chi B(\chi))}\right] Z \equiv 1 - \Xi^*(\chi)Z, \tag{7.32b}$$

with

$$\mathrm{Bi}_{\mathrm{conv}} = B(\chi) \quad \text{and} \quad \Xi^*(\chi) \equiv \left[\frac{B(\chi)}{(1 + \chi B(\chi))}\right]. \tag{7.32c}$$

With (7.32b, c), instead of the upper free-surface condition for $\mathbf{V}_0$, (7.30c), we derive the following condition:

$$\left.\frac{\partial \mathbf{V}_0}{\partial Z}\right|_{Z=\chi} = R^* M \left[\chi\left(\frac{\mathrm{d}\Xi^*(\chi)}{\mathrm{d}\chi}\right) + \Xi^*(\chi)\right](\mathbf{D}\chi) \equiv R^* M \Lambda(\chi)(\mathbf{D}\chi), \tag{7.33a}$$

where

$$\Lambda(\chi) = \frac{[B(\chi) + \chi(\mathrm{d}B(\chi)/\mathrm{d}\chi)]}{[1 + \chi B(\chi)]^2} \tag{7.33b}$$

when we take into account (7.32b), the right-hand side of (7.33a) being only a function of $\chi(T, X, Y, Z)$.

The problem of the determination of function $\mathbf{V}_0(T, X, Y, Z)$ is thereby reduced to resolution of the model (nonlinear parabolic) problem:

$$\frac{D\mathbf{V}_0}{DT} - \left(\frac{1}{R^*}\right)\frac{\partial^2 \mathbf{V}_0}{\partial Z^2} = -\left(\frac{G}{R^*}\right)\mathbf{D}\chi + W^*\mathbf{D}(\mathbf{D}^2\chi), \tag{7.34a}$$

subject to two boundary conditions:

$$\mathbf{V}_0|_{Z=0} = 0 \quad \text{and} \quad \left.\frac{\partial \mathbf{V}_0}{\partial Z}\right|_{Z=\chi} = R^* M \Lambda(\chi)(\mathbf{D}\chi), \tag{7.34b}$$

where $\Lambda(\chi)$ is a given function when $B(\chi)$ is known. In the material derivative D/DT for the vertical component of the velocity we have

$$W_0 = -\int_0^Z (\mathbf{D}\cdot\mathbf{V}_0)\,\mathrm{d}Z.$$

On the right-hand side of equation (7.34a) for $\mathbf{V}_0$ we have two effects, first the gravity effect ($G > 0$) and second the Weber (W^*) effect. A third Marangoni (M) effect, is present in the upper, free-surface, condition (7.34b).

7.4 Lubrication Evolution Equations for the Dimensionless Thickness of the Film

The case of $R^* \to 0$ ($\delta \to 0$, with Re fixed and O(1)) in equation (7.34a) for the horizontal velocity vector $\mathbf{V}_0$, allows us to determine the function $\mathbf{V}_0$ via the following simplifed linear problem:

$$\begin{aligned} &\frac{\partial^2 \mathbf{V}_0}{\partial Z^2} = -G\mathbf{D}\chi + W^{**}\mathbf{D}(\mathbf{D}^2\chi), \\ &\mathbf{V}_0|_{Z=0} = 0, \\ &\left.\frac{\partial \mathbf{V}_0}{\partial Z}\right|_{Z=\chi} = M^* \Lambda(\chi)(\mathbf{D}\chi), \end{aligned} \tag{7.35a}$$

when the following two constraints are assumed: a large modified (see (7.26b)) Weber number (W^*) and a large Marangoni number (M), such that we can write

$$W^{**} = R^* W^* = \mathrm{O}(1) \quad \text{and} \quad M^* = R^* M = \mathrm{O}(1) \tag{7.35b}$$

when $R^* \to 0$, both W^{**} and M^* are assumed to be fixed.

The reduced problem (7.35a) is a classical problem in the 'lubrication theory', which assumes that $R^* \ll 1$ in the long-wave approximation, a consequence of $R^* \ll 1$ being the following constraint on the length λ:

$$\lambda \gg \left(\frac{U_c}{\nu_A}\right) d^2. \tag{7.36}$$

The solution for $\mathbf{V}_0$ of the linear problem (7.35a), when we take into account the formula (7.33b) for $\Lambda(\chi)$, is given by

$$\begin{aligned} \mathbf{V}_0(\chi, Z) &= M^* \left\{ \left[B(\chi) + \chi \left(\frac{\mathrm{d}B(\chi)}{\mathrm{d}\chi} \right) \right] \frac{(\mathbf{D}\chi)}{[1 + \chi B(\chi)]^2} \right\} Z \\ &\quad + (1/2)\{G\mathbf{D}\chi - W^{**}\mathbf{D}(\mathbf{D}^2\chi)\}[Z^2 - 2\chi Z]. \end{aligned} \tag{7.37}$$

Then, from evolution equation (7.29d) after integration of $\mathbf{V}_0(\chi, Z)$, given by (7.37) from $Z = 0$ to $Z = \chi$, we derive the following 'lubrication equation' for the thickness $\chi(T, X, Y)$:

$$\begin{aligned} &\frac{\partial \chi}{\partial T} + (1/3)\mathbf{D}\Bigg\{ \chi^3[W^{**}\mathbf{D}(\mathbf{D}^2\chi) - G\mathbf{D}\chi] \\ &\quad + M^* \left[B(\chi) + \chi \left(\frac{\mathrm{d}B(\chi)}{\mathrm{d}\chi} \right) \right] \left[\frac{1}{[1 + \chi B(\chi)]^2} \right] \chi^2(\mathbf{D}\chi) \Bigg\} = 0. \end{aligned} \tag{7.38}$$

We observe from (7.38) an interesting feature of this new lubrication equation, with a variable convective Biot number, a function of the thickness $\chi(T, X, Y)$.

Namely, the Marangoni effect is coupled with two 'Biot effects':

$$M^* B(\chi)(\mathbf{D}\chi) \left[\frac{\chi^2}{[1 + \chi B(\chi)]^2} \right], \tag{7.39a}$$

and

$$M^* \left(\frac{\mathrm{d}B(\chi)}{\mathrm{d}\chi} \right) (\mathbf{D}\chi) \left[\frac{\chi^3}{[1 + \chi B(\chi)]^2} \right]. \tag{7.39b}$$

In the case of a vanishing convective Biot number, $B(\chi) \to 0$, the first effect, (7.39a), is not present in the lubrication equation (7.38) but, the second effect, (7.39b) is not necessarily zero, and the influence of a large Manrangoni effect is operative!

Usually, in a derived classical 'lubrication equation' (see, for instance, the survey paper on the 'long-scale evolution of thin liquid film' by Oron et al. [11]), with a vanishing Biot number, the Marangoni effect disappears. This non-physical consequence is practically always encountered in all derived lubrication equations, with thermocapillarity effect, by various authors; see, for instance, in addition to [11], also the two recent papers [9, 10]. In [9, section 4], the reader can find a short discussion concerning the small (single) Biot number. Indeed, again in [9], the Biot number (Bi) in the upper free-surface condition for the convection dimensionless temperature is, in fact, the same as the conduction Biot number which allows us to determine the adverse conduction temperature gradient β_S.

It is opportune at this point to observe that in a short paper by VanHook and Swift [12], it is clearly mentioned that:

> ... the Pearson result has two Biot numbers (one for the conduction state and the perturbation) while ...

and

> ... the distinction between the two Biot numbers has not been made in some experimental papers [13]; a theoretical analysis, however, should preserve the distinction!

On the contrary, in our lubrication equation (7.38), thanks to the term $\mathrm{d}B(\chi)/\mathrm{d}\chi \neq 0$, when $B(\chi) \to 0$ we obtain the following reduced lubrication equation, where the Marangoni effect remains:

$$\frac{\partial \chi}{\partial T} + (1/3)\mathbf{D}\cdot\left\{\chi^3\left[W^{**}\mathbf{D}(\mathbf{D}^2\chi) + \left[M^*\left(\frac{\mathrm{d}B}{\mathrm{d}\chi}\right) - G\right]\mathbf{D}\chi\right]\right\} = 0. \quad (7.40)$$

In the unsteady one-dimensional case (T, X), instead of equation (7.38) we obtain the following equation for the thickness, $\chi(T, X)$:

$$\frac{\partial \chi}{\partial T} + \left(\frac{W^{**}}{3}\right)\frac{\partial}{\partial X}\left[\chi^3\frac{\partial^3\chi}{\partial X^3}\right] + \left(\frac{M^*}{3}\right)\frac{\partial}{\partial X}\left[\chi^3\left(\frac{\mathrm{d}B}{\mathrm{d}\chi}\right)\frac{\partial \chi}{\partial X}\right] - \left(\frac{G}{3}\right)\frac{\partial}{\partial X}\left[\chi^3\frac{\partial \chi}{\partial X}\right] = 0. \quad (7.41)$$

Linearization of the dimensionless equation (7.41) around the basic state

$$\chi = 1 + \eta h(T, X) \quad \text{with } \eta \ll 1$$

gives, at the order η, a linear equation for the thickness $h(T, X)$:

$$\frac{\partial h}{\partial T} + \left(\frac{W^{**}}{3}\right)\frac{\partial^4 h}{\partial X^4} + \left(\frac{M^*}{3}\right)\left(\frac{\mathrm{d}B}{\mathrm{d}\chi}\right)_1 \frac{\partial^2 h}{\partial X^2} - \left(\frac{G}{3}\right)\frac{\partial^2 h}{\partial X^2} = 0, \quad (7.42)$$

where the term $(\mathrm{d}B/\mathrm{d}\chi)_1$ is a constant; the value of $\mathrm{d}B/\mathrm{d}\chi$ at $\chi = 1$.

The simple solution of (7.42), $h(T, X) = \exp[\sigma T + ikX]$, yields the characteristic equation

$$3\sigma = \left\{-G + M^*\left(\frac{\mathrm{d}B}{\mathrm{d}\chi}\right)_1 - W^{**}k^2\right\}k^2, \quad (7.43)$$

and the dimensionless cutoff wave number k_c (when $k > k_c$ there is a linear instability) is given in this case by

$$k_c = \left\{\left(\frac{\mathrm{d}B}{\mathrm{d}\chi}\right)_1\left[\frac{M^*}{W^{**}}\right] - \left(\frac{G}{W^{**}}\right)\right\}^{1/2}. \quad (7.44)$$

The characteristic equation (7.43) shows first that (in (7.40), the term $-(G/3)\mathbf{D}\cdot(\chi^3\mathbf{D}\chi)$ (which is linked with the gravity effect), has certainly a stabilizing effect in evolution of the free surface, in the Bénard convection problem, heated from below, and it is clear (if we return to a dimensional form in (7.40)) that the thicker the film, the stronger the gravitational stabilization.

Then the term proportional to W^{**}, linked with a Weber constant (large) surface tension effect, has also a stabilizing effect. On the contrary, the term proportional to M^*, linked with the thermocapillarity (large Marangoni effect) has a destabilizing effect on the free surface (if $\mathrm{d}B/\mathrm{d}\chi > 0$). This effect is well observed in [11, pp. 944–945]:

> thermocapillary destabilization is explained by examining the fate of an initial corrugated free surface in the linear temperature field by a thermal condition. Where the free surface is depressed, it lies in a region of higher temperature than its neighbors. Hence, if surface tension is a decreasing function of temperature, free surface stresses[3] drive liquid on the free surface away from the depression thus, because the liquid is viscous, causing the depression to deepen further. Hydrostatic and capillary forces cannot prevent this deepening.

---

[3] See, for example, the upper free-surface condition (2.35) or the above two conditions (7.21a, b).

In [11], the authors assume systematically that the Reynolds number of the flow (defined by (7.12) is not too large and use the analogy with Reynold's theory of lubrication – a general nonlinear evolution equation (as our above equation (7.32)) is derived for various particular cases (but, unfortunately, again, in an *ad hoc* non-rational manner).

When we start from the nonlinear problems (7.30a–c)–(7.31a–c) – with $\mathrm{Pr} = \mathrm{O}(1)$ and $R^* = \mathrm{O}(1)$ fixed – then, as a consequence of the limiting process $R^* \to 0$, time derivatives are dropped, first in equation (7.30a) and then in equation (7.31a) and we recover the two linear steady-state problems, (7.32a–c) for Ξ_0 and (7.35a) with (7.35b) for $\mathbf{V}_0$, where the time variable plays the role of a parameter via the thickness $\chi(T, X, Y)$! Again, one may ask: what is the order of magnitude of time necessary for establishing the velocity $\mathbf{V}_0$ (given by (7.37)) and the dimensionless temperature Ξ_0 (given by (7.32b)) which are both associated with the thickness $\chi(T, X)$?

The 'simple' answer is that: 'this time for the unsteady adjustment is $\mathrm{O}(R^*)$ and that the rate is exponential, the only interesting issue with this point being that one need not be anxious about any oscillations which might persist without attenuation after the $\mathrm{O}(\mathrm{Re}^*)$ period'.

Finally, we observe that the linear equation (7.42) for $h(T, X)$ is a linear Kuramoto–Sivashinsky (KS) equation for small wave amplitude in the long-wavelength theory and strong surface tension for a relatively large Marangoni number. In Section 7.5, we derive a nonlinear KS equation and also an extended KS equation that includes a dispersive term $(\partial^3 h/\partial x^3)$, a so-called KS–KdV equation. But first we derive an evolution equation *à la* Benney [14], discovered in 1966, that proved to be succesful in describing the initial evolution of nonlinear waves.

7.5 Benney, KS, KS–KdV, IBL Model Equations Revisited

In this section we consider mainly the BM thermocapillary convection down a free-falling vertical thin liquid, two-dimensional film, since most experiments and theories are linked precisely with such a configuration (see Figure 7.2), the wave dynamics on the free surface of a thin liquid layer along an inclined plan being quite analogous.

For the case of a convection down a free-falling vertical thin incompressible, two-dimensional, liquid film, we work with the dimensionless functions

$$u = \frac{u^*}{U_c}, \quad w = \frac{w^*}{\delta U_c}, \quad p - p_A = \frac{(p - p_A)^*}{\rho_c U_c^2} \quad \text{and} \quad \Theta = \frac{(T - T_A)}{(T_w - T_A)}$$

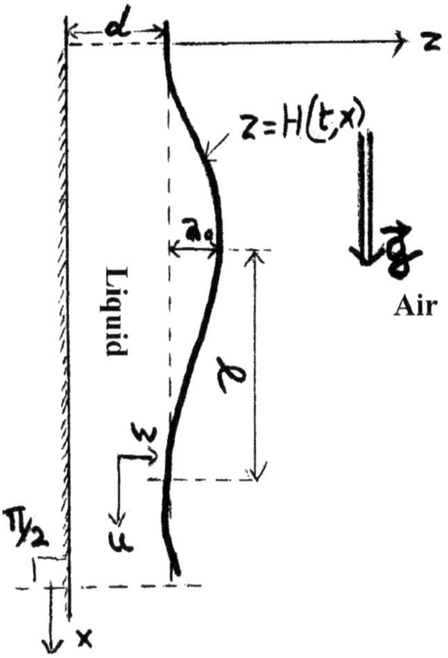

Fig. 7.2 Sketch of a free-falling thin film down along a vertical plane.

and the dimensionless time-space coordinates

$$t = \frac{t^*}{(\lambda/U_c)}, \quad x = \frac{x^*}{\lambda}, \quad z = \frac{z^*}{d}, \quad H(t,x) = \frac{H^*}{d},$$

where the $*$ is relative to quantities with dimensions. In terms of these dimensionless quantities, the dimensionless system of equations and boundary equations, governing the BM convection problem for a free-falling vertical thin liquid film, becomes

$$\frac{\partial u}{\partial x} + \frac{\partial w}{\partial z} = 0; \tag{7.45a}$$

$$\frac{Du}{Dt} + \frac{\partial p}{\partial x} - \frac{1}{\delta F^2} = \left(\frac{1}{\delta\,\mathrm{Re}_d}\right)\left[\frac{\partial^2 u}{\partial z^2} + \delta^2\frac{\partial^2 u}{\partial x^2}\right]; \tag{7.45b}$$

$$\delta^2\frac{Dw}{Dt} + \frac{\partial p}{\partial z} = \left(\frac{\delta}{\mathrm{Re}_d}\right)\left[\frac{\partial^2 w}{\partial z^2} + \delta^2\frac{\partial^2 w}{\partial x^2}\right]; \tag{7.45c}$$

$$\mathrm{Pr}\frac{D\Theta}{Dt} = \left(\frac{1}{\delta,\,\mathrm{Re}_d}\right)\left[\frac{\partial^2 \Theta}{\partial z^2} + \delta^2\frac{\partial^2 \Theta}{\partial x^2}\right]; \tag{7.45d}$$

- at $z = 0$:

$$u = w = 0 \tag{7.46a}$$

and

$$\Theta = 1; \tag{7.46b}$$

- at $z = H(t, x)$:

$$\frac{\partial u}{\partial z} = -\delta\,\mathrm{Ma}\left[\frac{\partial\Theta}{\partial x} + \left(\frac{\partial H}{\partial x}\right)\frac{\partial\Theta}{\partial z}\right] - \delta^2\left[\frac{\partial w}{\partial x} + 4\left(\frac{\partial H}{\partial x}\right)\frac{\partial w}{\partial z}\right]; \tag{7.47a}$$

$$\begin{aligned} p &= p_A + 2\left(\frac{\delta}{\mathrm{Re}_d}\right)\left[\frac{\partial w}{\partial z} - \left(\frac{\partial H}{\partial x}\right)\frac{\partial u}{\partial z}\right] - \delta^2\mathrm{We}\frac{\partial^2 H}{\partial x^2} \\ &\quad + \delta^2\left(\frac{\mathrm{Ma}}{\mathrm{Re}_d}\right)\left(\frac{\partial^2 H}{\partial x^2}\right)\Theta; \end{aligned} \tag{7.47b}$$

$$\frac{\partial\Theta}{\partial z} = -\mathrm{Bi}_{\mathrm{conv}}\Theta + \delta^2\left[\left(\frac{\partial H}{\partial x}\right)\frac{\partial\Theta}{\partial x} - (1/2)\mathrm{Bi}_{\mathrm{conv}}\left(\frac{\partial H}{\partial x}\right)^2\Theta\right]; \tag{7.47c}$$

$$w = \frac{\partial H}{\partial t} + u\frac{\partial H}{\partial x}; \tag{7.47d}$$

where

$$\delta = \frac{d}{\lambda}, \quad \mathrm{Re}_d = \frac{U_c d}{\nu_c}, \quad \mathrm{Pr} = \frac{\nu_c}{\kappa_c}, \quad \kappa_c = \frac{k_c}{\rho_c C_{pc}}, \quad F^2 = \frac{U_c^2}{gd}, \tag{7.48a}$$

$$\mathrm{We} = \frac{\sigma_A}{\rho_c d U_c^2}, \quad \mathrm{Ma} = \left(-\frac{\mathrm{d}\sigma}{\mathrm{d}T}\right)_A \frac{(T_w - T_A)}{\nu_c \rho_c U_c}, \quad \mathrm{Bi}_{\mathrm{conv}} = \frac{q_{\mathrm{conv}} d}{k_c}. \tag{7.48b}$$

We note that the upper, deformable free-surface conditions (at $z = H(t, x)$), (7.47a–c), are written with an error of $O(\delta^3)$. Below, we first consider the derivation of a Benney type equation.

When $\delta \to 0$, we assume that

$$\delta^2\mathrm{We} = \mathrm{We}^*$$

and

$$\frac{\mathrm{Re}_d}{F^2} = 1 \Rightarrow U_c = \frac{gd^2}{\nu_c}, \tag{7.48c}$$

and in $\mathrm{Bi}_{\mathrm{conv}}$ we take into account the dependence of the thickness H

$$\mathrm{Bi}_{\mathrm{conv}} = B(H)$$

and

$$\Xi^*(H) \equiv \left[\frac{B(H)}{(1 + HB(H))}\right]. \tag{7.49}$$

As in Benney [14], we write

$$U = [u, w, p, \Theta]^T = U_0 + \delta U_1 + \mathrm{O}(\delta^2) \quad \text{when } \delta \to 0, \tag{7.50}$$

but for the moment we do not expand the thickness of the film, $H(t, x) = [1 + \eta h(t, x)]$. The leading-order solution for U_0 is

$$u_0 = -z[(1/2)z - H]; \tag{7.51a}$$

$$w_0 = -(1/2)\left(\frac{\partial H}{\partial x}\right) z^2; \tag{7.51b}$$

$$p_0 = p_A - \mathrm{We}^* \left(\frac{\partial^2 H}{\partial x^2}\right); \tag{7.51c}$$

$$\Theta_0 = 1 - \Xi^*(H) z. \tag{7.51d}$$

We get also

$$\frac{\partial H}{\partial t} + H^2 \left(\frac{\partial H}{\partial x}\right) = \mathrm{O}(\delta), \tag{7.51e}$$

which is a consequence of the evolution equation (see (7.23) and also (4.65))

$$\frac{\partial H}{\partial t} + \mathbf{D} \cdot \left(\int_0^{Z=H} u \, \mathrm{d}Z\right) = 0, \tag{7.52}$$

but written for u_0.

Using $u_0, u_1, u_2, \ldots$, we may compute q_0 and q_1 in the expansion of

$$q(t, x) \equiv \int_0^{H(t,x)} u(t, x, z) \, \mathrm{d}z = q_0 + \delta q_1 + \delta^2 q_2 + \mathrm{O}(\delta^3). \tag{7.53}$$

Concerning

$$q_0 = (1/3) H^3, \tag{7.54}$$

this has already been taken into account with (7.51a), the relation (7.51e), which is a consequence of (7.52) with u_0 instead of u, being satisfied at the leading order. Concerning q_1, we get for u_1 a problem similar to problem (7.35a), considered in Section 7.4:

$$\frac{\partial^2 u_1}{\partial z^2} = \mathrm{Re}_d \left\{\frac{\partial u_0}{\partial t} + u_0 \frac{\partial u_0}{\partial x} + w_0 \frac{\partial u_0}{\partial z} + \frac{\partial p_0}{\partial x}\right\},$$

$$u_1 = 0 \quad \text{at } z = 0,$$

$$\left.\frac{\partial u_1}{\partial z}\right|_{z=H(t,x)} = -\mathrm{Ma}\left[\frac{\partial \Theta_0}{\partial x} + \left(\frac{\partial H}{\partial x}\right)\frac{\partial \Theta_0}{\partial z}\right]$$
$$= \mathrm{Ma}\Lambda(H)\frac{\partial H}{\partial x}, \tag{7.55a}$$

where (see (7.33b))

$$\Lambda(H) = \left\{B(H) + H\frac{[\mathrm{d}B(H)/\mathrm{d}H]}{[1 + HB(H)]^2}\right\}. \tag{7.55b}$$

If we want to take into account the influence of the Prandtl number Pr, assumed O(1), then it is necessary to consider for u_2 a problem similar to (7.55a, b):

$$\frac{\partial^2 u_2}{\partial z^2} = \mathrm{Re}_d\left\{\frac{\partial u_1}{\partial t} + u_0\frac{\partial u_1}{\partial x} + u_1\frac{\partial u_0}{\partial x} + w_0\frac{\partial u_1}{\partial z} + w_1\frac{\partial u_0}{\partial z} + \frac{\partial p_1}{\partial x}\right\},$$
$$u_2 = 0 \quad \text{at } z = 0,$$
$$\left.\frac{\partial u_2}{\partial z}\right|_{z=H(t,x)} = -\mathrm{Ma}\left[\frac{\partial \Theta_1}{\partial x} + \left(\frac{\partial H}{\partial x}\right)\frac{\partial \Theta_1}{\partial z}\right]$$
$$-\left[\frac{\partial w_0}{\partial x} + 4\left(\frac{\partial H}{\partial x}\right)\frac{\partial w_0}{\partial z}\right], \tag{7.56a}$$

but also, for p_1 take first into account the problem

$$\frac{\partial p_1}{\partial z} = \left(\frac{1}{\mathrm{Re}_d}\right)\frac{\partial^2 w_0}{\partial z^2},$$
$$p_1 = \left(\frac{2}{\mathrm{Re}_d}\right)\left[\frac{\partial w_0}{\partial z} - \left(\frac{\partial H}{\partial x}\right)\frac{\partial u_0}{\partial z}\right] \quad \text{at } z = H(t,x), \tag{7.56b}$$

and then, for Θ_1, the following problem:

$$\frac{\partial^2 \Theta_1}{\partial z^2} = \mathrm{Re}_d\,\mathrm{Pr}\left[\frac{\partial \Theta_0}{\partial t} + u_0\frac{\partial \Theta_0}{\partial x} + w_0\frac{\partial \Theta_0}{\partial z}\right],$$
$$\Theta_1 = 0 \quad \text{at } z = 0 \quad \text{and} \quad \frac{\partial \Theta_1}{\partial z} = -\mathrm{Bi}_{\mathrm{conv}}\Theta_1 \quad \text{at } z = H(t,x). \tag{7.56c}$$

We note that $\mathrm{Re}_d\,\mathrm{Pr} = \mathrm{Pé}$ is the Péclet number.

Obviously the determination of the solution of the above second-order problem (7.56a) for u_2, with (7.56b, c), which allows us to take into account the (large) Prandtl (Péclet) number effect in a second-order, non-isothermal, *à la* Benney equation, is rather lengthy, but also especially tedious; it can, however, be used as a good exercise for a 'plodder' reader!

The solution for u_1 is obtained from (7.55a) when we use the leading-order solution (7.51b–c) and also relation (7.55b) and equation (7.51e) for H, which gives at the leading order: $\partial H/\partial t = -H^2(\partial H/\partial x)$.

Namely we first obtain for u_1:

$$\begin{aligned} u_1 &= \mathrm{Re}_d \left\{ \mathrm{We}^* \frac{\partial^3 H}{\partial x^3}[Hz - (1/2)z^2] + (1/24)H\left(\frac{\partial H}{\partial x}\right)z^4 \right. \\ &\quad \left. - (1/6)H^2\left(\frac{\partial H}{\partial x}\right)z^3 + (1/3)H^4\left(\frac{\partial H}{\partial x}\right)z \right\} \\ &\quad + \mathrm{Ma}\,\Lambda(H)\left(\frac{\partial H}{\partial x}\right)z, \end{aligned} \tag{7.57a}$$

and then

$$\begin{aligned} q_1 &= (1/3)\mathrm{Re}_d \left\{ \mathrm{We}^* H^3\left(\frac{\partial^3 H}{\partial x^3}\right) + (2/5)H^6\left(\frac{\partial H}{\partial x}\right) \right\} \\ &\quad + (1/2)\mathrm{Ma}\, H^2\Lambda(H)\frac{\partial H}{\partial x}. \end{aligned} \tag{7.57b}$$

Finally, with an error of $O(\delta^2)$, we derive the following *à la* Benney equation for our thermocapillary convection problem with a variable convection Biot number :

$$\begin{aligned} &\frac{\partial H}{\partial x} + H^2\frac{\partial H}{\partial x} + \delta\frac{\partial}{\partial x}\left\{ (1/3)\mathrm{Re}_d\,\mathrm{We}^*\,H^3\left(\frac{\partial^3 H}{\partial x^3}\right) \right. \\ &\quad + (2/15)H^6\left(\frac{\partial H}{\partial x}\right) + (1/2)\mathrm{Ma}\left[\frac{B(H)}{[1+HB(H)]^2}\right]H^2\left(\frac{\partial H}{\partial x}\right) \\ &\quad \left. + (1/2)\mathrm{Ma}\left[\frac{[\mathrm{d}B(H)/\mathrm{d}H]}{[1+HB(H)]^2}\right]H^3\left(\frac{\partial H}{\partial x}\right) \right\} = 0. \end{aligned} \tag{7.58}$$

In a recent paper by Trevelyan and Kalliadasis [15] concerning inclusion of the Péclet number Pé in an extended *à la* Benney equation, the authors assumed that the Péclet number on the right-hand side of the equation for solution Θ_1 of problem (7.56c) is large so that the convective heat transport effects are included at a lower relevant order. More specifically, they assumed Pé $\sim \mathrm{O}(1/\delta^{2/3})$, and then carried out an expansion in δ up to $\mathrm{O}(\delta^{4/3})$, neglecting terms of $\mathrm{O}(\delta^2)$ and higher. According to the authors, 'this level of truncation allows the derivation of a relatively simple evolution equation for the free surface as the $O(\delta^2)$ terms are rather lengthy', the pressure and temperature being both expanded up to $O(\delta^{1/3})$. The reader can find the derived free-surface evolution equation (for $h(t,x)$ [15, p. 184, eq. (9)]) and we observe (in spite of fact that the authors write; 'the seventh term is of $O(\delta^{4/3})$') that, on the one hand, this seventh term in their equation (9) appears as proportional to δ^2 ($= \delta^{4/3}\delta^{2/3}$) and, on the other hand, instead of

$$\text{Pé}^* = \delta^{2/3}\,\text{Pé} = O(1), \tag{7.59}$$

in a term proportional to $\delta^{4/3}$ as is the case in a rational theory when a similarity rule, such as (7.59) between small δ and large Pé, is written – in this seventh term appears again the usual Pé as a term proportional to δ^2. However, we observe that, on the contary, in equation (9) of [15], among the four terms proportional to δ we have as the sixth term of this equation (9): $(2/3)\delta^2\,\text{We}\,h^3\partial^3 h/\partial x^3$, and this is justified because, at the start, the authors assume that $\text{We} = O(\delta^2)$ and $\delta^2\,\text{We} = O(1)$. These above various 'mistakes' introduce 'confusion' and it is not clear if the rescaled equation (10) obtained in [15] is correct, the explanation after this equation being not at all comprehensible. It is clear that the Benney equation obtained when, at start, Pé is assumed to be a large parameter is certainly not equivalent to a derived (with a Pé fixed – O(1)) extended Benney equation after which Pé is assumed to be large!

Without the two last terms, proportional to Ma, which take into account the Marangoni and also the Biot (convective, via $B(H)$) effects, the reduced (isothermal) Benney equation,

$$\frac{\partial H}{\partial x} + H^2\frac{\partial H}{\partial x} + \delta\frac{\partial}{\partial x}\left\{(1/3)\text{Re}_d\,\text{We}^* H^3\left(\frac{\partial^3 H}{\partial x^3}\right)\right\} = 0, \tag{7.60}$$

has been extensively studied over several decades; see, for instance, [16].

This Benney equation (7.60) was also numerically investigated as a partial first-order differential equation in [17]. However, along with the success of this Benney model equation in describing the dynamics of falling liquid film, there is a serious drawback. It turns out that there exists a subdomain in parameter space, where the Benney equation exhibits solutions whose amplitude grows without bound and loses its physical relevance (see [18, 19]). In a recent paper by Oron and Gottlieb [20], the authors have carried out a bifurcation analysis of the first-order Benney equation (7.60) and also of the second-order (in the form given by Lin [21], with several terms proportional to δ^2) Benney equation. Recently, alternative and more efficacious approaches that avoid the solutions blow-up in the model's equations, have been introduced [22, 23], which are refinements of the Shkadov isothermal IBL method [24] and below we discuss this IBL approach in the non-isothermal case.

Now, it is necessary to change our derived Benney equation (7.58) to a KS equation. First, we observe that equation (7.58) contains the small parameter δ, and it does not seem to be asymptotically coherent! Indeed, this is due to fact that the thickness $H(t,x) = 1 + \eta h'(t,x)$ has not been expanded (when $\eta \ll 1$ and assumed, $\eta = \mathrm{O}(\delta)$) as it should be in a fully consistent asymptotic modelling approach through an expansion with respect to δ.

Of course, we may expand the function $H(t,x)$ in different ways and we shall investigate below the same kind of phenomenon as the one which led to the KS equation. In order to obtain a KS equation from the Benney equation (7.58), we first put there the following change for horizontal coordinate x (where we consider a moving reference frame):

$$x \Rightarrow \xi = x - t$$

such that

$$\frac{\partial h'}{\partial t} \Rightarrow \frac{\partial h'}{\partial t} - \frac{\partial h'}{\partial x} \quad \text{and} \quad \frac{\partial h'}{\partial x} \equiv \frac{\partial h'}{\partial \xi}, \tag{7.61}$$

and as a consequence, for the function $h'(t,\xi)$, instead of (7.58), the following approximate equation is derived:

$$\begin{aligned}
&\eta\frac{\partial h'}{\partial t} + 2\eta^2 h'\frac{\partial h'}{\partial \xi} + \delta\eta[(1/3)\mathrm{Re}_d\,\mathrm{We}^*]\frac{\partial^4 h'}{\partial \xi^4} \\
&\quad + \delta\eta\left\{(2/15)\mathrm{Re}_d + (1/2)\mathrm{Ma}\,[1+B(1)]\left[B(1) + \left(\frac{\mathrm{d}B(H)}{\mathrm{d}H}\right)_{H=1}\right]\right\}\frac{\partial^2 h'}{\partial \xi^2} \\
&\quad = 0.
\end{aligned} \tag{7.62}$$

Finally, rescaling the time $\tau = \eta t$ and assuming that $\eta \equiv \delta$, we see that in equation (7.62) all terms contain δ^2, and hence we derive, for the thickness $h'(\tau,\xi)$ the KS equation, associated with (7.58),

$$\frac{\partial h'}{\partial \tau} + 2h'\frac{\partial h'}{\partial \xi} + [\beta + \gamma]\frac{\partial^2 h'}{\partial \xi^2} + \alpha\frac{\partial^4 h'}{\partial \xi^4} = 0, \tag{7.63a}$$

where

$$\alpha = (1/3)\mathrm{Re}_d\,\mathrm{We}^*, \quad \beta = (2/15)\mathrm{Re}_d, \tag{7.63b}$$

$$\gamma = (1/2)\mathrm{Ma}\,[1+B(1)]\left[B(1) + \left(\frac{\mathrm{d}B(H)}{\mathrm{d}H}\right)_{H=1}\right], \tag{7.63c}$$

with an error of $O(\delta)$.

An important remark is that, in coefficient γ, given by (7.63c) which takes into account in the KS equation (7.63a) the influence of the Marangoni effect but also the influence of the Biot effect, we have the constant value of $B(H)$ and $dB(H)/dH$, both at $H = 1$! It seems obvious that $B(1)$ must be identified with the constant value of the conduction Biot, Bi_s, number (see, for instance, Section 4.4), the constant value of $dB(H)/dH$ at $H = 1$ being a sequel of our starting hypothesis, $Bi_{conv} = B(H)$. As a consequence we do not have the usual (paradoxical) problem related to the cancelling of the Marangoni effect for a vanishing Biot number, since $B(1)$ is certainly different from zero because it is related to the conduction state! This result justifies our approach based on a variable convective number and resolves the (false) so-called 'vanishing Biot problem'!

With

$$A(\tau, \xi) = 2h'(\tau, \xi), \tag{7.64}$$

we obtain, instead of (7.63a), the following canonical KS equation:

$$\frac{\partial A}{\partial \tau} + A\frac{\partial A}{\partial \xi} + (\beta + \gamma)\frac{\partial^2 A}{\partial \xi^2} + \alpha\frac{\partial^4 A}{\partial \xi^4} = 0. \tag{7.65}$$

When $\alpha = 0$ and $(\beta + \gamma) = 0$, we obtain the well-known equation

$$\frac{\partial A}{\partial \tau} + A\frac{\partial A}{\partial \xi} = 0, \tag{7.65a}$$

and along characteristics (defined by $d\xi/d\tau = A(\tau, \xi)$) the solution $A(\tau, \xi(\tau))$ is constant.

When $\alpha = 0$ ($We^* = 0$), the surface tension term is removed and (7.65) reduces to Burgers' equation

$$\frac{\partial A}{\partial \tau} + A\frac{\partial A}{\partial \xi} + (\beta + \gamma)\frac{\partial^2 A}{\partial \xi^2} = 0. \tag{7.65b}$$

In this case, the Cole–Hopf transformation further reduces it to the heat equation. Since $\alpha > 0$ the Cole–Hopf transformation produces a heat equation backward in time and an initial disturbance will then grow without limit.

Below, we shall include the surface tension term and discuss the full canonical KS equation (7.65) or (7.63a), when both $\alpha > 0$ and $\beta + \gamma > 0$. This full KS equation (7.65) is capable of generating solutions in the form of irregular fluctuating quasi-periodic waves. The KS model equation provides a mechanism for the saturation of an instability, in which the energy in long-wave instabilities is transferred to short-wave modes which are then damped by surface tension.

In the full KS equation (7.65), the two terms $\partial A/\partial\tau + A\partial A/\partial\xi$, lead to steepening and wave breaking in the absence of stabilizing terms. The term $(\beta+\gamma)\partial^2 A/\partial\xi^2$ destabilizes shorter wave-length modes preferentially and therefore aggravates wave steepening (since β and γ are both positive). The Biot effect being usually small, it has no serious influence on this destabilization via Marangoni effect. Finally, the term $\alpha\partial^4 A/\partial\xi^4$ is required for saturation. Unfortunately, explicit analytic solutions of the KS equation are not available!

A very naïve linear stability analysis shows that, for the KS equation (7.63a), there exists a cutoff wave number. Indeed, if

$$h'(\xi) \sim \exp[\omega\tau + ik\xi],$$

then for ω we derive the dispersion relation

$$\omega - (\beta+\gamma)k^2 + \alpha k^4 = 0. \tag{7.66a}$$

The curve $\omega = 0$ corresponds to neutral linear stability; and in this case the phase velocity, $\omega/k = c = 0$, where the wavenumber k is assumed to be real, and as a consequence we obtain a 'cutoff wavenumber' k^* such that

$$(k^*)^2 = \frac{(\beta+\gamma)}{\alpha} \tag{7.66b}$$

$$= \left(\frac{1}{\mathrm{We}^*}\right)\left\{(2/5) + (3/2)\mathrm{Ma}\,[1+B(1)]\left[B(1) + \left(\frac{\mathrm{d}B(H)}{\mathrm{d}H}\right)_{H=1}\right]\right\}.$$

The linear dispersion relation (7.66a) shows that short waves are stable and long waves are unstable. The critical wavenumber is $k^* = [(\beta+\gamma)/\alpha]^{1/2}$ which ought to be small for the long-wave analysis to make sense. The maximum growth rate is $(\beta+\gamma)^2/4\alpha)$ and occurs at $k^*/\sqrt{2}$.

It is anticipated that the effect of the nonlinear $(A\partial A/\partial\xi)$ term in the canonical KS equation (7.65) will be to allow energy exchange between a wave with wavenumber k and its harmonics with the end result being nonlinear saturation.

The final state may be either chaotic oscillatory motion or a state involving only a few harmonics. The energy equation, corresponding to (7.65), is obtained by multiplying (7.65) by A and integrating by parts, assuming A is periodic, with period $2L$; namely we obtain

$$(1/2)\frac{\partial}{\partial\tau}\left[\int_0^{2L} A^2\,\mathrm{d}\xi\right] = \int_0^{2L}\left[(\beta+\gamma)\left(\frac{\partial A}{\partial\xi}\right)^2 - \alpha\left(\frac{\partial A^2}{\partial\xi^2}\right)^2\right]\mathrm{d}\xi. \tag{7.66c}$$

The minimization of the right-hand side of (7.66c), over all periodic functions, shows that this right-hand side will be negative for $\pi/L > k^*$ and, therefore, the nonlinear KS equation (7.65) is globally stable for an initial condition with a wavenumber satisfying the linear stability criterion.

In other words, if we put in an initial disturbance (e.g., $\sin(k\xi)$) with a wavenumber k' greater than k^*, then the nonlinear term in (7.65) creates higher harmonics, but it will not create waves with wavenumbers smaller than k', so there will be stability.

If we want to generate a component with a wavenumber in the unstable region, we have to put in an initial condition with a wavenumber less than k^*. Hence, we need to consider only the case $k < k^*$. The periodic boundary conditions allow A to be written as the Fourier series

$$A = \sum_{-\infty}^{+\infty} A_n(\tau)\exp(ink\xi), \quad A_{-n} = A_n^*, \tag{7.67}$$

where A_n^* is the complex conjugate of A_n. Since $A_0 = \text{const}$, we may put $A_0 = 0$ and substitution of (7.67) into the KS equation (7.65) gives the following system of equations:

$$\frac{\partial A_n}{\partial \tau} - \sigma_n A_n + inkB_n = 0, \tag{7.67a}$$

where

$$B_n = \sum_{r=1}^{\infty} A_r^* A_{r+n} + (1/2)\sum_{r=1}^{n-1} A_r A_{n-r}, \tag{7.67b}$$

and

$$\sigma_n = (\beta + \gamma)(nk)^2 - \alpha(nk)^4. \tag{7.67c}$$

The significant feature of the above system of equations (7.67a) with (7.67b, c), is that

for any given k, only a finite number of Fourier modes, say A_1, A_2, ..., are unstable ($\sigma_1 > 0$, $\sigma_2 > 0$, ...), and all higher modes are stable.

Note that the nth mode has a critical wavenumber of k^*/n, and a maximum growth rate of $[(\beta+\gamma)^2/4\alpha]$ – independent of n – at $k^*/(n\sqrt{2})$. This implies that unstable modes will be stabilized by energy transfer to higher harmonics.

The simplest case amenable to some theoretical rational analysis is when

$$\frac{k^*}{2} < k < k^*.$$

Only the $n = 1$ mode is unstable in this case and in the following it is assumed that it is sufficient to consider just the interaction between $n = 1$ and $n = 2$ modes. The approximate version of system of equations (7.67a), when we take into account (7.67b), is then (only two equations):

$$\frac{\partial A_1}{\partial \tau} - \sigma_1 A_1 + ikA_2A_1^* = 0, \tag{7.68a}$$

$$\frac{\partial A_2}{\partial \tau} - \sigma_2 A_2 + ik(A_1)^2 = 0, \tag{7.68b}$$

and we note that A_1 is unstable ($\sigma_1 > 0$) but A_2 ($\sigma_2 < 0$) is stable. Note also that the following relation is satisfied:

$$\sigma_1|A_1|^2 + \sigma_2|A_2|^2 = 0,$$

reflecting the required energy balance in the approximate version of

$$\frac{\partial}{\partial \tau}(\Sigma_n|A_n|^2) = 2\Sigma\sigma_n|A_n|^2, \quad n = 1, \dots, \infty,$$

as a consequence of (7.66c) with (7.67).

Equation (7.68a) has the steady solution

$$|A_1| = \left[-\left(\frac{1}{k^2}\right)\sigma_1\sigma_2\right]^{1/2}, \tag{7.69a}$$

since from (7.68b),

$$A_2 = \left(\frac{ik}{\sigma_2}\right)(A_1)^2. \tag{7.69b}$$

Here, A_1 is growing and A_2 is stabilizing. However, as k is decreased, the hypothesis that only two modes are involved becomes more suspect! Indeed, as k is decreased, the steady-state solution of the system of two equations, (7.68a, b), given approximately by (7.69a, b), is at first modified by the presence of a small correction due to A_3 and then when

$$\frac{k^*}{3} < k < \frac{k^*}{2} \quad \text{(i.e. } \sigma_2 > 0\text{, but } \sigma_3 < 0\text{)},$$

is replaced by another 'two-mode equilibrium' in which A_2 and A_4 are the dominant components. Further decrease in k then leads to a succession of states, alternating between 'two-mode equilibria' and 'bouncy states'. If the steady solution for A_2, given by (7.69b), is substituted into equation (7.68a) then a Landau–Stuart (LS) equation is obtained for A_1:

$$\frac{\partial A_1}{\partial \tau} = \sigma_1 A_1 + \left(\frac{k^2}{\sigma_2}\right) |A_1|^2 A_1, \tag{7.70}$$

and this LS equation (7.70) is, in fact, valid only for k *close to* k^*. If in (7.70) we assume that $A_1 = |A_1| \exp(i\varphi)$, then $\varphi = \text{const}$ and for $|A_1|$ we derive a classical Landau equation:

$$\frac{\partial |A_1|}{\partial \theta} = \sigma_1 |A_1| + \lambda |A_1|^3, \tag{7.71}$$

with $\lambda = (k^2/\sigma_2) < 0$, since $\sigma_2 < 0$. The solution of (7.71) is

$$|A_1| \sim A_1^0 \exp(\sigma_1 \tau) \quad \text{as } \tau \to -\infty, \tag{7.72a}$$

where A_1^0 is the initial value at $\tau = 0$ and $\sigma_1 > 0$, which decays like the linearized theory. However,

$$|A_1|^2 \to -\left(\frac{2\sigma_1}{\lambda}\right) \quad \text{as } \tau \to +\infty, \tag{7.72b}$$

for all values of A_1^0; this case is called supercritical stability.

If now we introduce a small perturbation parameter κ, defined by

$$\kappa^2 \mu = k^2 \left[\frac{(\beta + \gamma)}{\alpha} - k^2\right] > 0, \tag{7.73}$$

and a slow time scale $T = \kappa^2 \tau$, then for the slowly varying amplitude of the fundamental wave $H(T)$ such that $|A_1| = \kappa H$, from Landau's equation (7.71), with (7.67c) for σ_1 and σ_2, we derive the following canonical Landau equation for $H(T)$:

$$\frac{\partial H}{\partial T} = \gamma \mu H - \lambda H^3, \tag{7.74a}$$

where the (positive) Landau constant is:

$$\lambda = 1/16\gamma \left[k^2 - \left(\frac{(\beta + \gamma)}{4\alpha}\right)\right] > 0. \tag{7.74b}$$

In the above derivation of the Benney equation (7.58) the Reynolds number Re_d and also the Marangoni number Ma have been assumed both O(1) and fixed, when in long-wave approximation, the small parameter $\delta \to 0$.

Below we consider another case, linked with the low Reynolds and low Marangoni numbers, which leads to a KS–KdV evolution equation with a dispersive additional term, $\varphi \partial^3 h' / \partial \xi^3$.

Namely we assume that

$$\text{Re}_d \ll 1 \quad \text{and} \quad \text{Ma} \ll 1, \tag{7.75a}$$

such that we have three small parameters and as a consequence we write two similarity rules:

$$\frac{\text{Re}_d}{\delta} = \text{Re}^* \quad \text{and} \quad \frac{\text{Ma}}{\delta} = \text{Ma}^*, \tag{7.75b}$$

with R^* and M^* both O(1) and fixed when $\delta \to 0$.

Again we assume that

$$\delta^2\,\text{We} = \text{We}^* \quad \text{and} \quad \frac{\text{Re}_d}{F^2} = 1. \tag{7.75c}$$

This case, (7.75a–c), leads, instead of (7.58), to an *à la* Benney evolution equation, but with some additional terms and then to a modified KS equation with an additional dispersive term, a so-called KS–KdV equation. With the two new constraints (7.75a, b) in the full problem (7.45a–d), (7.46a, b) and (7.47a–d), we obtain the following new problem:

$$\frac{\partial u}{\partial x} + \frac{\partial w}{\partial z} = 0, \tag{7.76a}$$

$$\frac{\partial^2 u}{\partial z^2} + 1 = \delta^2\,\text{Re}^*\left[\frac{Du}{Dt} + \frac{\partial p}{\partial x} - \left(\frac{1}{\text{Re}^*}\right)\frac{\partial^2 u}{\partial x^2}\right], \tag{7.76b}$$

$$\frac{\partial p}{\partial z} - \left(\frac{1}{\text{Re}^*}\right)\frac{\partial^2 w}{\partial z^2} = \delta^2\left[\left(\frac{1}{\text{Re}^*}\right)\frac{\partial^2 w}{\partial x^2} - \frac{Dw}{Dt}\right], \tag{7.76c}$$

$$\frac{\partial^2 \Theta}{\partial z^2} = \delta^2\left[\text{Pr}\,\text{Re}^*\frac{D\Theta}{Dt} - \frac{\partial^2 \Theta}{\partial x^2}\right]. \tag{7.76d}$$

- At $z = 0$:

$$u = w \tag{7.77a}$$

$$w = 0 \tag{7.77b}$$

and

$$\Theta = 1. \tag{7.77c}$$

- At $z = H(t, x)$:

$$\frac{\partial u}{\partial z} = -\delta^2\,\text{Ma}^*\left[\frac{\partial \Theta}{\partial x} + \left(\frac{\partial H}{\partial x}\right)\frac{\partial \Theta}{\partial z}\right] - \delta^2\left[\frac{\partial w}{\partial x} + 4\left(\frac{\partial H}{\partial x}\right)\frac{\partial w}{\partial z}\right] + \text{O}(\delta^4), \tag{7.78a}$$

$$p = p_A - \mathrm{We}^* \frac{\partial^2 H}{\partial x^2} + \left(\frac{2}{\mathrm{Re}^*}\right)\left[\frac{\partial w}{\partial z} - \left(\frac{\partial H}{\partial x}\right)\frac{\partial u}{\partial z}\right] + \mathrm{O}(\delta^2), \quad (7.78b)$$

$$\frac{\partial \Theta}{\partial z} = -\mathrm{Bi}_{\mathrm{conv}}\Theta + \mathrm{O}(\delta^2), \quad (7.78c)$$

$$w = \frac{\partial H}{\partial t} + u\frac{\partial H}{\partial x}. \quad (7.78d)$$

Obviously in this case, the formal Benney expansion in δ is modified. Here, it is necessary to write

$$U = (u, w, p, \Theta)^T = U_0 + \delta^2 U_2 + \cdots \quad \text{when } \delta \to 0. \quad (7.79)$$

The solution U_0 is obtained in a straightforward way, when $\delta \to 0$ in the problem, (7.76a–d), (7.77a, b) and (7.78a–d):

$$u_0 = -(1/2)z^2 + Hz, \quad w_0 = -(1/2)\left(\frac{\partial H}{\partial x}\right)z^2, \quad (7.80a)$$

$$p_0 = p_A - \mathrm{We}^* \frac{\partial^2 H}{\partial x^2} - \left(\frac{1}{\mathrm{Re}^*}\right)\left(\frac{\partial H}{\partial x}\right)(H + z), \quad (7.80b)$$

$$\Theta_0 = 1 - \Xi^*(H)z. \quad (7.80c)$$

We obtain again

$$\frac{\partial H}{\partial t} + H^2\frac{\partial H}{\partial x} = \mathrm{O}(\delta^2), \quad (7.80d)$$

since $q_0 = (1/3)H^3$, but valid with an error of $\mathrm{O}(\delta^2)$.

Writing out the set of equations and boundary conditions at order δ^2, from (7.76b), (7.77a) and (7.78a), with (7.79), for u_2, and assuming that $H(t, x)$ is not yet expanded, we may obtain an awkward expression for u_2 that may be integrated with respect to z in order to obtain an explicit expression for q_2 in

$$\frac{\partial H}{\partial t} + H^2\frac{\partial H}{\partial x} + \delta^2\frac{\partial q_2}{\partial x} = \mathrm{O}(\delta^4).$$

The final result is analogous to (7.58), but with *two additional terms*:

$$\frac{\partial H}{\partial t} + H^2\frac{\partial H}{\partial x} + \varepsilon^2\frac{\partial}{\partial x}\left\{(1/3)H^3\left[\mathrm{Re}^*\,\mathrm{We}^*\frac{\partial^3 H}{\partial x^3} + \underline{7\left(\frac{\partial H}{\partial x}\right)^2}\right]\right.$$
$$\left. + \underline{H^4\frac{\partial^2 H}{\partial x^2}} + (2/15)H^6\frac{\partial H}{\partial x} + (1/2)\mathrm{Ma}^*\,H^2\Lambda(H)\left(\frac{\partial H}{\partial x}\right)\right\} = 0. \quad (7.81)$$

where the function (of H), $\Lambda(H)$, is given above by (7.55b). The evolution equation (7.81) above, for $H(\tau, x)$, is valid with an error of $O(\delta^4)$. Now, with this evolution equation (7.81) we intend to play the same game as the one considered for the reduction of (7.58) to a KS equation (7.63a).

Thus we use:

$$\tau = \delta t, \quad \xi = x - t, \quad H = 1 + \left(\frac{1}{\varphi}\right)\delta^2 h'(\tau, \xi) + \cdots, \tag{7.82a}$$

with

$$\eta = \left(\frac{1}{\varphi}\right)\delta^2, \tag{7.82b}$$

where φ is the dispersive similarity parameter. Carrying out again the limiting process $\delta \to 0$, we find instead of (7.81) an equation which combines the features of the KdV equation on the one hand and the KS equation on the other hand:

$$\frac{\partial h'}{\partial \tau} + 2h'\frac{\partial h'}{\partial \xi} + (\beta^* + \gamma^*)\frac{\partial^2 h'}{\partial \xi^2} + \varphi\frac{\partial^3 h'}{\partial \xi^3} + \alpha^*\frac{\partial^4 h'}{\partial \xi^4}, \tag{7.83a}$$

where

$$\beta^* = \frac{2}{15}\varphi\,\mathrm{Re}^*, \quad \alpha^* = \frac{1}{3}\varphi\,\mathrm{Re}^*\,\mathrm{We}^*, \tag{7.83b}$$

$$\gamma^* = \frac{1}{2}\varphi\,\mathrm{Ma}^*[1 + B(1)]\left[B(1) + \left(\frac{\mathrm{d}B(H)}{\mathrm{d}H}\right)_{H=1}\right], \tag{7.83c}$$

The evolution KS–KdV equation (7.83a) is again a significant model equation valid for large time with an error of $O(\delta)$. The coefficients α^*, γ^*, β^* and φ are all positive constants characterizing dissipation (via Re^*), instability (via Ma^*), and dispersion (via φ), respectively.

As a consequence of the derivation of the KS–KdV equation (7.83a), valid for low Reynolds and Marangoni numbers, we conclude that the features of a thin film for a strongly viscous liquid are quite different: the dispersive term, $\varphi(\partial^3 h'/\partial\xi^3)$, changes the behavior of the thickness of the liquid film $h'(t, \xi)$ in space and in time.

The above derivation of the KS–KdV model equation was first published in 1995 [25]. When the dispersion term in equation (7.83a) is zero, this above equation reduces to a self-exciting dissipative KS equation which exhibits turbulent (chaotic) behavior. On the other hand, in the limiting case when Re^* and Ma^* both tend to zero – a non-viscous liquid film, without the Marangoni effect – equation (7.83a) reduces to the classical KdV equation, well known in the theory of 'nonlinear long surface waves in shallow water', and known to admit soliton solutions instead of chaos!

Thus, in the general case of non-zero α^*, $\gamma^* + \beta^*$ and φ, increasing the value of φ is expected to change the character of the solution of equation (7.83a) from an irregular wave train to a regular row of solitons (a row of pulses of equal amplitude). The trend is amplified at larger values of φ, and the asymptotic state of the solution for large φ takes the form of a row of the KdV solitons. There seems to exist a critical value of about unity for the dimensionless parameter:

$$\mu = \frac{\varphi}{[\alpha^*(\beta^* + \gamma^*]^{1/2}}$$

which represents the relative importance of dispersion corresponding to the transition from an irregular wave train to a regular row of solitons.

We note that the complicated evolution of solutions of (7.83a) is described by the weak interaction of pulses, each of which is a steady-state solution of (7.83a) and when the dispersion is strong, pulse interactions become repulsive, and the solutions tend, in fact, to form stable lattices of pulses.

The linear dispersion relation of the KS–KdV equation (7.83a) for the wave,

$$h'(\tau, \xi) \approx \exp[ik\xi + \sigma\tau]$$

is expressed as

$$\sigma = (\beta^* + \gamma^*)k^2 - \alpha^* k^4 + i\varphi k^3. \tag{7.84a}$$

For Real$(\sigma) > 0$ we have instability, for Real$(\sigma) < 0$ stability and

$$\text{Real}(\sigma) = 0 \quad \text{if} \quad k = k_c = \left[\frac{(\beta^* + \gamma^*)}{\alpha^*}\right]^{1/2}. \tag{7.84b}$$

Consequently, the cut-off wavenumber for the KS–KdV equation (7.83a) satisfies the relation

$$k_c^2 = (2/5\,\text{We}^*) + (3/2)\frac{\text{Ma}^*}{\text{Re}^*\,\text{We}^*}[1 + B(1)]\left[B(1) + \left(\frac{\mathrm{d}B(H)}{\mathrm{d}H}\right)_{H=1}\right]. \tag{7.84c}$$

Thus waves of small wavenumber are amplified while those of large wavenumber are damped.

To demonstrate the competition between the stationary waves and the non-stationary (possibly chaotic) attractors of the KS–KdV equation (7.83a), we convert this equation, with $\alpha^* = \beta^* + \gamma^* = 1$ (with the help of a judicious change of function and space-time coordinates) into a finite-dimensional dynamical system by the Galerkin projection in a periodic medium with wavelength $2\pi/k$:

$$h'(\tau,\xi) = (1/2)\Sigma A_p(\tau)\cos(pk\xi) + B_p(\tau)\sin(pk\xi), \quad p > 1. \quad (7.85a)$$

For a qualitative analysis of projections of the chaotic phase trajectory onto the plane it seems sufficient to consider a dynamical system truncated at the third harmonics (as in the Lorenz case). This system can easily be written in an explicit form. First we make a simple linear transformation of the coordinates: $k\xi \to x$, $k\tau \to t$, with the initial condition $h'(0,\xi) = h'^0(\xi)$, and periodic boundary conditions

$$h'(\tau,\xi) = h'(\tau,\xi + 2\pi/k);$$

the spatial period (wavelength) of the equation is then equal to 2π. Next, substituting (7.85a) into the KS-KdV equation (7.83a), we derive for the amplitudes $A_1(\tau)$, $B_1(t)$ and $B_2(t)$, the following reduced dynamical system:

$$\begin{aligned} \frac{dA_1}{dt} &= \sigma_1 A_1 + k^2\varphi B_1 - 2A_1B_2, \\ \frac{dB_1}{dt} &= \sigma_1 B_1 - k^2\varphi A_1 + 2B_1B_2, \\ \frac{dB_2}{dt} &= 2\sigma_2 B_2 + 2[(A_1)^2 - (B_1)^2], \end{aligned} \quad (7.85b)$$

where

$$\sigma_1 = k(1-k^2) \quad \text{and} \quad \sigma_2 = 2k(1-4k^2). \quad (7.85c)$$

The phase flow of the above dynamical system (7.85b) is dissipative if the following relation is satisfied:

$$\sigma_1 + \sigma_2 < 0,$$

and because of this dissipative effect, the corresponding strange attractors have zero phase volume and dimensionality smaller than 3 (when time t tends to infinity) for the wavenumber k such that $0.58 < k < 1$. This three-amplitude DS (7.85b) can be studied qualitatively and numerically.

A final comment concerning the Benney type single evolution equation, which leads in various cases to a non-physical finite-time blow-up (see, for instance, [18]). The Ooshida regularization procedure [26] of the Benney expansion leads to a single evolution equation for the free surface h that does not exhibit this severe drawback – nevertheless, the Ooshida equation fails to describe accurately the dynamics of the film, for moderate Reynolds numbers, as its solitary wave solutions exhibit unrealistically small amplitudes and speeds. Another single evolution equation including the second-order dissipation effects was recently introduced by Panga and Balakotaiah [27]; in fact, the inertial terms appearing in the model equations offered by both Ooshida and Panga and Balakotaiah can be shown to be equivalent to each other by using the lowest-order expression $\partial h/\partial t = -h^2 \partial h/\partial x$ provided by the flat-film solution and the mass conservation equation. This simple procedure was shown to cure the non-physical loss of the solitary wave solutions and thus to avoid the occurence of finite-time blow-up (see the recent paper by Ruyer-Quil and Manneville [28]).

Concerning the IBL approach, in the isothermal case, this method combines the assumption of a self-similar parabolic viscocity profile beneath the film with the Kármán–Polhausen averaging method used in classical boundary-layer theory. It seems that this IBL approach was first suggested by Petr Leonidovitch Kapitza, at the end of the 1940s, to describe stationary waves and later was extended by Shkadov and coworkers to non-stationary and three-dimensional films (see [24, 30, 31]). As this is well observed in [9]:

> The IBL model does not suffer from the shortcomings of Benney's expansion and performs well in a region of moderate Reynolds numbers and without any singularities for the solitary wave solution branch.

For the non-isothermal case, Zeytounian [5, 8] first derived an IBL model consisting of three averaged equations in terms of the local film thickness (h), flow rate (q) and a function (Σ) related to the mean temperature across the layer. Later another (more effective) non-isothermal IBL model was proposed by Kaliadasis et al. [29]. Finally, more recently, Ruyer-Quil et al. [9] considered the modelling of the thermocapillary flow by using a gradient expansion combined with a Galerkin projection with polynomial test function for both velocity and temperature fields and obtained a system of equations for h, q and Θ which is the temperature at the free surface $z = h$. In [9], a model consistent at second order is also derived. In [10] the reader can find various results of the numerical computation and, in particular, the analysis of the effects of Reynolds, Prandtl and Marangoni numbers on the shape of waves, flow patterns and temperature distribution in a film.

In our 1998 paper [8], instead of the upper free-surface condition (7.47c) for Θ, we used (like other research workers investigating the liquid film flow problem) the generally accepted, but in fact controversial, Davis [41] condition for the dimensionless temperature $\theta = (T - T_d)/T_w - T_d)$, namely,

$$\frac{\partial \theta}{\partial z} = -(1 + \mathrm{Bi}\,\theta) + \mathrm{O}(\delta^2) \quad \text{at } z = H(t, x);$$

but in fact this has no influence since below we assume that Bi $= 0$ – the Biot effect being neglected in our 1998 non-isothermal IBL system of three averaged equations. For this, with the above θ, instead of (7.48b), we have

$$\mathrm{We} = \frac{\sigma_d}{\rho_c d U_c^2}, \quad \mathrm{Ma} = \left(-\frac{\mathrm{d}\sigma}{\mathrm{d}T}\right)_d \frac{(T_w - T_d)}{\nu_c \rho_c U_c} \quad \text{and} \quad \mathrm{Bi} = \frac{qd}{k_c},$$

these three parameters (Weber, Marangoni and Biot) being assumed constant in the convection regime.

Below we consider, again, the situation corresponding to a long wave, $\delta = d/\lambda \ll 1$ and assume $\mathrm{Re}_d \gg 1$, such that $\delta\,\mathrm{Re}_d = R^* = \mathrm{O}(1)$. In this case, from equation (7.45c), we obtain the limiting equation $\partial p/\partial z = 0$, when $\delta \to 0$ and, according to (7.47b), where instead of $\delta^2\,\mathrm{We}$ we have $\mathrm{We}^* = \mathrm{O}(1)$, we can write, with an error of $\mathrm{O}(\delta^2)$,

$$p = -\mathrm{We}^* \frac{\partial^2 H}{\partial x^2}. \tag{7.86}$$

With ($\delta\,\mathrm{Re}_d = R^*$, large Reynolds Re_d number)

$$R^*\,\mathrm{We}^* = \frac{1}{K^*} = \mathrm{O}(1), \quad \frac{\mathrm{Re}_d}{F^2} = 1 \quad \text{and} \quad \delta\,\mathrm{Ma} = M^{**} = \mathrm{O}(1), \tag{7.87}$$

considering also the large Froude (since $\mathrm{Re}_d = F^2$) and Marangoni numbers, when $\delta \to 0$, we derive at the leading order a reduced 'boundary layer' (BL) type two-dimensional problem:

$$\frac{\partial u}{\partial x} + \frac{\partial w}{\partial z} = 0, \tag{7.88a}$$

$$R^* \frac{Du}{Dt} - \frac{\partial^2 u}{\partial z^2} - \frac{1}{\delta F^2} + \left(\frac{1}{K^*}\right) \frac{\partial^3 H}{\partial x^3}, \tag{7.88b}$$

$$R^*\,\mathrm{Pr} \frac{D\theta}{Dt} = \frac{\partial^2 \theta}{\partial z^2}, \tag{7.88c}$$

with (when $\delta \to 0$) as boundary conditions,

$$u = w = 0 \quad \text{and} \quad \theta = 1 \quad \text{at } z = 0, \tag{7.89a}$$

$$\frac{\partial u}{\partial z} = -M^{**} \left[\frac{\partial \theta}{\partial x} + \left(\frac{\partial H}{\partial x}\right) \frac{\partial \theta}{\partial z}\right] \quad \text{at } z = H(t, x), \tag{7.89b}$$

$$\frac{\partial \theta}{\partial z} = -(1 + \mathrm{Bi}\,\theta) \quad \text{at } z = H(t, x), \tag{7.89c}$$

$$w = \frac{\partial H}{\partial t} + u\frac{\partial H}{\partial x} \quad \text{at } z = H(t,x), \tag{7.89d}$$

which remains a complicated BL problem very similar to BM long-wave reduced model problem (7.29a–d) formulated in Section 7.3, but formulated here for a two-dimensional free-falling vertical film.

When $M^{**} = 0$, and in this case the thermal field is decoupled from the dynamical one, Shkadov in 1967 [24], using the integral method, reduced the above problem to a system of two averaged equations for $H(t,x)$ and $q(t,x) = \int_0^H u(t,x,z)\,\mathrm{d}z$, to use the self-similarity assumption for u,

$$u(t,x,z) = \left[\frac{U(t,x)}{H}\right]\{z - (1/2H)z^2\}, \tag{7.90a}$$

where $U(t,x)$ is an arbitrary unknown function, but related to q by

$$U(t,x) = \left[\frac{3}{H(t,x)}\right]q(t,x). \tag{7.90b}$$

In a more general case when $M^{**} \neq 0$ but $\mathrm{Bi} = 0$ (or $\mathrm{Bi}/\delta = \mathrm{Bi}^* = \mathrm{O}(1)$, the Biot number being usually very small) we can derive two averaged IBL model equations for $q(t,x)$ and also for $Q(t,x)$, a function linked with the thermal field

$$Q(t,x) = \int_0^H [\theta(t,x,z) - 1 + z)]\,\mathrm{d}z. \tag{7.91}$$

Namely we obtain

$$R^*\left\{\frac{\partial q}{\partial t} + \frac{\partial}{\partial x}\left[\int_0^H u^2\,\mathrm{d}z\right]\right\}\left(\frac{\partial u}{\partial z}\right)_{z=0}$$
$$= -M^{**}\left[\frac{\partial \theta}{\partial x} + \left(\frac{\partial H}{\partial x}\right)\frac{\partial \theta}{\partial z}\right] + \left(\frac{1}{K^*}\right)H\frac{\partial^3 H}{\partial x^3} + H; \tag{7.92a}$$

$$R^*\,\mathrm{Pr}\left\{\frac{\partial Q}{\partial t} + \frac{\partial}{\partial x}\left[\int_0^H u[\theta(t,x,z) - 1 + z)]\,\mathrm{d}z\right] - \int_0^H w\,\mathrm{d}z\right\}$$
$$+ \left(\frac{\partial \theta}{\partial z}\right)_{z=0} = 0. \tag{7.92b}$$

With the averaged equation for $H(t,x)$, see (7.52),

$$\frac{\partial H}{\partial t} + \frac{\partial q}{\partial x} = 0, \tag{7.92c}$$

an IBL system of three averaged equations (7.92a–c) for three functions H, q and Q, the Biot number Bi being assumed zero, but the influence of the Prandtl number being taken into account, with $\Pr = O(1)$.

Now, with the relations

$$u(t,x,z) = \left[\frac{U(t,x)}{H}\right]\{z - (1/2H)z^2\} + M^{**}\left(\frac{\partial \Sigma}{\partial x}\right)z, \tag{7.93a}$$

$$\theta(t,x,z) - 1 + z = 2\left[\frac{\Sigma(t,x)}{H(t,x)}\right]\{z - (1/2H)z^2\}, \tag{7.93b}$$

where the function $Q(t,x)$, defined by (7.91), is related to $\Sigma(t,x)$ and H by

$$Q = (2/3)H(H - \Sigma), \tag{7.93c}$$

instead of the averaged equations (7.92a) and (7.92b), we derive our two IBL equations for $q(t,x)$ and $\Sigma(t,x)$:

$$\begin{aligned} R^*\left\{\frac{\partial q}{\partial t} + (6/5)\frac{\partial(q^2/H)}{\partial x} + (1/20)M^{**}\frac{\partial}{\partial x}\left[qH\frac{\partial \Sigma}{\partial x}\right]\right\} + \left(\frac{3}{H^2}\right)q \\ = \left(\frac{1}{K^*}\right)H\frac{\partial^3 H}{\partial x^3} + H + (3/2)M^{**}\frac{\partial \Sigma}{\partial x} \\ - (1/120)R^*(M^{**})^2\frac{\partial}{\partial x}\left[H^3\left(\frac{\partial \Sigma}{\partial x}\right)^2\right]; \end{aligned} \tag{7.94a}$$

$$\begin{aligned} \frac{\partial \Sigma}{\partial t} + \left[2 - \left(\frac{\Sigma}{H}\right)\right]\frac{\partial q}{\partial x} + (6/5H)\frac{\partial}{\partial x}[q(\Sigma - H)] + (3/2)\left(\frac{q}{H}\right)\frac{\partial H}{\partial x} \\ - (9/16H)\frac{\partial (Hq)}{\partial x} + M^{**}(1/32H)\frac{\partial}{\partial x}\left[H^3\frac{\partial \Sigma}{\partial x}\right] \\ + M^{**}(1/40H)\frac{\partial}{\partial x}\left[H^2(\Sigma - H)\frac{\partial \Sigma}{\partial x}\right] + 3\left(\frac{1}{R^*\Pr}\right)\left[\frac{(\Sigma - H)}{H^2}\right], \end{aligned} \tag{7.94b}$$

and with

$$\frac{\partial H}{\partial t} + \frac{\partial q}{\partial x} = 0, \tag{7.94c}$$

we obtain our 1998 IBL non-isothermal system of three averaged equations.

In the linear case, when we introduce the perturbations h, ψ and ζ, such that

$$H = 1 + \delta h(t,x), \quad q = (1/3) + \delta \psi(t,x), \quad \Sigma = 1 + \delta \zeta(t,x), \quad (7.95)$$

from (7.94a–c) we derive, with $\partial\psi/\partial x = -\partial h/\partial t$, for the perturbations, h and ζ, the following linear system of two equations:

$$\frac{\partial^2 h}{\partial t^2} + (4/5)\frac{\partial^2 h}{\partial t \partial x} + (2/15)\frac{\partial^2 h}{\partial x^2} + \mathrm{We}^* \frac{\partial^4 h}{\partial x^4} + \left(\frac{3}{R^*}\right)\left[\frac{\partial h}{\partial t} + \frac{\partial h}{\partial x}\right] + (3/2)\left(\frac{M^{**}}{R^*}\right)\frac{\partial^2 \zeta}{\partial x^2} - \left(\frac{M^{**}}{60}\right)\frac{\partial^3 \zeta}{\partial x^3} = 0, \quad (7.96a)$$

$$\frac{\partial \zeta}{\partial t} - (7/16)\frac{\partial h}{\partial t} + (2/5)\frac{\partial \zeta}{\partial x} - (7/80)\frac{\partial h}{\partial x} + \left(\frac{M^{**}}{32}\right)\frac{\partial^2 \zeta}{\partial x^2} + \left(\frac{3}{R^* \mathrm{Pr}}\right)(\zeta - h) = 0. \quad (7.96b)$$

In Section 7.6, devoted to various aspects of the linear and weakly nonlinear stability analysis of the thermocapillary convection, the above system (7.96a, b) is investigated in the framework of infinitesimal disturbances.

7.6 Linear and Weakly Nonlinear Stability Analysis

As our first example we consider a linear stability analysis of the classical BM two-dimensional (as in [32]) problem when, again, instead of the dimensionless temperature Θ, we use the dimensionless temperature $\theta = (T - T_d)/T_w - T_d)$. In this linear case, from our 'correct' upper, free-surface dimensionless condition for θ (1.24c), at $z = H = 1 + \eta h(t,x)$, we can write

$$\frac{\partial \theta}{\partial n'} + \left[\frac{\mathrm{Bi}_{\mathrm{conv}}}{\mathrm{Bi}_s(T_d)}\right]\{1 + \mathrm{Bi}_s(T_d)\theta\} = 0,$$

when $Q_0 = 0$. Because in the conduction case, $\theta_s(z) = 1 - z$, we write:

$$\theta = 1 - z + \eta\theta' + \cdots \quad \text{and} \quad \mathrm{Bi}_{\mathrm{conv}} = B(H) = B(H) + \eta h \left(\frac{dB}{dH}\right)_{H=1},$$

with $B(H) \equiv \mathrm{Bi}_s$ as the conduction Biot number.

From the above upper free-surface condition, written at $z = 1$, and the expansion for θ and $\mathrm{Bi}_{\mathrm{conv}}$, we first obtain

$$-1 + \frac{\partial \theta'}{\partial z} + \left[1 + \eta \left(\frac{1}{\mathrm{Bi}_s}\right)\left(\frac{\mathrm{d}B}{\mathrm{d}H}\right)_{H=1} h\right]\{1 + \eta\,\mathrm{Bi}_s[\theta' - h]\} = 0,$$

and as a consequence at the order η, we derive the following linearized upper condition (at $z = 1$):

$$\left.\frac{\partial \theta'}{\partial z}\right|_{z=1} + \mathrm{Bi}_s[h - \theta'_{z=1}] + \left(\frac{1}{\mathrm{Bi}_s}\right)\left(\frac{\mathrm{d}B}{\mathrm{d}H}\right)_{H=1} h = 0. \tag{7.97}$$

If the last (third) term on the left-hand side of (7.97) can be assumed to be zero, this is not the case for the second term proportional to conduction constant Biot number Bi_s.

Below, if we want to use the Takashima classical linear theory [35], then as condition we choose, neglecting the term $(1/\mathrm{Bi}_s)(\mathrm{d}B/\mathrm{d}H)_{H=1}h$,

$$\left.\frac{\partial \theta'}{\partial z}\right|_{z=1} + \mathrm{Bi}_s[h - \theta'_{z=1}] = 0, \tag{7.98}$$

which is the Takashima condition but with $\mathrm{Bi}_s \neq 0$ as Biot number. Obviously, the general case (with the term $(1/\mathrm{Bi}_s)(\mathrm{d}B/\mathrm{d}H)_{H=1}h$) deserves consideration. With (7.98) the various results of Takashima are correct, for $\mathrm{Bi}_s \neq 0$; on the contrary, the Takashima results for (the Takashima) Biot number = 0 are questionable! As a basic steady-state solution we have

$$\mathbf{u}_{\mathrm{BM}} = 0, \quad \pi_{\mathrm{BM}} = 0, \quad \theta_{\mathrm{BM}} = 1 - z \quad \text{and} \quad H = 1.$$

For a two-dimensional case, Takashima assumes that in the linearized ($\eta \ll 1$) BM problem, the perturbations u', w', π' and θ' are decomposed in terms of normal modes (with $h^0 = \text{const}$):

$$(u', w', \pi', \theta', h) = [U(z), W(z), P(z), T(z), h^0]\exp[\sigma t + ikx]. \tag{7.99}$$

As in Takashima's example, two linear OD equations are derived:

$$\left\{\sigma - \left[\frac{\mathrm{d}^2}{\mathrm{d}z^2} - k^2\right]\right\}\left[\frac{\mathrm{d}^2 W}{\mathrm{d}z^2} - k^2 W\right] = 0, \tag{7.100a}$$

$$\left\{\mathrm{Pr}\,\sigma - \left[\frac{\mathrm{d}^2}{\mathrm{d}z^2} - k^2\right]\right\} T = \mathrm{Pr}\,W, \tag{7.100b}$$

with linear conditions

$$\left.\frac{\mathrm{d}W}{\mathrm{d}z}\right|_{z=0} = 0, \quad W(0) = 0, \quad T(0) = 0; \tag{7.100c}$$

$$\left.\frac{\mathrm{d}^2W}{\mathrm{d}z^2}\right|_{z=1} + k^2W(1) + k^2\,\mathrm{Ma}[T(1) - h^0] = 0; \tag{7.100d}$$

$$[\sigma + 3k^2]\left.\frac{\mathrm{d}W}{\mathrm{d}z}\right|_{z=1} - \frac{\mathrm{d}^3W}{\mathrm{d}z^3} + k^2\left[\left(\frac{1}{\mathrm{Fr}^2_{Ad}}\right) + k^2\mathrm{We}\right]h^0 = 0; \tag{7.100e}$$

$$\left.\frac{\mathrm{d}T}{\mathrm{d}z}\right|_{z=1} + \mathrm{Bi}_s[T(1) - h^0] = 0; \tag{7.100f}$$

$$W(1) = \sigma h^0. \tag{7.100g}$$

The problem (7.100a–g) is our eigenvalue (linear) problem, and Takashima considered two cases: (1) $\sigma = 0$, when the neutral state is a stationary one, and (2) $\sigma \neq 0$, when we have 'overstability'. In case (1), the general solution of the two equations (7.100a, b), with $\sigma = 0$, for $W(z)$ and $T(z)$, can easily be obtained through the $\sinh(kz)$ and $\cosh(kz)$; when these solutions are substituted into the boundary conditions (7.100c–g), then we derive the following eigenvalue relationship:

$$\mathrm{Ma} = \frac{8k[\sinh(k)\cosh(k) - k][k\cosh(k) + \mathrm{Bi}_s\sinh(k)](Bd + k^2)}{8\mathrm{Cr}^* k^5\cosh(k) + (Bd + k^2)[\sinh^3(k) - k^3\cosh(k)]} \tag{7.101}$$

which is a result derived in [32]. In [33] the growth rate of disturbances for the non-zero mode is also studied. For fixed values of Bi_s, $Bd = \mathrm{Pr}\,\mathrm{Cr}^*/\mathrm{Fr}^2_{Ad}$ and $\mathrm{Cr}^* = 1/\mathrm{Pr}\,\mathrm{We}$, relation (7.101) enables us to plot the stability curve in the $(k; \mathrm{Ma})$-plane (see the figures in [32, pp. 2748, 2749]). In particular, when $Bd > 0$ (the case when the upside of the liquid layer is a free surface), the values of Cr^* are given in [32, fig. 1, p. 2748]: since the region below each curve represents a stable state, the lowest point of each curve gives the critical Marangoni number Ma_c and the corresponding critical wave number k_c. It follows that Ma has a minimum value at $k = 0$. The values of Ma_c and k_c, when $\mathrm{Cr}^* < Bd/120$, are almost independent of Bd and Cr^* and are almost the same as those obtained by Pearson [34] in 1958. In such a case the free surface deformation is not important and therefore the assumption of a non-deformable free surface is valid. The condition under which the free-surface deformation becomes important can be expressed as (see also our condition (1.11)),

$$d < d^* = \left[120\frac{\nu\kappa}{g}\right]^{1/3}. \tag{7.102}$$

For water, $d^* = 0.012$ cm. We observe also that for $k_c = 0$ we have, from problem (7.100a–g), that $W = 0$ and there can be no motion. In practice, however, the presence of lateral boundaries will impose a non-zero lower bound on the horizontal wave number, and the minimum Marangoni number required to cause convection will be raised (see, for instance [35]).

In case (2), considered in [32, pp. 2751–2756], again the general solutions of problem (7.100a, b), with $\sigma \neq 0$, can be obtained and these, when substituted into (7.100c–g), yield a time-dependent eigenvalue relationship of the form

$$\mathrm{Ma} = L(\mathrm{Bi}_s, Bd, \mathrm{Cr}^*, \mathrm{Pr}, k, \sigma), \tag{7.103}$$

where L is a real-valued function of parameters in parentheses. The neutral stability curves for the onset of overstability (the alternative possibility of the instability setting in as oscillations) lie only in the negative region of Ma and, contrary to the case of a stationary mode, the region above each curve in [32, fig. 2, p. 2754] represents a stable state. The highest point of each curve gives the critical Marangoni number Ma_c for the onset of overstability and the corresponding critical wave number k_c. In fact, the free surface deformation must be taken into account when (for $Bd > 0$) $\mathrm{Ma} > 0$ and $120\mathrm{Cr}^* > Bd > 0$, or $\mathrm{Ma} < 0$ and $Bd > 0$, and when $\mathrm{Ma} < 0$ the liquid layer can become unstable via a marginal state of purely oscillatory motions. It is therefore concluded that, when the upside of the liquid layer is a free surface ($Bd > 0$), instability is possible for heating in either direction, but the manner of the onset of instability depends on the direction of heating.

Concerning now the KS-KdV equation (7.83a), another way for the derivation of a three-amplitude DS (different from (7.85b)) is the Fourier series. In this case we assume that

$$h(\tau, \xi) = (1/2)\Sigma A_n(\tau) \exp[in(\omega_1^0 \tau - k_1 \xi)], \tag{7.104a}$$

where $A_n(\tau)$ is the complex amplitude of the n-th spatial harmonic, and where ω_1^0 is the linear angular frequency (in fact, the angular frequency of the fundamental harmonic, with k_1 as wavenumber, at the first stage of its growth). It must be stressed that, if the wavenumber $k_n = nk_1$ is the actual wavenumber of the n-th harmonic, the frequency $n\omega_1^0$, on the contrary, cannot be considered as its actual frequency ω_n; the latter may vary a little, owing to possible small dispersive effects. The slow variation $\psi_n(\tau)$ of the phase corresponding to this small frequency shift is taken into account in the complex amplitude

$$A_n(\tau) = |A_n(\tau)| \exp[i\psi_n(\tau)]. \tag{7.104b}$$

Inserting (7.104a) into equation (7.83a), for the first three harmonics we derive the following three-amplitude DS (again, with $\alpha^* = 1$ and $\beta^* + \gamma^* = 1$):

$$\frac{dA_1}{d\tau} = \gamma_1 A_1 + ik_1 A_1^* A_2, \tag{7.105a}$$

$$\frac{dA_2}{d\tau} = (\gamma_2 - 6ik_1^3\varphi)A_2 + ik_1 A_1^2, \tag{7.105b}$$

$$\frac{dA_3}{d\tau} = (\gamma_3 - 24ik_1^3\varphi)A_3 + 3ik_1 A_1 A_2, \tag{7.105c}$$

where

$$\gamma_n = (nk_1)^2[1 - (nk_1)^2], \quad n = 1, 2, 3. \tag{7.105d}$$

Near criticality, where the mode A_1 is the only unstable mode, while the others are linearly strongly damped, the dynamics are controlled by the marginally unstable mode A_1, to which the other two modes are slaved. As a consequence, from (7.105b), we have that the dynamics of the harmonics A_2 is slaved to the dynamics of the fundamental harmonic, A_1, according to

$$A_2 = -\left[\frac{ik_1}{(\gamma_2 - 6ik_1^3\varphi)}\right] A_1^2. \tag{7.106}$$

From (7.105a), with (7.106), the fundamental harmonic A_1 obeys the following Stuart–Landau type equation:

$$\frac{dA_1}{d\tau} = \gamma_1 A_1 + \lambda A_1^* A_1^2, \tag{7.107a}$$

where

$$\lambda = \left(\frac{\gamma_2 k_1^2}{a^2}\right)\left[\frac{1 + i(6k_1^3\varphi)}{\gamma_2}\right], \tag{7.107b}$$

with $a^2 = \gamma_2^2 + 36k_1^6\varphi^2$, and $\gamma_2 > 0$.

The real part (positive) of the complex Landau constant λ corresponds to nonlinear dissipation, and its imaginary part to nonlinear frequency correction (due to dispersive effects). The dispersive character of the waves plays a crucial role (via the parameter φ in (7.83a)) in the occurence of amplitude collapses and frequency locking.

This may be understood within the framework of the above DS (7.105a–c) after separation of modulus and phase of the complex amplitudes (according to (7.104b)).

For the simple case of $|A_1(\tau)|$ and $|A_2(\tau)|$ and phase difference $\Theta(\tau) = \psi_2 - 2\psi_1$, we derive the following DS of three equations instead of (7.105a–c):

$$\frac{\mathrm{d}|A_1|}{\mathrm{d}\tau} = \gamma_1|A_1| - k_1|A_1|\,|A_2|\sin\Theta, \tag{7.108a}$$

$$\frac{\mathrm{d}|A_2|}{\mathrm{d}\tau} = \gamma_2|A_2| + k_1|A_1|^2\sin\Theta, \tag{7.108b}$$

$$\frac{\mathrm{d}\Theta}{\mathrm{d}\tau} = -6(k_1)^3\varphi + k_1\left\{\frac{[|A_1|^2 - 2|A_2|^2]}{|A_2|}\right\}\cos\Theta, \tag{7.108c}$$

which is a particular case of the more complicated DS considered in [36] and deserves a further careful numerical investigation.

We return to the system of two linear equations (7.96a, b) which are deduced from our IBL system (7.94a–c) in Section 7.5. These two equations also certainly deserve further careful investigation. Here we will consider only a particular case when Pr is vanishing, $\mathrm{Pr} \to 0$, and in a such case the equation (7.96b) leads to $\zeta = h$. As a consequence, in this particular case, from (7.96a) we obtain the following single linear equation for the thickness of the film $h(t, x)$:

$$\begin{aligned} &\frac{\partial^2 h}{\partial t^2} + (4/5)\frac{\partial^2 h}{\partial t\partial x} + \left[(2/15) + (3/2)\left(\frac{M^{**}}{R^*}\right)\right]\frac{\partial^2 h}{\partial x^2} \\ &\quad - \left(\frac{M^{**}}{60}\right)\frac{\partial^3 h}{\partial x^3} + \mathrm{We}^*\frac{\partial^4 h}{\partial x^4} + \left(\frac{3}{R^*}\right)\left[\frac{\partial h}{\partial t} + \frac{\partial h}{\partial x}\right] = 0. \end{aligned} \tag{7.109}$$

This evolution equation, with the parameters M^{**}, R^* and We^* is, in fact, an evolution equation for the deformation of the free surface which generalizes the KS–KdV classical equation.

When

$$h(t, x) \approx \exp[ik(x - ct)],$$

we obtain as dispersion equation

$$\begin{aligned} &\left(\frac{3}{R^*}\right)[c - 1] - ik[c_2 - (4/5)c + (2/5)] + ik^3\mathrm{We}^* \\ &\quad = \left(\frac{M^{**}}{2}\right)\left[(1/30)k_2 + i\left(\frac{3}{R^*}\right)k\right], \end{aligned} \tag{7.110}$$

and if $c = c_r + ic_i$, with k real, we obtain two equations for real and imaginary parts:

$$\left(\frac{3}{R^*}\right)[1 - c_r] + kc_i[(4/5) - 2c_r] + \left(\frac{M^{**}}{60}\right)k^2 = 0, \tag{7.111a}$$

$$\left(\frac{3}{R^*}\right) c_i - \left[c_r^2 - c_i^2 - (4/5)c_r + (2/15) + (3/2)\left(\frac{M^{**}}{R^*}\right)\right] k + k^3\,\mathrm{We}^* = 0. \tag{7.111b}$$

If $c_i > 0$, the disturbances are amplified, while if $c_i < 0$, the disturbances are vanishing. When $c_i = 0$ in (7.111a), then for the phase velocity corresponding to a neutral disturbance, we obtain the following relationship:

$$c_r^* \equiv c^* = 1 + (1/180) M^{**} R^* k^2. \tag{7.112}$$

As a consequence, when $M^{**} \neq 0$, it appears that the infinitesimal perturbances are dispersives. If we introduce

$$B^* \equiv (1/180) M^{**} R^*,$$

then from (7.111b), when $c_i = 0$ and with (7.112) we obtain for the determination of the cut-off (neutral) wavenumber k_c ($\neq 0$) the following equation:

$$B^* (k_c^2)^2 + [(6/5)B^* - \mathrm{We}^*] k_c^2 + (1/3) + (3/2)\left(\frac{M^{**}}{R^*}\right) = 0, \tag{7.113}$$

and when $M^{**} \neq 0$, $B^* \neq 0$ and if

$$\mathrm{We}^* \geq (6/5)B^* + 2B^* \left[(1/3) + (3/2)\left(\frac{M^{**}}{R^*}\right)\right]^{1/2}, \tag{7.114a}$$

then we have, for k_c^2, one or two positive values. A particular case which leads to a single value for k_c^2 is

$$k_c^2 = \left(\frac{1}{B^*}\right)\left[(1/3) + (3/2)\left(\frac{M^{**}}{R^*}\right)\right]^{1/2}, \tag{7.114b}$$

and this is the case, when in space parameters (We^*, M^{**}, R^*) the following relationship is justified:

$$\left(\frac{30}{M^{**} R^*}\right) \mathrm{We}^* = (1/5) + (1/3)\left[(1/3) + (3/2)\left(\frac{M^{**}}{R^*}\right)\right]^{1/2}. \tag{7.114c}$$

When $k > k_c$, the disturbances are vanishing and for $k < k_c$, the disturbances are unstable. We observe that from the relation (7.113) we obtain a relationship for M^{**} as a function of We^* and R^*:

$$M^{**} = \frac{k_c^2\,\mathrm{We}^* - (1/3)}{(3/2)(1/R^*) + (1/180) R^* k_c^2 [k_c^2 + (6/5)]} \tag{7.115}$$

and $M^{**} > 0$ if

$$k_c > (1/3\,\mathrm{We}^*)^{1/2}. \tag{7.116a}$$

When $M^{**} = 0$, the condition for a neutral stability, $c_i = 0$ (and in a such case also $c_r \equiv 1$), is

$$k_c = (1/3\,\mathrm{We}_c^*)^{1/2} \tag{7.116b}$$

and we have linear stability ($k > k_c$) for

$$\mathrm{We}^* > \mathrm{We}_c^* \equiv (1/3k_c^2). \tag{7.116c}$$

When $k < k_c$ we have linear instability and in this case, $c_i > 0$ and $c_r \equiv 1$.

If we consider both equations (7.96a) and (7.96b) for the functions $\zeta(t, x)$ and $h(t, x)$ and assume that

$$h(t, x) \approx h^0 \exp[ik(x - ct)] \quad \text{and} \quad \zeta(t, x) \approx \zeta^0 \exp[ik(x - ct)],$$

then instead of the single equation (7.110) we derive the following dispersion relation:

$$\begin{aligned}\left(\frac{3}{R^*}\right)[c - 1] - ik[c^2 - (4/5)c + (2/5)] + ik^3\mathrm{We}^* \\ = \left[\frac{B}{A}\right]\left(\frac{M^{**}}{2}\right)\left[(1/30)k^2 + i\left(\frac{3}{R^*}\right)k\right],\end{aligned} \tag{7.117a}$$

with

$$\left[\frac{B}{A}\right] = \frac{3(1/\mathrm{Pr}\,R^*) + (7/16)ik[c - (1/5)]}{3(1/\mathrm{Pr}\,R^*) - (1/32)M^{**}k^2 - ik[c - (2/5)]}, \tag{7.117b}$$

and unless the case of $\mathrm{Pr} = 0$ (and in a such case $[B/A] = 1$), the ratio $[B/A]$ is a complex function of k and c and the dispersion relation (7.117a) is very complicated, a numerical computation certainly being necessary.

The above very concise linearized stability theory gives very limited results concerning instability, because the amplitude of the disturbance is found to grow exponentially in time for values of certain flow parameters above a critical value.

In reality, such disturbances do not grow exponentially without limit, and an at least weakly nonlinear stability (WNS) analysis is obviously a necessary task! In this WNS theory the Landau–Stuart (LS) equation plays an important role. As a simple example, we consider the Shkadov IBL isothermal model with two equations (7.92c) and (7.92a) for h and q but with $M^{**} = 0$; for the details of the derivation the reader can turn to our survey paper [8].

The main small parameter, using a multiscale technique, is η which is a measure for the deformation of the upper free-surface $z = H$ ($\equiv 1 + \eta h(t, x, y)$) and as bifurcation parameter we choose the modified Weber number We* such that

$$\mathrm{We}^* = \mathrm{We}_c^* + S\eta^2, \tag{7.118a}$$

where S is a scalar assumed O(1); as $\eta \ll 1$, we are interested in a weakly nonlinear stability analysis, near neutral stability. For the phase velocity we write

$$c_r = c_r^* + c_2\eta^2 \quad \text{with } c_r^* = 1, \tag{7.118b}$$

and we introduce slow coordinates

$$\xi_k = \eta^k \xi, \quad \xi = (x - c_r t), \quad k = 0, 1, 2, \ldots \tag{7.118c}$$

In reality, if we want to derive the LS envelope equation, it is sufficient to assume that the amplitude of the wave packet envelope is only a function of $\xi^2 \equiv X$. Now, for the functions $h(t, x)$ and $q(t, x)$ we consider the expansion:

$$h = h'(\xi, X; \eta) = 1 + \eta h_1 + \eta^2 h_2 + \eta^3 h_3 + \cdots, \tag{7.119a}$$

$$q = q'(\xi, X; \eta) = (1/3) + \eta q_1 + \eta^2 q_2 + \eta^3 q_3 + \cdots. \tag{7.119b}$$

According to linear theory,

$$h_1(\xi, X) = A(X)E(\xi) + A^*(X)E(-\xi), \tag{7.120a}$$

with $E(\pm\xi) = \exp(\pm i k_c \xi)$, where k_c is the neutral (cut-off) wavenumber and A^* the complex conjugate of the amplitude A ($AA^* = |A|^2$).

For the derivation of the LS equation for the amplitude $A(X)$ of the wave-packet envelope, it is necessary to eliminate the secular terms in the equations for $h_3(\xi, X)$ and $q_3(\xi, X)$, assuming that the asymptotic expansions (7.119a, b) are uniformly valid with respect to the coordinate ξ. Next, taking into account the relations

$$\frac{\partial}{\partial t} = -c_r\left(\frac{\partial}{\partial \xi} + \eta^2 \frac{\partial}{\partial X}\right) \quad \text{and} \quad \frac{\partial}{\partial x} = \frac{\partial}{\partial \xi} + \eta^2 \frac{\partial}{\partial X}, \tag{7.120b}$$

we substitute the expansions (7.119a, b) and (7.117a, b) for We* and c_r into equations (7.92c) and (7.92a), for h and q, but with $M^{**} = 0$. By identification of the terms in different orders of η, up to 3, we derive a sequence of differential equations. The solution of these differential equations is straightforward. If we introduce the operator

$$\Lambda(h,q) \equiv -(1/5)\frac{\partial q}{\partial \xi} - (2/15)\frac{\partial h}{\partial \xi} + \left(\frac{3}{R^*}\right)[q-h] - (1/3k_c^2)\frac{\partial^3 h}{\partial \xi^3}, \quad (7.121)$$

then we derive successively the following solutions: first

$$h_1(\xi, X) = A(X)E(\xi) + A^*(X)E(-\xi), \quad (7.122a)$$

$$q_1(\xi, X) = A(X)E(\xi) + A^*(X)E(-\xi), \quad (7.122b)$$

then

$$h_2(\xi, X) = q_2(\xi, X) - 2|A(X)|^2, \quad (7.122c)$$

$$q_2(\xi, X) = \beta|A(X)|^2 E(2\xi) + \beta^*|A^*(X)|^2 E(-2\xi), \quad (7.122d)$$

with

$$E(\pm 2\xi) = \exp(\pm 2ik_c\xi)$$

and

$$\beta = -[(7/10) + i(3/2k_c R^*)], \quad \beta^* = -[(7/10) - i(3/2k_c R^*)].$$

At the third order, there appears first the equation

$$\Lambda(h_3, q_3) = \gamma|A^*(X)|^3 E(3\xi) + \left[-(2/3)\frac{dA}{dX} + \lambda A(X) - \mu A(X)|A^*(X)|^2\right] E(\xi) + \text{c.c.}$$

and also, for h_3, the relation

$$h_3 = q_3 - c_2 h_1.$$

The solution for q_3 is of the form

$$q^3 = D(A)E(3\xi) - \left\{ \underline{(3/2)\left[\kappa c_2 A + (2/3)\frac{dA}{dX} - \lambda A(X) + \mu A(X)|A(X)|^2\right] E(\xi)} \right\}\xi + \text{c.c.}, \quad (7.122e)$$

and, in (7.122e), the term underlined in { } before ξ, is a 'secular term', which is very large with increasing ξ!

As a consequence, this term $\eta^3 q_3$, in (7.119b) may not be small relative to the term $\eta^2 q_2$! Finally, by elimination of this secular term, we derive our LS equation:

$$(2/3)\frac{\mathrm{d}A}{\mathrm{d}X} + \nu A + \mu A(X)|A(X)|^2 = 0. \tag{7.123}$$

In the LS equation (7.123), the coeficients ν and μ are given by

$$\nu = \kappa c_2 - \lambda = \left(\frac{3}{R^*}\right) c_2 + i[-(6/5)k_c c_2 + Sk_c^3], \tag{7.124a}$$

$$\mu = \left(\frac{93}{10R^*}\right) + i\left[(31/50)k_c + \left(\frac{9}{k_c}\right)\left(\frac{1}{R^*}\right)^2\right]. \tag{7.124b}$$

For the determination of c_2, we can assume that the coeficient ν is real and, in such a case,

$$c_2 = (5/6)Sk_c^2, \tag{7.124c}$$

and, instead of (7.123), we obtain for the amplitude $A = A(X)$, the LS equation

$$-\frac{\mathrm{d}A}{\mathrm{d}X} = \alpha SA + (3/2)\mu A|A|^2 \tag{7.125a}$$

with

$$\alpha = (15/4)\left(\frac{1}{R^*}\right) k_c^2. \tag{7.125b}$$

From (7.125a) it is judicious to derive a Landau classical equation for the amplitude A, with real coefficient αS and $\mu_r = (93/10R^*)$. For this the complex amplitude $A(X)$, a solution of the above LS equation (7.125), where $\mu = \mu_r + i\mu_i$, is written as

$$A(X) = |A(X)| \exp[i\Psi(X)], \tag{7.126a}$$

and for $|A(X)|$ we derive, from (7.125a), the Landau equation

$$\frac{\mathrm{d}|A|(X)|}{\mathrm{d}X} = -\alpha S|A(X)| - (3/2)\mu_r|A|^3. \tag{7.126b}$$

For the phase $\Psi(X)$ we obtain the relation

$$\frac{\mathrm{d}\Psi(X)}{\mathrm{d}X} = -(3/2)\mu_i|A(X)|^2, \tag{7.126c}$$

with

$$\mu_i = (31/50)k_c + \left(\frac{9}{k_c}\right)\left(\frac{1}{R^*}\right)^2. \tag{7.126d}$$

The above Landau equation (7.126b) implies that the solution $|A| = 0$ is an equilibrium solution which is stable if

$$\mathrm{We}^* < \mathrm{We}_c^*,$$

or unstable if

$$\mathrm{We}^* > \mathrm{We}_c^*,$$

respectively, and $|A| \to |A|_c$ is a new equilibrium value as $X \to -\infty$. The branching of the curve of the equilibrium solution $|A| = 0$ at $\mathrm{We}^* = \mathrm{We}_c^*$ is called the Landau bifurcation. When $\mathrm{We}^* > \mathrm{We}_c^*$, then $S > 0$ so that

$$-\frac{\mathrm{d}|A(X)|}{\mathrm{d}X} > 0 \text{ and } |A|(X) \textit{increases exponentially} \text{ as } X \to -\infty.$$

This case corresponds to a rapid transition to turbulence (chaos). The Landau equation (7.125b) is, in fact, an OD linear equation,

$$\frac{\mathrm{d}L(X)}{\mathrm{d}X} - 2\alpha S L(X) = 3\mu_r, \tag{7.127a}$$

for the function

$$L(X) = \left[\frac{1}{|A(X)|^2}\right], \tag{7.127b}$$

and as a consequence, we may solve (7.127a) explicitly. Using this explicit solution we may estimate the critical rupture $X = X_c$, corresponding to $L(X_c) = 0$. Concerning the nonlinear rupture of the films, the reader can find various useful information in a paper by Erneux and Davis [40].

Obviously, it is necessary to consider such a theory as above, for the non-isothermal case when, instead of Shkadov's IBL system (7.92c), (7.92a), we investigate the weakly nonlinear stability of the IBL system with three equations, (7.92c) with (7.96a) and (7.96b). Obtaining such an amplitude equation from these non-isothermal three equations is a good working example, and can also be performed for a non-isothermal system of three equations derived, respectively, in [29], [9] and [10]; in Section 10.4, we discuss the derivation of some non-isothermal systems of three equations obtained in the papers cited above. On the other hand, in Section 10.5, the reader can find a discussion concerning a paper published by Golovin et al. (in 1994) [37] relative to 'Interaction between short-scale Marangoni thermocapillary convection and long-scale deformational instability', with some comments on the numerical results of Kazhdan [38] obtained via a three amplitude DS.

Finally, we observe that for the lubrication equation (7.40) derived in Section 7.4, as in [39] for example, we can (at least in the unsteady one-dimensional case, see (7.41) use the concept of a Liapounov function and derive the corresponding conservation law; on the other hand the existence of a Liapounov function (decreasing function along any trajectory) provides a start for any meaningful nonlinear stabity analysis of equation (7.41). We recall that *if a system has a Liapounov functional bounded from below (a free energy functional), then any initial data evolve into a steady state.*

7.7 Some Complementary Remarks

We begin with the formulation of a second (modified) model problem for the BM thermocapillary convection, for the function

$$\mathbf{u}', \pi \text{ and } \theta = \frac{(T - T_d)}{(T_w - T_d)} \tag{7.128}$$

instead of

$$\mathbf{u}', \pi \text{ and } \Theta \left[= \frac{(T - T_A}{(T_w - T_A)} \right],$$

which are solutions of the model problem (7.6a–c), (7.7a, b) and (7.8a–e) in Section 7.2.

In such a case, according to the discussion in Chapter 1 relative to the upper free-surface condition (1.38), for the dimensionless temperature θ, we take into account, instead of (7.8d), just this condition (1.38) which seems similar to the Davis condition derived in [41], but different owing to the fact that in (7.38) we have the convection Biot number $\mathrm{Bi}_{\mathrm{conv}}$ instead of Bi_s, the conduction Biot number in the Davis condition.

We have again (with ε, the expansibility parameter given by (1.10a)), $\mathrm{Fr}^2_{d,} \approx 1$ and, as a consequence, $\mathrm{Gr} \approx \varepsilon \ll 1$. For the functions $\mathbf{u}'$, π and θ as model equations, we obtain (unless prime for $\mathbf{u}_{\mathrm{BM}}$):

$$\nabla \cdot \mathbf{u}_{\mathrm{BM}} = 0, \tag{7.129a}$$

$$\frac{\mathrm{d}\mathbf{u}_{\mathrm{BM}}}{\mathrm{d}t} + \nabla \pi_{\mathrm{BM}} = \boldsymbol{\Delta} \mathbf{u}_{\mathrm{BM}}, \tag{7.129b}$$

$$\mathrm{Pr}\frac{\mathrm{d}\theta_{\mathrm{BM}}}{\mathrm{d}t} = \Delta \theta_{\mathrm{BM}}, \tag{7.129c}$$

where

$$\mathrm{Pr} = \frac{\nu(T_d)}{\kappa(T_d)} \quad \text{with} \quad \kappa(T_d) = \frac{k(T_d)}{\rho(T_d) C_v(T_d)}.$$

At $z = 0$, we have

$$\mathbf{u}_{\mathrm{BM}}|_{z=0} = 0 \tag{7.130a}$$

and

$$\theta_{\mathrm{BM}}|_{z=0} = 1. \tag{7.130b}$$

For the leading-order BM equations (7.129a–c), as associated leading-order upper conditions for $\pi_{\mathrm{BM}}|_{z=H}$ we have:

$$\begin{aligned}
\pi_{\mathrm{BM}} = &\left(\frac{1}{\mathrm{Fr}_d^2}\right)(H-1) + \left(\frac{2}{N}\right)\left\{\left(\frac{\partial u_{1\mathrm{BM}}}{\partial x}\right)\left(\frac{\partial H}{\partial x}\right)^2 + \left(\frac{\partial u_{2BM}}{\partial y}\right)\left(\frac{\partial H}{\partial y}\right)^2\right. \\
&+ \frac{\partial u_{3\mathrm{BM}}}{\partial z} + \left[\frac{\partial u_{1\mathrm{BM}}}{\partial y} + \frac{\partial u_{2\mathrm{BM}}}{\partial x}\right]\left(\frac{\partial H}{\partial x}\right)\left(\frac{\partial H}{\partial y}\right) \\
&\left. - \left[\frac{\partial u_{1\mathrm{BM}}}{\partial z} + \frac{\partial u_{3\mathrm{BM}}}{\partial x}\right]\left(\frac{\partial H}{\partial x}\right) - \left[\frac{\partial u_{2\mathrm{BM}}}{\partial z} + \frac{\partial u_{3\mathrm{BM}}}{\partial y}\right]\left(\frac{\partial H}{\partial y}\right)\right\} \\
&- \left(\frac{1}{N}\right)^{3/2} [\mathrm{We} - \mathrm{Ma}\,\theta_{\mathrm{BM}}]\left\{N_2\left(\frac{\partial^2 H}{\partial x^2}\right)\right. \\
&\left. - 2\left(\frac{\partial H}{\partial x}\right)\left(\frac{\partial H}{\partial y}\right)\left(\frac{\partial^2 H}{\partial x \partial y}\right) + N_1\left(\frac{\partial^2 H}{\partial y^2}\right)\right\} \quad \text{at } z = H(t, x, y),
\end{aligned} \tag{7.131a}$$

where, according to (1.18a, b),

$$\begin{aligned}
\mathrm{We} &= \frac{\sigma_d d}{\rho_d \nu_d^2}, \\
\mathrm{Ma} &= \frac{\gamma_\sigma d \Delta T}{\rho_d \nu_d^2},
\end{aligned}$$

with $\Delta T = T_w - T_d$, and we observe that in the first BM model (see (7.8a), in the definition of Ma, instead of ΔT, we have the temperature difference $(T_w - T_A)$, this is also the case for the expansibility small parameter ε!

Then, instead of the two tangential upper free-surface conditions (7.8b, c), with Marangoni effect, we have

$$\begin{aligned}
&\left[\frac{\partial u_{1\mathrm{BM}}}{\partial x} - \frac{\partial u_{3\mathrm{BM}}}{\partial z}\right]\left(\frac{\partial H}{\partial x}\right) + (1/2)\left[\frac{\partial u_{1\mathrm{BM}}}{\partial y} + \frac{\partial u_{2\mathrm{BM}}}{\partial x}\right]\left(\frac{\partial H}{\partial y}\right) \\
&+ (1/2)\left[\frac{\partial u_{2\mathrm{BM}}}{\partial z} + \frac{\partial u_{3\mathrm{BM}}}{\partial y}\right]\left(\frac{\partial H}{\partial x}\right)\left(\frac{\partial H}{\partial y}\right) \\
&- (1/2)\left[1 - \left(\frac{\partial H}{\partial x}\right)^2\right]\left[\frac{\partial u_{3\mathrm{BM}}}{\partial x} + \frac{\partial u_{1\mathrm{BM}}}{\partial z}\right]
\end{aligned}$$

$$= \left(\frac{N^{1/2}}{2}\right) \mathrm{Ma} \left[\frac{\partial \theta_{\mathrm{BM}}}{\partial x} + \left(\frac{\partial H}{\partial x}\right) \frac{\partial \theta_{\mathrm{BM}}}{\partial z}\right] \quad \text{at } z = H(t, x, y), \tag{7.131b}$$

and

$$\left[\frac{\partial u_{1\mathrm{BM}}}{\partial x} - \frac{\partial u_{2\mathrm{BM}}}{\partial y}\right]\left(\frac{\partial H}{\partial y}\right)\left(\frac{\partial H}{\partial x}\right)^2 + \left[\frac{\partial u_{2\mathrm{BM}}}{\partial y} - \frac{\partial u_{3\mathrm{BM}}}{\partial z}\right]\left(\frac{\partial H}{\partial y}\right)$$
$$+ \left[\frac{\partial u_{1\mathrm{BM}}}{\partial z} + \frac{\partial u_{3\mathrm{BM}}}{\partial x}\right]\left(\frac{\partial H}{\partial x}\right)\left(\frac{\partial H}{\partial y}\right)$$
$$+ (1/2)\left[1 + \left(\frac{\partial H}{\partial x}\right)^2 - \left(\frac{\partial H}{\partial y}\right)^2\right]\left(\frac{\partial u_{1\mathrm{BM}}}{\partial y} + \frac{\partial u_{2\mathrm{BM}}}{\partial x}\right)\left(\frac{\partial H}{\partial x}\right)$$
$$- (1/2)\left[1 - \left(\frac{\partial H}{\partial x}\right)^2 - \left(\frac{\partial H}{\partial y}\right)^2\right]\left[\frac{\partial u_{2\mathrm{BM}}}{\partial z} + \frac{\partial u_{3\mathrm{BM}}}{\partial y}\right]$$
$$= \left(\frac{N^{1/2}}{2}\right) \mathrm{Ma} \left\{-\left(\frac{\partial H}{\partial x}\right)\left(\frac{\partial H}{\partial y}\right)\frac{\partial \Theta_{\mathrm{BM}}}{\partial x} + \left[1 + \left(\frac{\partial H}{\partial x}\right)^2\right]\frac{\partial \Theta_{\mathrm{BM}}}{\partial y}\right.$$
$$\left. + \left(\frac{\partial H}{\partial y}\right)\frac{\partial \Theta_{\mathrm{BM}}}{\partial z}\right\} \quad \text{at } z = H(t, x, y). \tag{7.131c}$$

Next, instead of (7.8d) for θ_{BM} as upper free-surface condition we have (see (2.48)):

$$\frac{\partial \theta}{\partial z} + N^{1/2}(1 + \mathrm{Bi}_{\mathrm{conv}}\theta_{\mathrm{BM}})$$
$$= \frac{\partial \theta_{\mathrm{BM}}}{\partial x}\left(\frac{\partial H}{\partial x}\right) + \frac{\partial \theta_{\mathrm{BM}}}{\partial y}\left(\frac{\partial H}{\partial y}\right) \quad \text{at } z = H(t, x, y), \tag{7.131d}$$

and the dimensionless temperature $\theta_S(z)$, for the steady motionless conduction regime, is $\theta_s(z) = 1 - z$, while the kinematic condition is unchanged:

$$u_{3\mathrm{BM}} = \frac{\partial H}{\partial t} + u_{1\mathrm{BM}}\left(\frac{\partial H}{\partial x}\right) + u_{2\mathrm{BM}}\left(\frac{\partial H}{\partial y}\right) \quad \text{at } z = H(t, x, y). \tag{7.131e}$$

A fundamental question is linked with the real 'significance' of these three model problems related to three different upper free-surface conditions and two definitions of the dimensionless temperature: (1) with the Davis upper condition (1.25), where in fact (the Davis [41]) $B = \mathrm{Bi}_s$; then (2) with upper condition (7.8d) for Θ_{BM}, or (3) with (7.131d) for θ_{BM}. Here we have only to bring attention to this question and we do not have to give any definitive answer!

Still the fact remains that, personally, I think the model problem with Θ_{BM} (case (2)) for the BM thermocapillary convection (considered in Section 7.2) is a fundamentally rational model, where the difference between conduction regime (with $\mathrm{Bi}_s(T_A)$ and $\Theta_S(z)$) and convection regime (with a variable $\mathrm{Bi}_{\mathrm{conv}}$ and the dimensionless temperature Θ_{BM}) is clearly taken into account! But, on the other hand, the model problem (case (3) with θ_{BM}) for the BM thermocapillary convection allows us to use classical results derived with the help of the Davis (case (1)) upper condition, changing B $(= \mathrm{Bi}_s)$ to $\mathrm{Bi}_{\mathrm{conv}}$.

However, here it seems necessary to note a 'negative' aspect of the BM model with Θ_{BM} (see Section 2.5) linked with the fact that, when we use the dimensionless temperature Θ, when the difference of temperature $T_w - T_A$ appears, then we are constrained to write for the temperature-dependent surface tension $\sigma(T)$ the approximate relation

$$\sigma(T) = \sigma(T_A)\left[1 - \left(\frac{\mathrm{Ma}}{\mathrm{We}}\right)\Theta_{\mathrm{BM}}\right],$$

with

$$\mathrm{Ma} = \left[-\frac{\mathrm{d}\sigma(T)}{\mathrm{d}T}\right]_A d\frac{(T_w - T_A)}{\rho(T_A)\nu(T_A)^2}$$

where in Ma we have also $T_w - T_A$, and in reality

$$\sigma(T) = \sigma(T_A) + \left[-\frac{\mathrm{d}\sigma(T)}{\mathrm{d}T}\right]_A (T - T_A)$$

is related to the T_A instead of T_d.

This fact does not express very well the physical nature of the Marangoni (temperature-dependent surface tension along the free surface) effect, because this is equivalent to assuming that in the conduction regime, in fact $T_d = T_A$, and, in such a case, from (1.21a) we obtain $\beta_s = 0$, when $\mathrm{Bi}_s = \mathrm{O}(1)$; only when

$$\mathrm{Bi}_s \uparrow \infty \quad \text{and} \quad (T_d - T_A) \downarrow 0$$

can we assume that $\beta_s = \mathrm{O}(1)$ – the flat, $z = d$, free surface in conduction state at that time being assumed to be a perfect conductor! Rigorously speaking, it is necessary to use relation (1.17a) for $\sigma(T)$, and then in (1.17a) to replace $(T - T_d)$ by $[(T - T_A) - d\beta_s/\mathrm{Bi}_s]$.

With this second BM model problem with θ_{BM} as lubrication equation, instead of (7.38), in a similar way we derive the following equation:

$$\frac{\partial \chi}{\partial T} + (1/3)\mathbf{D}\cdot\Big\{\chi^3[A\mathbf{D}(\mathbf{D}^2\chi) - G\mathbf{D}\chi] + B(1+\mathrm{Bi}_{\mathrm{conv}})\left[\frac{1}{[1+\chi\,\mathrm{Bi}_{\mathrm{conv}}]^2}\right]\chi^2(\mathbf{D}\chi)\Big\} = 0, \quad (7.132)$$

where $A = \sigma_d/\rho_d\lambda^2 g$ and $B = \gamma_\sigma\beta_s/\rho_d dg$.

On the other hand, in the KS equation (7.63a), the coefficient $[\beta+\gamma]$ in front of $\partial^2 h'/\partial\xi^2$ is

$$[\beta+\gamma] = (2/15)\mathrm{Re}_d + (1/2)\frac{\mathrm{Ma}}{(1+\mathrm{Bi}_{\mathrm{conv}})}. \quad (7.133)$$

In the case when $\chi(T,x,y) \Rightarrow H(t,x)$, instead of (7.132) we write a one-dimensional nonlinear evolution equation (when $G \equiv 1$):

$$\frac{\partial H}{\partial t} + \frac{\partial}{\partial x}\left\{\left(\frac{B}{3}\right)\frac{(1+\mathrm{Bi}_{\mathrm{conv}})H^2}{(1+\mathrm{Bi}_{\mathrm{conv}}H)^2} - \frac{H^3}{3}\frac{\partial H}{\partial x}\right\} + \left(\frac{A}{3}\right)\frac{\partial}{\partial x}\left[H_3\left(\frac{\partial^3 H}{\partial x^3}\right)\right] = 0. \quad (7.134)$$

Concerning the KS equation, a more convenient reduction of the KS equation (for instance (7.63a)) is obtained if we introduce the new function $H(t,x)$ and new variables, t and x, by the relations

$$h' = 2[\beta+\gamma]\left[\frac{(\beta+\gamma)}{\alpha}\right]^{1/2} H, \quad \tau = \frac{\alpha}{[\beta+\gamma]^2}t, \quad \xi = \frac{\alpha}{[\beta+\gamma]}x; \quad (7.135a)$$

in this case we obtain the following reduced KS equation for the amplitude $H(t,x)$:

$$\frac{\partial H}{\partial t} + 4H\frac{\partial H}{\partial x} + \frac{\partial^2 H}{\partial x^2} + \frac{\partial^4 H}{\partial x^4} = 0. \quad (7.135b)$$

Transforming (7.135b) to a moving coordinate system with speed C and integrating once, one obtains for $H^*(\xi_C)$,

$$\frac{\partial^3 H^*}{\partial \xi_C^3} + \frac{\partial H^*}{\partial \xi_C} - CH^* + 2H^{*2} = Q, \quad (7.136a)$$

where $Q = \langle 2H^{*2}\rangle$ is the deviation flux in the moving frame obtained by invoking the constant-thickness condition

$$\langle H^*\rangle = 0, \quad (7.136b)$$

and $\langle\,\rangle$ denotes averaging over one wavelength in the scaled moving, ξ_C – coordinate. If however, the constant-flux condition is imposed, $Q = 0$, equation (7.136a) reduces to

$$\frac{\partial^3 H'}{\partial \xi_C^3} + \frac{\partial H'}{\partial \xi_C} - CH' + 2H'^2 = 0, \tag{7.137}$$

and constraint (7.136b) is unnecessary and no longer holds, $\langle H' \rangle \neq 0$.

We observe that the constant-flux equation has one parameter less and involves only one equation (7.137), whereas the two equations (7.136a, b) must be solved for the constant-thickness approach and two parameters Q and C are involved; for a detailed discussion of the properties of these equations (above, companion to KS), see [42]. There are myriad infinite wave families; in particular, we observe the solitary-wave regime and note that some families of waves are traveling-wave solutions which have unique solitary-wave shapes. It is important to note that after the solitary-wave regime, the wave breaks into non-stationary, three-dimensional patterns. This implies that 3D stationary waves either do not exist or have very short lifetimes, thus being insignificant. The final transition to interfacial 'turbulence' must then be analyzed with an entirely different approach. In Section 10.8 we return to the discussion of solitary wave phenomena in a convection regime.

We observe that various authors (see, for instance, Parmentier et al. [43]) were interested in a weakly nonlinear analysis of coupled surface-tension and gravitational-driven instability in a thin layer. Unfortunately, the existence of such a coupled Bénard–Rayleigh–Marangoni model problem on the basis of a rational analysis and asymptotic modelling was not demonstrated in this publication. Yet, such an approach certainly deserves further investigation! Obviously when buoyancy is the single responsibility of convection, only rolls will be observed. As soon as capillary effects are present, the situation is more complex; however, a general tendency is observed and it appears that a hexagonal structure is preferred at the linear threshold. The more the thermocapillary forces are dominant with respect to the buoyancy forces, the larger the size of the region where hexagons are stable. The influence of the Prandtl number has received particular attention from Parmentier et al. [43]!

Here it seems not superfluous to observe that the quantity β_S (> 0) is defined as minus the vertical temperature gradient that would appear in a purely conductive steady state (see, for instance, (1.19a, b)). Since in the pure heat conducting state, the temperature at the upper (flat) free surface is uniform, there is no ambiguity in determining experimentally $\beta_s d$, which is related to the difference between the temperature at the lower rigid plate (T_w) and the temperature of the air surmounting the liquid layer (T_A) (see (1.21b)), where the Biot number (Bi_s) is defined with a constant heat transfer coefficient $q_s(T_d)$, this assumption being strictly satisfied only when the temperature at the upper free surface is uniform – such a condition is met in pure buoyancy-driven (RB) convection, i.e., when the upper free surface is rigid (a simple Rayleigh problem considered in Chapter 3); if we except the reference heat conductive steady case (1.19a), this is no longer true in Marangoni's instability (in BM thermocapillary convection) as the temperature at the upper deformed (by the convection) free surface varies from point to point – the heat transfer coefficient q_{conv} (or convection Biot number $\mathrm{Bi}_{\mathrm{conv}}$) is then not a constant! In a convection regime, when the fluid is set in motion, β_S is no longer the temperature gradient in the fluid layer since convection induces a non-zero mean perturbative temperature at the upper free fluid surface. As a consequence the dimensionless numbers of Marangoni and Rayleigh must be experimentally evaluated with, as given by (1.21b),

$$\beta_s = \frac{(T_w - T_A)}{[(k/qs) + d]}, \tag{7.138}$$

with k the thermal conductivity (assumed a constant) of the fluid layer. Namely (the subscript '0' is relative to the 'room' temperature):

$$\rho \approx \rho_0[(T - T_0)] \quad \text{with} \quad \alpha_0 = -\left(\frac{1}{\rho_0}\right)\left[\frac{\partial \rho}{\partial T}\right]_0,$$

$$\sigma(T) \approx \sigma_0\left[1 - \left(\frac{\gamma_0}{\sigma_0}\right)(T - T_0)\right] \quad \text{with} \quad \gamma_0 = -\left[\frac{\partial \sigma(T)}{\partial T}\right]_0,$$

$$\mathrm{Ra} = \frac{g\alpha_0\beta_s d^4}{k\nu},$$

$$\mathrm{Ma} = \frac{\gamma_0\beta_s d^2}{\rho_0 k\nu}. \tag{7.139}$$

In [43], as an alternative to the Marangoni (Ma) and Rayleigh (Ra) numbers, Parmentier et al. define two new dimensionless numbers α and λ by the relations

$$\alpha\lambda = \frac{\mathrm{Ra}}{\mathrm{Ra}^0} \quad \text{and} \quad \lambda(1-\alpha) = \frac{\mathrm{Ma}}{\mathrm{Ma}^0}, \tag{7.140}$$

where Ra^0 and Ma^0 are two arbitrary constants – namely, Ra^0 is the critical Ra for pure buoyancy and Ma^0 is the critical Ma for pure thermocapillarity. According to Parmentier et al. [43], we observe that, in physical situations, the main control parameter is neither Ma nor Ra, but the temperature gradient β_S defined above by (7.138). The use of α and λ is motivated by the fact that α is a combination of the relevant physical parameters, while λ is the quantity directly proportional to the control temperature difference. From (7.140), because

$$(1-\alpha)\left[\frac{\text{Ra}}{\text{Ra}^0}\right]=\alpha\left[\frac{\text{Ma}}{\text{Ma}^0}\right], \tag{7.141a}$$

we see that α can be considered as the percentage of buoyancy effect with regard to thermocapillary effect – it takes values between zero and one: $\alpha = 0$ corresponds to pure thermocapillarity and $\alpha = 1$ to pure buoyancy. On the other hand, from the obvious relation

$$\lambda=\frac{\text{Ra}}{\text{Ra}^0}+\frac{\text{Ma}}{\text{Ma}^0}, \tag{7.141b}$$

we see that λ is directly proportional to the temperature gradient; in weakly nonlinear problems λ remains close to 1. A challenging problem, linked with the above situation, is the formulation of a 'correct' dominant dimensionless model, as this has been demonstrated here in Chapter 4 and the use, then, of an IBL averaged approach! The analytical weakly nonlinear approach, given in [43], via the derivation of a set of Ginzburg–Landau amplitude equations and, then, to a consideration of a reduced system of three amplitude equations, is interesting from a mathematical point of view (in particular, concerning the construction of a complete basis of eigenfunctions with the incorporation a mode with a zero wave number in the nonlinear development) but, rather complicated to keep in mind the results obtained? Obviously the main interest of the approach realized in [43] is the investigation related to the competition between hexagonal and roll patterns, which are the most current patterns observed near the threshold (the relative distance from the threshold being given by $\varepsilon = (\lambda - \lambda_c)/\lambda_c$, with as critical lambda, $\lambda_c = \min(\alpha, \text{Bi}, k)$, relative to the set $]0, \infty[$ of admissible values of the wave number k; the k corresponding to λ_c is the critical $k \to k_c$. Except for pure buoyancy instability, when $\alpha = 1$, the convective patterns that appear at the linear threshold are always formed with hexagons – below this threshold, a subcritical region where hexagons can be stable is also found. Hexagonal patterns are unstable when buoyancy is the only factor of instability. When the temperature gradient is increased, a region where rolls and hexagons coexist is diplayed; at still higher temperature gradients, rolls are expected. We note that, when buoyancy is the single responsibility of the convection, only rolls will be observed. On the other hand, it appears that a hexagonal structure is preferred at the linear threshold. The more the thermocapillary forces are dominant with respect to the buoyancy forces, the larger the size of the region where hexagons are stable. The results of Parmentier et al. [43] very well exhibit the fact that, by increasing temperature-dependent surface tension, one promotes the hexagonal pattern and, on the contrary, in the limiting case (in a pure buoyancy thermal convection) of a negligible temperature dependence of the surface tension, only rolls are stable. But it is necessary to note as well that the (in the thermocapillary case) important effect of a deformable free surface is not taken into account in [43] and in, fact, the starting problem in [43] and in the Dauby and Lebon paper [44] are similar and are a consistent RB model problem which takes into account only partially the real effect linked with the thermocapillarity (see our discussion in Section 5.2). More precisely, the upper free surface condition for the dimensionless pressure π is not taken into account and just this condition poses a problem in a weakly expansible liquid subject to a temperature-dependent surface tension!

References

1. D.G. Crowdy, C.J. Lawrence and S.K. Wilson (Guest-Editors), *The Dynamics of Thin Liquid Films*. Special issue of *J. Engng. Math.* **50**(2–3), 95–341, November 2004.
2. H.-C. Chang, Wave evolution on a falling film. *Annu. Rev. Fluid Mech.* **26**, 103–136, 1994.
3. P. Colinet, J.C. Legros and M.G. Velarde, *Nonlinear Dynamics of Surface-Tension-Driven Instabilities*, 1st Edn. Wiley-VCH, 2001.
4. A.A. Nepomnyaschy, M.G. Velarde and P. Colinet, *Interfacial Phenomena and Convection*. Chapman & Hall/CRC, London, 2002.
5. M.G. Velarde and R.Kh. Zeytounian, *Interfacial Phenomena and the Marangoni Effect*. CISM Courses and Lectures, No. 428, CISM, Udine and Springer-Verlag, Wien/New York, 2002.
6. E. Guyon, J-P. Hulin and L. Petit, *Hydrodynamique physique*. Savoirs Actuels, InterEditions/Editions du CNRS, Paris, Meudon, 1991.
7. E. Guyon, J-P. Hulin, L. Petit and C.D. Mitescu, *Physical Hydrodynamics*, Oxford University Press, Oxford, 2001.
8. R.Kh. Zeytounian, The Bénard–Marangoni thermocapillary-instability problem. *Phys. Uspekhi* **41**, 241–267, 1988.
9. C. Ruyer-Quil, B. Scheid, S. Kalliadasis, M.G. Velarde and R.Kh. Zeytouian, Thermocapillary long waves in a liquid film fmow. Part 1. Low-dimensional formulation. *J. Fluid Mech.* **538**, 199–222, 2005.
10. B. Scheid, C. Ruyer-Quil, S. Kalliadasis, M.G. Velarde and R.Kh. Zeytouian, Thermocapillary long waves in a liquid film fmow. Part 2. Linear stability and nonlinear waves. *J. Fluid Mech.* **538**, 223–244, 2005.
11. A. Oron, S.H. Davis and S.G. Bankoff, Long-scale evolution of thin liquid films. *Rev. Modern Phys.* **69**(3), 931–980, 1997.
12. S.J. VanHook and J.B. Swift, *Phys. Rev. E* **56**(4), 4897–4898, October 1997.
13. M.F. Schatz, S.J. VanHook, W.D. McCormick, J.B. Swift and H.L. Swinney, *Phys. Rev. Lett.* **75**, 1938, 1995.
14. D.J. Benney, Long waves on liquid films. *J. Math. Phys.* (N.Y.) **45**, 150–155, 1966.

15. P.M.J. Trevelyan and S. Kalliadasis, *J. Engng. Math.* **50**(2-3), 177–208, 2004.
16. C. Nakaya, Waves of a viscous fluid down a vertical wall. *Phys. Fluids* **A1**, 1143–1154, 1989.
17. B. Sheid, A. Oron, P. Colinet, U. Thiele and J.C. Legros, *Phys. Fluids* **14**, 4130–4151, 2002. Erratum: *Phys. Fluids* **15**, 583, 2003.
18. A. Pumir, P. Manneville and Y. Pomeau, *J. Fluid Mech.* **135**, 27–50, 1983.
19. P. Rosenau, A. Oron and J.M. Hyman, *Phys. Fluids* **A4**, 1102–1104, 1992.
20. A. Oron and O. Gottlieb, Subcritical and supercritical bifurcations of the first- and-second-order Benney equations. *J. Engng. Math.* **50**(2–3), 121-140, 2004.
21. S.-P. Lin, Finite amplitude side-band stability of a viscous film. *J. Fluid Mech.* **63**, 417–429, 1974.
22. C. Ruyer-Quil and P. Manneville, Improved modeling of flows down inclined planes. *Eur. Phys. J.* **B15**, 357–369, 2000.
23. C. Ruyer-Quil and P. Manneville, *Phys. Fluids* **14**, 170–183, 2002.
24. V.Ya. Shkadov, Wave flow regimes of a thin layer of viscous fluid subject to gravity. *Izv. Akad. Nauk SSSR, Mekh. Zhidk. Gaza* **2**, 43–51, 1967 [English transl. in *Fluid Dyn.* **19**, 29–34, 1967.
25. R.Kh. Zeytounian, *Long-Waves on Thin Viscous Liquid Film/Derivation of Model Equations*. Lecture Notes in Physics, Vol. 442, Springer-Verlag, Berlin/Heidelberg, pp. 153–162, 1995.
26. T. Ooshida, *Phys. Fluids* **11**, 3247–3269, 1999.
27. M.K.R. Panga and V. Balakotaiah, *Phys. Rev. Lett.* **90**(15), 154501, 2003.
28. C. Ruyer-Quil and P. Manneville, *Phys. Rev. Lett.* **93**(19), 199401, 2004.
29. S. Kalliadasis, A. Kiyashko and E.A. Demekhin, *J. Fluid Mech.* **475**, 377–408, 2003.
30. E.A. Demekhin, M. Kaplan and V.Ya Shkadov, Mathematical models of the theory of viscous liquid films. *Izv. Akad. Nauk SSSR, Mekh. Zhidk Gaza* **6**, 73–81, 1987.
31. E.A. Demekhin and V.Ya Shkadov, *Izv. Akad. Nauk SSSR, Mekh. Zhidk Gaza* **5**, 21–27, 1984.
32. M. Takashima, *J. Phys. Soc. Japan* **50**(8), 2745–2750 and 2751–2756, 1981.
33. V.C. Regnier and G. Lebon, *Q. J. Mech. Appl. Math.* **48**(1), 57–75, 1995.
34. J.R.A. Pearson, On convection cells induced by surface tension. *J. Fluid Mech.* **4**, 489–500, 1958.
35. D.A. Niels, Surface tension and buoyancy effects in cellular convection. *J. Fluid Mech.* **19**, 341–352, 1964.
36. P. Barthelet, F. Charry and J. Fabre, *J. Fluid Mech.* **303**, 23, 1995.
37. A.A. Golvin, A.A. Nepomnyaschy and L.M. Pismen, *Phys. Fluids* **6**(1), 35–48, 1994.
38. D. Kashdan et al., Nonlinear waves and turbulence in Marangoni convection. *Phys. Fluids* **7**(11), 2679–2685, 1995.
39. A. Oron and Ph. Rosenau, Formation of patterns induced by thermocapillarity and gravity. *J. Phys. (France) II* **2**, 131–146, 1992.
40. E. Erneux and S.H. Davis, Nonlinear rupture of free films. *Phys. Fluids* **A5**, 1117–1122, 1993.
41. S.H. Davis, *Annu. Rev. Fluid Mech.* **19**, 403–435, 1987.
42. H.-C. Chang, E.A. Demekhin and D. Kopelevitch, Nonlinear evolution of waves on a vertically falling fluid. *J. Fluid Mech.* **250**, 433–480, 1993.
43. P.M. Parmentier, V.C. Regnier and G. Lebon, *Phys. Rev. E* **54**(1), 411–423, 1996.
44. P.C. Dauby and G. Lebon, *J. Fluid Mech.* **329**, 25–64, 1996.

Chapter 8
Summing Up the Three Significant Models Related with the Bénard Convection Problem

8.1 Introduction

One year ago, in reply to my proposal concerning the 'possible' publication of the present book, with a clear emphasis on rational analysis and asymptotic modelling in derivation of model equations for the main kinds of convective flows, Professor René Moreau, Series Editor of FMIA, wrote to me:

> ... However, from a certain point of view, you know that some readers do not care much for this rigor and just want to know what are the relevant model equations for their problem ...

and later:

> May I ask you a question? Could you imagine to add, in a kind of general conclusion, a sort of table made on the following idea, which, in my opinion, would significantly increase the potential sales.

This short chapter, with a summing up of Chapters 3, and 5 to 7, is, to a certain extent, a reponse to the above suggestion from Professor Moreau. Nevertheless, I hope that the preceding seven chapters have captured the interest of the majority of my readers and that, for them, this chapter will serve as only a concentrated review, at least concerning the sections devoted to RB, deep and BM convections. In the preceding chapters, our main objective was a rational clarification of the various steps which lead to now well-known approximate leading-order, Rayleigh–Bénard (shallow-thermal), Zeytounian (deep-thermal) and Bénard–Marangoni (thermocapillarity-surface tension) convections. As a starting physical phenomemon we chose the simplest Bénard problem of a liquid layer heated from below, in the absence of rotation, magnetic field, porous-medium or two-component fluid (for definitions of such convections, see Chapter 10).

In a horizontal liquid layer, an adverse temperature gradient (β_S) is maintained by heating the underside (a lower horizontal rigid, $z = 0$, heated plane at temperature, T_w). The occurence of the phenomena seems to be associated with cooling of the liquid at its deformable (free at the reference, conduction, level $z = d$) surface (where exposed to the air, at temperature T_A), when the layer of the liquid is at a temperature somewhat above that assumed by a thin superficial film.

A very slight excess of temperature in the layer of the liquid above that of the surrounding air is sufficient to institute the 'tesselated' changing structure (according to Thompson [1], as this was noted by Lord Rayleigh in 1916). More precisely, the conduction adverse temperature gradient in liquid, β_S, is directly determined by the difference of temperature $(T_w - T_A)$, via a Newton's cooling law of heat transfer with a unit constant thermal surface conductance q_s:

$$\beta_s = \frac{(T_w - T_A)}{[(k/qs) + d]}$$

where k is the thermal conductivity of the liquid and d the thickness of the layer, both constant in a conduction motionless state.

In the simplest Bénard problem of a liquid layer heated from below, with β_S, we have four main driving effects:

1. the *buoyancy* directly related to the thermal shallow convection,
2. the *temperature-dependent surface tension* which is responsible for the thermocapillary convection,
3. the *viscous dissipation* which leads to consideration to deep thermal convection, and
4. the effect related to the influence of the *deformable free surface*.

These four effects affect mainly the Bénard convection phenomenon and it is necessary from the start of formulation of the full Bénard problem to take into account these four driving forces. In Figure 8.1 we have sketched (with pecked lines) the significant interconnections between these three main facets of Bénard convection.

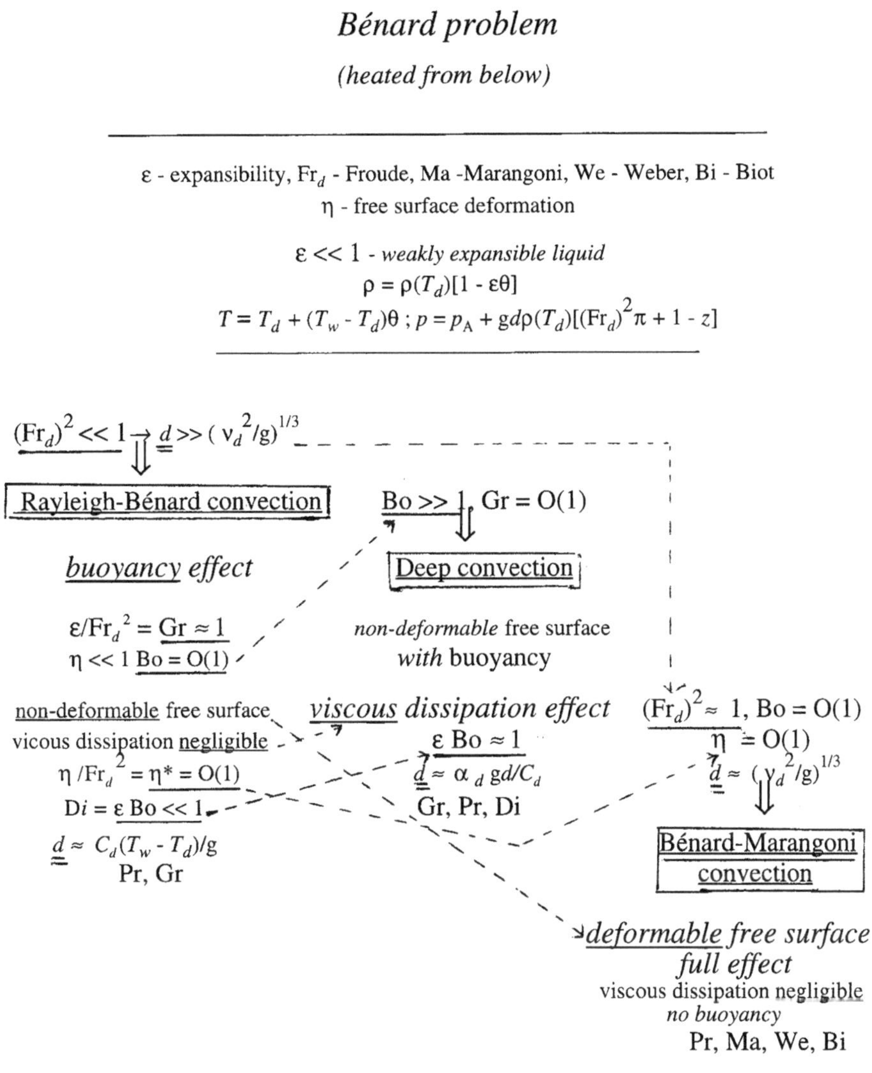

Fig. 8.1 Three main facets of Bénard convection.

8.2 A Rational Approach to the Rayleigh–Bénard Thermal Shallow Convection Problem

In Chapter 3, in the framework of the simple Rayleigh thermal convection problem, the reader was initiated into our rational analysis and asymptotic modelling approach. The keys to such an approach are based, from the beginning, on consideration of a problem formulation with the four main driving forces, mentioned above, and careful analysis of the influence of the various parameters which govern these driving forces! In the case of the classical Bénard, heated from below, thermal convection, the main driving force (in particular when, as in Chapter 3, the liquid layer is between two rigid horizontal planes) is the buoyancy, and the Grashof, Gr, number (or Rayleigh, Ra, number) governs the RB shallow thermal convction, when we assume that the liquid is weakly expansible.

We first define the expansibility parameter by

$$\varepsilon = \alpha(T_d)\Delta T, \tag{8.1a}$$

where, with $\rho = \rho(T)$, the influence of the pressure being negligible at the leading order,

$$\alpha(T_d) = -\left(\frac{1}{\rho_d}\right)\left[\frac{\mathrm{d}\rho(T)}{\mathrm{d}T}\right]_d \quad \text{and} \quad \Delta T = T_w - T_d \tag{8.1b}$$

with T_w, the temperature at the lower horizontal rigid plane ($z = 0$) and T_d, the temperature at the upper horizontal rigid plane or the reference level of the deformable free surface ($z = d$). Then, the square of the Froude number relative to the constant conduction thickness d of the liquid layer (where ν_d is the kinematic viscosity, at temperature T_d) being

$$\mathrm{Fr}_d^2 = \frac{(\nu_d/d)^2}{gd}, \tag{8.1c}$$

we discovered that the usual Grashof Gr number is simply the ratio of ε to Fr_d^2!

$$\mathrm{Gr} = \frac{\varepsilon}{\mathrm{Fr}_d^2} = g\left[-\frac{\mathrm{d}\rho(T)}{\mathrm{d}T}\right]_d \frac{(T_w - T_d)d^3}{\rho_d \nu_d^2}. \tag{8.1d}$$

The definition of the Grashof number as the ratio of the small expansibility parameter (ε) to the square of the Froude number (Fr_d^2), is the first main key step to a rational derivation of the RB model problem. On the other hand, when the Prandtl number (at $T = T_d$)

$$\mathrm{Pr} = \frac{\nu_d}{\kappa_d} = \mathrm{O}(1) \tag{8.1e}$$

is not very low or very large ($\nu_d \approx \kappa_d$ where κ_d is the thermal diffusivity) the Rayleigh number

$$\mathrm{Ra} \equiv \mathrm{Pr}\,\mathrm{Gr} = g\left[-\frac{\mathrm{d}\rho(T)}{\mathrm{d}T}\right]_d \frac{(T_w - T_d)d^3}{\rho_d \nu_d \kappa_d}$$

plays a similar role as Gr; but both cases, when Pr $\ll 1$ or Pr $\gg 1$, deserve particular attention (see some references in Chapter 10).

The second key step in rational, asymptotic derivation of the RB equations for the shallow thermal convection, emerging from $T_w > T_d$ or more precisely, from

$$\frac{\Delta T}{d} = \beta_s \left(\equiv -\frac{\mathrm{d}T_s(z)}{\mathrm{d}z} \right) > 0, \tag{8.2a}$$

where $T_s(z) = T_w - \beta_s z$, is the conduction temperature in a steady motionless state, is related to the introduction of a dimensionless temperature for a convection regime:

$$\theta = \frac{(T - T_d)}{d\beta_s}, \tag{8.2b}$$

and its companion dimensionless pressure

$$\pi = \left(\frac{1}{\mathrm{Fr}_d^2}\right) \left\{ \left[\frac{(p - p_d)}{g d \rho_d} \right] + z - 1 \right\}. \tag{8.2c}$$

With $\rho = \rho(T)$ we write, according to (8.2a, b),

$$\rho = \rho(T = T_d + \Delta T \theta) \approx \rho_d (1 - \varepsilon\theta), \tag{8.2d}$$

with an error of $\mathrm{O}(\varepsilon^2)$.

In a third key step, when the limiting (*à la* Boussinesq) process

$$\varepsilon \to 0 \quad \text{and} \quad \mathrm{Fr}_d^2 \to 0, \text{ simultaneously} \equiv \lim_{\text{Boussinesq}} \tag{8.3a}$$

is performed, such that

$$\mathrm{Gr} = \frac{\varepsilon}{\mathrm{Fr}_d^2} = \mathrm{O}(1) \text{ is fixed}, \tag{8.3b}$$

from the dimensionless starting full Navier–Stokes and Fourier equations for an expansible liquid we derive, first, as continuity equation, the divergence-free constraint

$$\nabla \cdot \mathbf{u} = \varepsilon \frac{\mathrm{d}\theta}{\mathrm{d}t} \Rightarrow \nabla \mathbf{u}_{\mathrm{RB}} = 0, \tag{8.3c}$$

for the limiting value of the dimensionless velocity

$$\mathbf{u}_{\mathrm{RB}} = \lim_{\text{Boussinesq}} \mathbf{u}.$$

On the other hand, from the dimensionless momentum equation for $\mathbf{u}$, as a consequence of the Boussinesq limiting process (8.3a), when we take into account (8.2b–d), the following limit leading-order equation for the above $\mathbf{u}_{RB}$ is derived:

$$\frac{\mathbf{d}\mathbf{u}_{RB}}{\mathrm{d}t} + \nabla\pi_{RB} - \mathrm{Gr}\,\theta_{RB}\mathbf{k} = \mathbf{\Delta}\mathbf{u}_{RB}, \tag{8.3d}$$

which is asymptotically correct in the leading order, with an error of $O(\varepsilon)$.

Now, it is necessary (as a fourth key step) to derive also a limit equation for

$$\theta_{RB} = \lim_{\text{Boussinesq}} \theta$$

and for this from the starting energy equation (for the specific internal energy, e) it is necessary to first obtain the corresponding equation for the temperature; in the case of the simple equation of state $\rho = \rho(T)$ this is an easy exercise, because in such a case $e = E(T)$ and

$$\frac{\mathrm{d}E}{\mathrm{d}t} = \left(\frac{\mathrm{d}E}{\mathrm{d}T}\right)\frac{\mathrm{d}T}{\mathrm{d}t}, \tag{8.4a}$$

where $\mathrm{d}E/\mathrm{d}T = C(T)$ is the specific heat for our expansible liquid.

Nevertheless, an essential problem is elucidation of the role of the viscous dissipation term in the dimensionless equation for temperature T. This viscous dissipation term, in a non-dimensional equation for the temperature T, is proportional to dissipation parameter Di, such that

$$(1/2\,\mathrm{Gr})\mathrm{Di} \equiv (1/2\,\mathrm{Gr})\varepsilon\,\mathrm{Bo}, \tag{8.4b}$$

where the so-called ‘Boussinesq number’ is given by

$$\mathrm{Bo} = \frac{gd}{C(T_d)\Delta T}. \tag{8.4c}$$

Gr is O(1) and fixed when we perform the above Boussinesq limiting process (8.3a). From (8.4b) we see that:

$$\text{if Bo} = \mathrm{O}(1) \text{ fixed, when (8.3a) is realized, then Di} \to 0.$$

In such a case, for the above dimensionless temperature θ_{RB}, we obtain from the full equation for the dimensionless temperature θ, defined by (8.2b), the model, leading-order equation

$$\frac{\mathrm{d}\theta_{RB}}{\mathrm{d}t} = \left(\frac{1}{\mathrm{Pr}}\right)\Delta\theta_{RB}. \tag{8.4d}$$

The condition $\mathrm{Bo} = \mathrm{O}(1)$ gives the following constraint for the thickness d:

$$d \approx \frac{C(T_d)\Delta T}{g},$$

and we obtain that the limiting, leading-order, shallow RB convection equations (8.3c, d) and (8.4d), for $\mathbf{u}_{\mathrm{RB}}$, π_{RB} and θ_{RB}, are valid in a liquid layer thickness d such that

$$1\ \mathrm{mm} \approx \left(\frac{\nu_d^2}{g}\right)^{1/3} \ll d \approx \frac{C(T_d)\Delta T}{g}. \tag{8.5}$$

For the *à la* Rayleigh model problem considered above, the boundary conditions are very simple because the convection operates (as a direct consequence of a low squared Froude number $\mathrm{Fr}_d^2 \ll 1$) inside a constant horizontal incompressible liquid layer, $z \in [0, 1]$, and in such a case,

$$\mathbf{u}_{\mathrm{RB}} = 0 \quad \text{and} \quad \theta_{\mathrm{RB}} = 1 \quad \text{on } z = 0;\ \mathbf{u}_{\mathrm{RB}} = 0 \text{ and } \theta_{\mathrm{RB}} = 0 \text{ on } z = 1. \tag{8.6}$$

When, as in Chapter 5, we assume that above the liquid layer exists an ambient atmosphere (*passive* air) and, as a consequence, it is necessary to take into account, in the starting formulation of the Bénard convection problem, heated from below, the presence of a deformable free surface, obviously the above conditions on $z = 1$ must be replaced by more complicated upper conditions! But, curiously, in the framework of the limiting process (8.3a), and in particular because $\mathrm{Fr}_d^2 \to 0$, these conditions are considerably simplified. The reason being that, for the dimensionless pressure π, defined by (8.2c), the associated dimensionless upper condition is written rigorously as:

$$\begin{aligned}\pi \ &= \frac{[H(t, x_1, x_2) - 1]}{\mathrm{Fr}_d^2} + \left(\frac{\partial u_i'}{\partial x_j'} + \frac{\partial u_j'}{\partial x_i'}\right) n_i' n_j' \\ &\quad + (\mathrm{We} - \mathrm{Ma}\theta)(\nabla_{\|} \cdot \mathbf{n}) - (2/3)\varepsilon \frac{\mathrm{d}\theta}{\mathrm{d}t}, \\ &\quad \text{on } z = H[= 1 + \eta h(t, x, y)],\end{aligned} \tag{8.7a}$$

and we observe that We and Ma are defined in Chapter 1 by (1.18a, b).

$\mathrm{Fr}_d^2 \to 0$ being the denominator, we can only assume that $H \to 1$ – as a consequence, in a rational theory it is necessary to assume the following similarity rule (assuming that the dimensionless deformation of the upper free surface $h = \mathrm{O}(1)$) between two small parameters η and Fr_d^2:

$$\frac{\eta}{\mathrm{Fr}_d^2} = \eta^* = \mathrm{O}(1) \quad \text{when } \mathrm{Fr}_d^2 \to 0 \text{ and } \eta \to 0 \text{ simultaneously.} \tag{8.7b}$$

From (8.7b) for the RB model system of shallow convection the upper, free-surface condition must be written (at least to leading order) on $z = 1$.

According to results of Section 5.2, we obtain, on $z = 1$,

$$\mathbf{u}_{\text{RB}} \cdot \mathbf{k} \equiv u_{3\text{RB}} = 0, \tag{8.7c}$$

$$\frac{\partial^2 u_{3\text{RB}}}{\partial z^2} = \text{Ma}\left[\frac{\partial^2 \theta_{\text{RB}}}{\partial x^2} + \frac{\partial^2 \theta_{\text{RB}}}{\partial y^2}\right], \tag{8.7d}$$

$$\frac{\partial \theta_{\text{RB}}}{\partial z} + [1 + \text{Bi}_{\text{conv}}\theta_{\text{RB}}] = 0. \tag{8.7e}$$

Finally, from (8.7a) we derive with (8.7b), when we assume that $\text{We} \gg 1$ and

$$\eta\,\text{We} = \text{We}^* = \text{O}(1), \tag{8.8a}$$

an equation for the free-surface deformation $h(t, x, y)$,

$$\frac{\partial^2 h}{\partial x^2} + \frac{\partial^2 h}{\partial y^2} - \left(\frac{\eta^*}{\text{We}^*}\right) h = -\left(\frac{1}{\text{We}^*}\right) \pi(t, x, y, z = 1). \tag{8.8b}$$

The reader has now, in the above few pages, all information linked to the full theoretical statement of the shallow Rayleigh–Bénard convection leading-order model problem. Further, if the reader wants to obtain the second-order model equations (with non-Boussinesq effects) associated with the leading-order shallow RB model equations, then this is an easy task as is shown in Sections 3.5 and 3.6 and also in Section 5.3.

In fact, 25 years ago, in October 1983, I first published a short note [1] in *Comptes Rendus de l'Académie des Sciences – Paris* entitled 'One Asymptotic Formulation of Rayleigh–Bénard's Problem via Boussinesq Approximation for the Expansible Liquid'.

A recent book by Getling [2] gives a pertinent account linked with the 'structure and dynamics' of the Rayleigh–Bénard convection. In a more recent review paper, by Bodensghatz et al. [3], the reader can find a modern account relative to theoretical and experimental investigations of the Rayleigh–Bénard convection problem. Finally, we observe that equation (8.8b) for the free-surface deformation $h(t, x, y)$ has been first derived thanks to our rational analysis and asymptotic modelling!

8.3 The Deep Thermal Convection with Viscous Dissipation

An attentive examination of the dominant equation for the dimensionless temperature θ shows that, as a consequence of the *à la* Boussinesq, limiting process (8.3a), two terms in this equation, which are proportional to ε Bo, have been neglected at leading order, because of the assumption that Bo $= O(1)$ is fixed during (8.3a)! As a consequence, in particular, if we want to take into account the viscous dissipation term in the limit leading-order equation for θ, derived from (8.3a), obviously it is necessary to assume now that

$$\text{Bo} \gg 1 \text{ such that, with } \varepsilon \ll 1\text{: } \varepsilon\,\text{Bo} = O(1) \text{ fixed, during (8.3a),} \tag{8.9}$$

which is the key for obtaining the 'deep thermal convection' equations.

According to (1.15) we have introduced the parameter

$$\text{Di} = \varepsilon\,\text{Bo} \tag{8.10}$$

and instead of (8.3a), in the case of deep convection, it is necessary to consider the following 'deep' limiting process:

$$\varepsilon \to 0 \text{ and } \text{Fr}_d^2 \to 0,\, \text{Bo} \to \infty \text{ simultaneously } \equiv \lim_{\text{deep}} \tag{8.11a}$$

such that

$$\text{Gr} = \frac{\varepsilon}{\text{Fr}_d^2} = O(1) \quad \text{and} \quad \text{Di} = \varepsilon\,\text{Bo} = O(1), \tag{8.11b}$$

Gr and Di both being fixed.

As a consequence, instead of the RB shallow convection equations (8.3c, d) and (8.4d), for $\mathbf{u}_{\text{RB}}$, π_{RB} and θ_{RB}, with the boundary conditions (8.6) on $z = 0$ and $z = 1$, we derive for the functions

$$\lim_{\text{deep}}(\mathbf{u}, \pi, \theta) = \mathbf{u}_D, \pi_D, \theta_D, \tag{8.11c}$$

the following dimensionless deep convection (DC) leading-order model equations:

$$\nabla \cdot \mathbf{u}_D = 0, \tag{8.12a}$$

$$\frac{d\mathbf{u}_D}{dt} + \nabla \pi_D - \text{Gr}\,\theta_D \mathbf{k} = \Delta \mathbf{u}_D, \tag{8.12b}$$

$$\{1 - \text{Di}[p_d + 1 - z]\}\frac{d\theta_D}{dt} = \left(\frac{1}{\text{Pr}}\right)\Delta\theta_D + (1/2\,\text{Gr})\text{Di}\left[\frac{\partial u_{\text{Di}}}{\partial x_j} + \frac{\partial u_{Dj}}{\partial x_i}\right]^2, \tag{8.12c}$$

where

$$\text{Di} = \frac{g\alpha(T_d)d}{C(T_d)} \tag{8.13}$$

is our 'depth' parameter, defined in 1989 in [4] by δ. For the DC equations (8.12a–c) as boundary conditions, we have again

$$\mathbf{u}_D = 0 \quad \text{and} \quad \theta_D = 1 \quad \text{on } z = 0;\ \mathbf{u}_D = 0 \text{ and } \theta_D = 0 \text{ on } z = 1. \tag{8.14}$$

But we can also consider these DC equations (8.12a–c) with the 'upper, non-deformable (on $z = 1$), free-surface' conditions, as in (8.12c–e) with (8.8a, b). I think that these DC equations with viscous dissipation term deserve various complementary investigations linked with a more detailed account of modifications introduced to the classical Rayleigh–Bénard shallow convection structure, dynamics and attractors via the route to chaos.

8.4 The Thermocapillary Convection with Temperature-Dependent Surface Tension

If we want to consider the full effect of a temperature-dependent tension on the upper, deformable free-surface conditions, then it is imperative to assume that the square of the Froude number Fr_d^2 is fixed and of order 1 when the expansibility parameter ε tends to zero. In such a case, at leading order, the effect of buoyancy is negligible and the limit (relative to $\varepsilon \to 0$) model equations are simply the usual Navier incompressible equations for

$$\lim_{\varepsilon \to 0}(\mathbf{u}, \pi) = (\mathbf{u}_{\mathrm{BM}}, \pi_{\mathrm{BM}}) \tag{8.15a}$$

and the *à la* Fourier usual temperature equation for

$$\lim_{\varepsilon \to 0} = \theta_{\mathrm{BM}}, \tag{8.15b}$$

namely,

$$\nabla \cdot \mathbf{u}_{\mathrm{BM}} = 0, \tag{8.16a}$$

$$\frac{\mathrm{d}\mathbf{u}_{\mathrm{BM}}}{\mathrm{d}t} + \nabla \pi_{\mathrm{BM}} = \boldsymbol{\Delta} \mathbf{u}_{\mathrm{BM}}, \tag{8.16b}$$

$$\mathrm{Pr} \frac{\mathrm{d}\theta_{\mathrm{BM}}}{\mathrm{d}t} = \Delta \Theta_{\mathrm{BM}}. \tag{8.16c}$$

However, in the framework of the Bénard-Marangoni (BM) model problem, for the model equations (8.16a–c), we have complicated upper, deformable free-surface conditions, because the similarity rule (8.7b) is unnecessary when $\mathrm{Fr}_d^2 = \mathrm{O}(1)$. Namely, we write on the upper-deformable free surface, $z = H(t, x, y)$:

$$\pi_{\mathrm{BM}} = \left(\frac{1}{\mathrm{Fr}_d^2}\right)(H-1) + \left(\frac{2}{N}\right)\left\{\left(\frac{\partial u_{1\mathrm{BM}}}{\partial x}\right)\left(\frac{\partial H}{\partial x}\right)^2 + \left(\frac{\partial u_{2\mathrm{BM}}}{\partial y}\right)\left(\frac{\partial H}{\partial y}\right)^2\right.$$

$$+ \frac{\partial u_{3\mathrm{BM}}}{\partial z} + \left[\frac{\partial u_{1\mathrm{BM}}}{\partial y} + \frac{\partial u_{2\mathrm{BM}}}{\partial x}\right]\left(\frac{\partial H}{\partial x}\right)\left(\frac{\partial H}{\partial y}\right)$$

$$\left. - \left[\frac{\partial u_{1\mathrm{BM}}}{\partial z} + \frac{\partial u_{3\mathrm{BM}}}{\partial x}\right]\left(\frac{\partial H}{\partial x}\right) - \left[\frac{\partial u_{2\mathrm{BM}}}{\partial z} + \frac{\partial u_{3\mathrm{BM}}}{\partial y}\right]\left(\frac{\partial H}{\partial y}\right)\right\}$$

$$- \left(\frac{1}{N}\right)^{3/2} [\mathrm{We} - \mathrm{Ma}\,\theta_{\mathrm{BM}}]\left\{N_2\left(\frac{\partial^2 H}{\partial x^2}\right)\right.$$

$$\left. - 2\left(\frac{\partial H}{\partial x}\right)\left(\frac{\partial H}{\partial y}\right)\left(\frac{\partial^2 H}{\partial x \partial y}\right) + N_1\left(\frac{\partial^2 H}{\partial y^2}\right)\right\}; \tag{8.17a}$$

$$\left[\frac{\partial u_{1\mathrm{BM}}}{\partial x} - \frac{\partial u_{3\mathrm{BM}}}{\partial z}\right]\left(\frac{\partial H}{\partial x}\right) + (1/2)\left[\frac{\partial u_{1\mathrm{BM}}}{\partial y} + \frac{\partial u_{2\mathrm{BM}}}{\partial x}\right]\left(\frac{\partial H}{\partial y}\right)$$

$$+ (1/2)\left[\frac{\partial u_{2\mathrm{BM}}}{\partial z} + \frac{\partial u_{3\mathrm{BM}}}{\partial y}\right]\left(\frac{\partial H}{\partial x}\right)\left(\frac{\partial H}{\partial y}\right)$$

$$- (1/2)\left[1 - \left(\frac{\partial H}{\partial x}\right)^2\right]\left[\frac{\partial u_{3\mathrm{BM}}}{\partial x} + \frac{\partial u_{1\mathrm{BM}}}{\partial z}\right]$$

$$= \left(\frac{N^{1/2}}{2}\right)\mathrm{Ma}\left[\frac{\partial \theta_{\mathrm{BM}}}{\partial x} + \left(\frac{\partial H}{\partial x}\right)\frac{\partial \theta_{\mathrm{BM}}}{\partial z}\right]; \tag{8.17b}$$

$$\left[\frac{\partial u_{1\mathrm{BM}}}{\partial x} - \frac{\partial u_{2\mathrm{BM}}}{\partial y}\right]\left(\frac{\partial H}{\partial y}\right)\left(\frac{\partial H}{\partial x}\right)^2 + \left[\frac{\partial u_{2\mathrm{BM}}}{\partial y} - \frac{\partial u_{3\mathrm{BM}}}{\partial z}\right]\left(\frac{\partial H}{\partial y}\right)$$

$$+ \left[\frac{\partial u_{1\mathrm{BM}}}{\partial z} + \frac{\partial u_{3\mathrm{BM}}}{\partial x}\right]\left(\frac{\partial H}{\partial x}\right)\left(\frac{\partial H}{\partial y}\right)$$

$$+ (1/2)\left[1 + \left(\frac{\partial H}{\partial x}\right)^2 - \left(\frac{\partial H}{\partial y}\right)^2\right]\left(\frac{\partial u_{1\mathrm{BM}}}{\partial y} + \frac{\partial u_{2\mathrm{BM}}}{\partial x}\right)\left(\frac{\partial H}{\partial x}\right)$$

$$- (1/2)\left[1 - \left(\frac{\partial H}{\partial x}\right)^2 - \left(\frac{\partial H}{\partial y}\right)^2\right]\left[\frac{\partial u_{2\mathrm{BM}}}{\partial z} + \frac{\partial u_{3\mathrm{BM}}}{\partial y}\right]$$

$$= \left(\frac{N^{1/2}}{2}\right)\mathrm{Ma}\left\{-\left(\frac{\partial H}{\partial x}\right)\left(\frac{\partial H}{\partial y}\right)\frac{\partial \Theta_{\mathrm{BM}}}{\partial x} + \left[1 + \left(\frac{\partial H}{\partial x}\right)^2\right]\frac{\partial \Theta_{\mathrm{BM}}}{\partial y}\right.$$

$$\left. + \left(\frac{\partial H}{\partial y}\right)\frac{\partial \Theta_{\mathrm{BM}}}{\partial z}\right\}. \tag{8.17c}$$

Next, for θ_{BM} as upper free-surface condition we have

$$\frac{\partial \theta_{\text{BM}}}{\partial z} + N^{1/2}(1 + \text{Bi}_{\text{conv}}\theta_{\text{BM}}) = \frac{\partial \theta_{\text{BM}}}{\partial x}\left(\frac{\partial H}{\partial x}\right) + \frac{\partial \theta_{\text{BM}}}{\partial y}\left(\frac{\partial H}{\partial y}\right), \quad (8.17\text{d})$$

while the kinematic condition is unchanged:

$$u_{3\text{BM}} = \frac{\partial H}{\partial t} + u_{1\text{BM}}\left(\frac{\partial H}{\partial x}\right) + u_{2\text{BM}}\left(\frac{\partial H}{\partial y}\right). \quad (8.17\text{e})$$

The full BM model problem for the phenomena of thermocapillarity is very complicated and necessitates a numerical computation, but this BM model problem is relative to thin films (of the order of the millimetre) and when the amplitude of the free surface deformation is moderate we have the possibility to consider, instead of the above BM problem (8.16a–c), (8.17a–e), a BM reduced long-wave model problem which has been discussed in Section 7.3, see (7.29a–d). Such a computation for the thermocapillary waves has been proposed recently, in [5, 6]. This computation has been described in [6] with the help of a regularized reduced model derived in [5]; see also Section 10.4.

In recent years, several books have been published relative to thermocapillary temperature-dependent, surface tension driven, Marangoni convection by Colinet et al. [7], Velarde and Zeytounian [8], and Nepomnyaschy et al. [9].

Obviously the present book, mainly devoted to an analytical approach to the full Bénard problem, heated from below, and its rational analysis and asymptotic modelling, have really, relative to purpose, very little overlap with the above cited books! Our fundamental concept is the possibility to extract, consistently, from this very complicated, but 'rich' Bénard, physical-mathematical problem, three relatively simple but significant approximate model problems.

On the other hand, to those familiar with the large panoply of films, liquid layers, and their behavior in its full complexity and variety, the approach and the derived model problems in this present book may seem abstract and remote. Nonetheless, I consider the field to have come a long way; at least now we are, from the start of formulation of the problem, devoted to modelling the main physical driving effects, whereas two decades ago a large number of papers were based on *ad hoc* pseudo-theoretical approaches and numerical computations. In the meantime, I hope that careful study of these seven preceding chapters in this volume will kindle new insights into the Bénard problem, and perhaps suggest fresh approaches.

References

1. R.Kh. Zeytounian, Sur une formulation asymptotique du problème de Rayleigh–Bénard, via l'approximation de Boussinesq pour les liqudes dilatables. *C. R. Acad. Sci., Paris Ser. I* **297**, 271–274, October 1983.
2. A.V. Getling, *Rayleigh–Bénard Convection: Structure and Dynamics*. World Scientific, Singapore, 1998.
3. E. Bodenschatz, W. Pesch and G. Ahlers, *Ann. Rev. Fluid Mech.* **32**, 709–778, 2000.
4. R.Kh. Zeytounian, The Bénard problem for deep convection: Rigorous derivation of approximate equations. *Int. J. Engng. Sci.* **27**(11), 1361–1366, 1989.
5. C. Ruyer-Quil, B. Scheid, S. Kalliadasis, M.G. Velarde and R.Kh. Zeytounian, Thermocapillary long-waves in a liquid film flow. Part 1. Low-dimensional formulation, *J. Fluid Mech.* **538**, 199–222, 2005.
6. B. Scheid, C. Ruyer-Quil, S. Kalliadasis, M.G. Velarde and R.Kh. Zeytounian, Thermocapillary long-waves in a liquid film flow. Part 2. Linear stability and nonlinear waves, *J. Fluid Mech.* **538**, 223–244, 2005.
7. P. Colinet, J. Legros and M.G. Velarde, *Nonlinear Dynamics of Surface-Tension-Driven-Instabilities*, 1st Edn. Wiley/VCH, 2001.
8. M.G. Velarde and R.Kh. Zeytounian, *Interfacial Phenomena and the Marangoni Effect*, 1st Edn. CISM Courses and Lectures, No. 428, Udine, Springer-Verlag, Wien/New York, 2002.
9. A.A. Nepomnyaschy, M.G. Velarde and P. Colinet, *Interfacial Phenomena and Convection*. Chapman & Hall/CRC, London, 2002.

Chapter 9
Some Atmospheric Thermal Convection Problems

9.1 Introduction

The thermal convection problems considered in this chapter are very different from the scales in time and space of the phenomena studied in the preceding chapters. As this is well and pertinently discussed in the paper by Velarde and Normand [1], published in July 1980 in *Scientific American*:

> In the Earth's atmosphere convection is observed at several scales of length. The temperature gradient between the Tropics and the poles drives a global circulation, which can be decomposed into at least three large convective cells in each hemisphere. Distortions of these patterns caused by the rotation of the Earth give rise to the trade winds of the Tropics and the prevailing westerlies of the temperate zones.
>
> Local heating of the atmosphere near the Earth's surface gives rise to smaller-scale convective flows, including those of most storms. Cumulus clouds, which form when warm air rises and cools and thereby becomes supersaturated with moisture, often mark the convective overturning of the atmosphere.
>
> A theoretical analysis of atmospheric convection must take into account the compressibility of air, which gives rise to a density gradient even when the temperature is constant with height.
>
> An accurate description of atmospheric circulation would also have to include the compressive heating of air when it sinks into a region of higher pressure. Viscosity and other properties of air also vary with pressure and temperature, and the presence of water vapor, which gives up heat when it condenses, adds still another level of complexity!

It is obvious that the theories that describe these convective processes require many simplifying assumptions, if they are to be of any practical use, Even then they are far from simple! But, it is therefore all the more remarkable that these few related theories, governed by a handful of dimensionless numbers, can account for phenomena that differ so greatly in scale.

It is interesting to note that Prandtl, in 1944, obtained an explicit solution of a local very simple thermal convection model problem on the assumption that the idealized mountain slope under consideration consists of an infinite, thermally homogeneous plane. Namely (see, for instance, chapter 7 in the book by Gutman [2]), in such a simple case, the thermal convection phenomenon is governed by the following two linear equations, for the temperature perturbation θ, and horizontal velocity u along the mountain slope with inclination angle α with respect to the horizontal direction:

$$\frac{\partial \theta}{\partial t} + Su \sin\alpha = k\frac{\partial^2 \theta}{\partial z^2}, \tag{9.1a}$$

$$\frac{\partial u}{\partial t} - \lambda\theta \sin\alpha = \nu\frac{\partial^2 u}{\partial z^2}. \tag{9.1b}$$

When the stratification parameter S and parameter of convection λ are both constant, with $k \equiv \nu =$ const, the steady-state solution of equations (9.1a, b), satisfying the conditions

$$z = 0: \quad u = 0, \quad \theta = \theta_0 = \text{const} \quad \text{and} \quad u \to 0, \quad \theta \to 0 \quad \text{when } z \to \infty, \tag{9.1c}$$

is of the following form (when $S > 0$):

$$u = \theta_0 \left[\left(\frac{\lambda}{S}\right)\right]^{1/2} \exp(-\xi)\sin\xi \quad \text{and} \quad \theta = \theta_0 \exp(-\xi)\cos\xi, \tag{9.2a}$$

where

$$\xi = z\left[(\lambda S)\sin^2\frac{\alpha}{4\nu^2}\right]^{1/4}. \tag{9.2b}$$

For the neutral or unstable stratification ($S \leq 0$) of the undisturbed atmosphere, equations (9.1a, b) do not have steady-state solutions which would satisfy conditions (9.1c). Obviously, the solution (9.2a) can be used as a test for convergence and for the accuracy of the more realistic numerical solution. In Section 9.5, the reader can find some complements which generalize the above simple Prandtl example.

Our intention in this chapter is not to give a full account of the various convective atmospheric phenomena (for this even a volume would not suffice), but rather to illustrate via some typical problems the rich variety of these convective atmospheric phenomena. The reader can find in the book by Emanuel [3] a relatively recent text devoted entirely to 'atmospheric convection'.

Concerning the Boussinesq approximation and the associated Boussinesq atmospheric equations which are a very good approximation for the slow (hyposonic, [4]) atmospheric motions, the reader can find in [5], and in the more recent paper [6] various information relative to the rational derivation of these Boussinesq atmospheric equations for dry air assumed to be a thermally perfect gas.

Namely, in this chapter I study only some particular (mainly meso or local) convection motions in the atmosphere. After this introduction, in Section 9.2, the formulation of the breeze problem via the Boussinesq hydrostatic approximation is considered and Section 9.3 is then devoted to the infuence of a local temperature field in an atmospheric Ekman layer via a triple deck asymptotic approach. In Section 9.4, a periodic, double-boundary layer thermal convection over a curvilinear wall is investigated. In Section 9.5, some other particular atmospheric convection problems are also briefly discussed.

9.2 The Formulation of the Breeze Problem via the Boussinesq Hydrostatic Approximation

We consider a first local atmospheric convection phenomenon, the free circulation-breeze problem. The modeling of this breeze problem is directly related to the justification of the Boussinesq hydrostatic approximation for atmospheric convection and, for this, it is necessary to start from the full, non-adiabatic and non-hydrostatic, viscous, unsteady, and compressible NSF equations and take into account a rotating system of spherical coordinates.

In a coordinate frame rotating with the Earth, we consider from the beginning the full NSF equations (for a thermally perfect gas) and take into account, first, the Coriolis force ($2\mathbf{\Omega} \wedge \mathbf{u}$), the gravitational acceleration (modified by the centrifugal force) $\mathbf{g}$, and also the effect of thermal radiation Q. The (relative) velocity vector is $\mathbf{u} = (\mathbf{v}, w)$, as observed in the Earth's frame rotating with the angular velocity $\mathbf{\Omega}$. The thermodynamic functions are again, ρ (atmospheric air density), p (atmospheric air pressure) and T (absolute temperature of dry thermally perfect air).

First, we assume that

$$\frac{\mathrm{d}R_s}{\mathrm{d}z_s} = \rho_s Q \quad \text{with } R_s = R_s(T_s(z_s)), \tag{9.3a}$$

and we note that $\rho_s(z_s)$ and $T_s(z_s)$, in relation (9.3a), are the density and temperature in the hydrostatic reference state (function only of the altitude z_s in the reference state). The heat source Q (i.e., thermal radiation) is assumed to be only a function of this hydrostatic reference state via the 'standard' temperature $T_s(z_s)$. Doing this, we consider only a so-called, mean, standard, distribution for Q and ignore variations from it for the perturbed atmosphere in motion (namely, in a 'convection' regime). The temperature $T_s(z_s)$, for the hydrostatic reference state (in a so-called 'standard atmosphere'), satisfies the following (balance) equation, written here in dimensional form:

$$k_c \frac{\mathrm{d}T_s}{\mathrm{d}z_s} + R_s(T_s(z_s)) = 0, \tag{9.3b}$$

where the conductivity k_c = const (in a simplified case). Now $\mathbf{\Omega}$ can be expressed as

$$\mathbf{\Omega} = \Omega^0 \mathbf{e} \quad \text{with } \mathbf{e} = \mathbf{k} \sin\varphi + \mathbf{j} \cos\varphi, \tag{9.4}$$

where Ω^0 = const and φ is the algebraic latitude of the point P^0 of the observation on the Earth's surface, around which the atmospheric convections (free circulations) are analyzed (in the Northern Hemisphere, $\varphi > 0$). The unit vectors directed to the east, north, and zenith (in the opposite direction from $\mathbf{g} = -g\mathbf{k}$), are denoted by $\mathbf{i}$, $\mathbf{j}$, and $\mathbf{k}$, respectively. It is helpful to employ spherical coordinates λ, φ, r, and in this case u, v, w denote again the corresponding relative velocity components in these directions, respectively, increasing azimuth (λ), latitude (φ), and radius (r). However, it is very convenient to introduce here the transformation

$$x = a_0 \cos\varphi^\circ, \quad y = a_0(\varphi - \varphi^\circ), \quad z = r - a_0, \tag{9.5}$$

where φ° is a reference latitude (for $\varphi^\circ \approx 45^\circ$ we have $a_0 \approx 6367$ km, which is the mean radius of the Earth), and the origin of this right-handed curvilinear coordinates system lies on the Earth's surface (for a flat ground, where $r = a_0$) at latitude φ° and longitude $\lambda = 0$.

We assume therefore that the atmospheric convection phenomenon occurs in a midlatitude, mesoscale region, distant from the equator ($\sin\varphi^\circ$, $\cos\varphi^\circ$, and $\tan\varphi^\circ$ are all of order unity), and in this case the sphericity parameter

$$\delta = \frac{L_0}{a_0} \text{ is assumed small.} \tag{9.6}$$

Although x and y are, in principle, new longitude and latitude coordinates in terms of which the NSF atmospheric equations may be rewritten without approximation (see, for instance, [5, chapter II]), they are obviously introduced in the expectation that (for small δ, in the leading order, with a sufficiently good approximation) will be the Cartesian coordinates of the so-called $f°$-plane (tangent) approximation.

In fact, in the breeze problem, it is necessary to take into account the influence of the Coriolis force as a main driving force, and also write a boundary condition for the temperature expressing the influence of a local thermal non-homogeneity on flat ground surface; such a condition has, for example, the following form:

$$T = T_s(0) + (\Delta T)_0 \Xi \quad \text{at } z = 0, \tag{9.7}$$

with Ξ is a given (known data) function of the time and the position on a bounded region D (of diameter L_0) on the flat ground surface. As a consequence, we have the possibility to introduce a vertical characteristic length scale (R is the perfect gas constant)

$$h_0 = \frac{R(\Delta T)_0}{g} \ll L_0, \tag{9.8a}$$

such that

$$\tau = \frac{(\Delta T)_0}{T_s(0)} \equiv \frac{g h_0}{R T_s(0)} \ll 1. \tag{9.8b}$$

On the one hand, the height $h^0 = 10^3$ m is significant for the breeze phenomenon and, on the other hand, when we take into account the Coriolis force in the dynamic atmospheric unsteady viscous equation, the horizontal characteristic length scale L_0 for this breeze phenomenon is $L_0 \approx 10^5$ m and as a consequence the Rossby number

$$\mathrm{Ro} = \frac{U_0}{f° L_0} \approx 1 \quad \text{with } f° = 2\Omega^0 \sin\varphi°, \tag{9.8c}$$

where U_0 can be evaluated after a judicious choice when we consider the breeze as a low Mach number,

$$M = \frac{U_0}{[\gamma R T_s(0)]^{1/2}} \ll 1, \tag{9.8d}$$

phenomenon, such that (γ = ratio of the specific heats)

$$\frac{\tau}{M} = \tau^* \approx 1 \Rightarrow U_0 \approx (\Delta T)_0 \left[\frac{\gamma R}{T_s(0)}\right]^{1/2}. \tag{9.8e}$$

Obviously, as long-wave parameter we must now consider the ratio

$$\lambda = \frac{h_0}{L_0} \ll 1, \tag{9.9a}$$

and for a high Reynolds number

$$\mathrm{Re} = \frac{U_0 L_0}{\nu_0} \gg 1, \tag{9.9b}$$

where the kinematic viscosity ν_0 is assumed constant, the following hydrostatic limit process must be performed:

$$\lambda \to 0 \text{ and } \mathrm{Re} \to \infty, \text{ simultaneously, such that } \lambda^2 \mathrm{Re} = \mathrm{Re}_\perp = \mathrm{O}(1). \tag{9.9c}$$

We observe that, when we use (9.8a), (9.8e), and (9.9a, b), with (9.9c) we derive the following value for the temperature rate $(\Delta T)_0$, in thermal boundary condition (9.7):

$$(\Delta T)_0 \approx \left(\frac{R}{g}\right)^{2/3} \left(\frac{1}{\nu_0 L_0}\right)^{1/3} \left[\frac{\gamma R}{T_s(0)}\right]^{1/6}. \tag{9.9d}$$

When dimensionless time-space variables are considered,

$$t' = \frac{t}{(1/\Omega^0)}, \quad x' = \frac{x}{L_0}, \quad y' = \frac{y}{L_0} \quad \text{and} \quad z' = \frac{z}{h_0} \tag{9.10a}$$

then, for the dimensionless horizontal velocity vector $\mathbf{v}'$, vertical velocity w', pressure p', density ρ', and temperature T', we write:

$$\mathbf{v}' = U_0 \mathbf{v}, \quad w' = \lambda U_0 w, \quad p' = p_s(0) p, \quad \rho' = \rho_s(0) \rho, \quad T' = T_s(0) T, \tag{9.10b}$$

where $\mathbf{v}'$, w', p', ρ', and T' are dependent on t', x', y', and z'.

Finally, with the Stokes relation, under the hydrostatic limit process (9.9c) we derive the following set of dimensionless quasi-hydrostatic dissipative (Q-HD) equations for $\mathbf{v}_{\mathrm{Q-HD}}$, $w_{\mathrm{Q-HD}}$, $p_{\mathrm{Q-HD}}$, $\rho_{\mathrm{Q-HD}}$, and $T_{\mathrm{Q-HD}}$:

$$S \frac{\mathrm{d}\rho_{\mathrm{Q-HD}}}{\mathrm{d}t'} + \rho_{\mathrm{Q-HD}} \left[\mathbf{D}' \cdot \mathbf{v}_{\mathrm{Q-HD}} + \frac{\partial w_{\mathrm{Q-HD}}}{\partial z'} \right] = 0; \tag{9.11a}$$

$$\rho_{\mathrm{Q-HD}} [S \frac{\mathrm{d}\mathbf{v}_{Q-HD}}{\mathrm{d}t'} + \left(\frac{1}{\mathrm{Ro}}\right) (\mathbf{k} \wedge \mathbf{v}_{\mathrm{Q-HD}}) + \left(\frac{\tau}{\gamma M^2}\right) \mathbf{D}' p_{\mathrm{Q-HD}}$$

$$= \left(\frac{1}{\mathrm{Re}_\perp}\right) \frac{\partial^2 \mathbf{v}_{\mathrm{Q-HD}}}{\partial z'^2}; \tag{9.11b}$$

$$\frac{\partial p_{\text{Q-HD}}}{\partial z'} + \tau \rho_{\text{Q-HD}} = 0; \tag{9.11c}$$

$$\rho_{\text{Q-HD}} S \frac{\mathrm{d}T_{\text{Q-HD}}}{\mathrm{d}t'} - \left[\frac{(\gamma-1)}{\gamma}\right] S \frac{\mathrm{d}p_{\text{Q-HD}}}{\mathrm{d}t'}$$
$$= \left(\frac{1}{\Pr \mathrm{Re}_\perp}\right)\left[\frac{\partial^2 T_{\text{Q-HD}}}{\partial z'^2} + (\gamma-1)\Pr M^2 \left|\frac{\partial \mathbf{v}_{\text{Q-HD}}}{\partial z'}\right|^2\right.$$
$$\left. + \tau^2 \Sigma_0 \frac{\mathrm{d}R'_S(T'_S(z'_S))}{\mathrm{d}z'_S}\right]; \tag{9.11d}$$

$$p_{\text{Q-HD}} = \rho_{\text{Q-HD}} T_{\text{Q-HD}}. \tag{9.11e}$$

In the quasi-hydrostatic dissipative equations (9.11a–e) we have

$$S\frac{\mathrm{d}}{\mathrm{d}t'} = S\frac{\partial}{\partial t'} + \mathbf{v}_{\text{Q-HD}} \cdot \mathbf{D}' + w_{\text{Q-HD}} \frac{\partial}{\partial z'},$$

where $S = L_0\Omega^0/U_0$ is the Strouhal number, $\Pr = C_p\mu_0/k_0$ the usual Prandtl number (with C_p the specific heat at pressure constant and $\mu_0 = \rho_s(0)\nu_0$, the constant dynamic viscosity) and $\mathbf{D}' = (\partial/\partial x', \partial/\partial y')$, $\mathbf{k}\cdot\mathbf{D}' = 0$. In equation (9.11d) for $T_{\text{Q-HD}}$ the dimensionless parameter Σ_0 is a measure for the (standard) heat source term. We observe that, if we assume that the Strouhal number $S \approx 1$, then the characteristic time is such that

$$\mathrm{Ro} = \frac{1}{(2\sin\varphi^\circ)} \approx 1, \tag{9.12}$$

and in such a case, unsteadiness and the Coriolis force are both operative in equation (9.11b) for the horizontal velocity vector in the convection problem – but in equation (9.11b) we have the small Mach number M. In standard atmosphere, from equation (9.11d), in particular we obtain for $T'_S(z'_S)$ (the dimensionless temperature in hydrostatic reference state), with the relation:

$$z'_S = \tau z', \tag{9.13}$$

(because $z' = z/h_0$ but $z'_S = z_S/H_S$, with $H_S = RT_s(0)/g$) according to (9.8b) the following dimensionless equation for $T_S(z'_S)$, instead of (9.3b).

$$\frac{\mathrm{d}T_S(z'_S)}{\mathrm{d}z'_S} + \Sigma_0 R'_S(T'_S(z'_S)) = 0. \tag{9.14}$$

In fact, (9.14) is an equation defining $T'_S(z'_Z)$ via $R'_S(T'_S(z'_S))$, when the judicious boundary conditions are assumed in z'_S. This determination of $T'_S(z'_S)$

is in reality a complicated problem; see, for instance, Kibel's book [7, section 1.4] – but, for our purpose here this problem has no influence, because z'_S tends to zero with τ tending to zero (when z' is fixed, see (9.13)), and $T'_S(0) \equiv 1$.

For equations (9.11a–e), from (9.7) in dimensionless form we write as thermal boundary condition at the flat ground:

$$T_{\mathrm{Q-HD}} = 1 + \tau\,\Xi(t', x', y'), \tag{9.15}$$

for $(x', y') \in D$ and $T' \equiv 1$, if $(x', y') \notin D$. Now, from the Q-HD equations (9.11a–e), with (9.15), we can derive via a Boussinesq limit:

$$\tau \to 0 \text{ and } M \to 0 \quad \text{with } \frac{\tau}{M} = \tau^* \approx 1, \tag{9.16}$$

with the following 'low Mach number asymptotics':

$$\begin{aligned}
\mathbf{v}_{\mathrm{Q-HD}} &= \mathbf{v}_{\mathrm{conv}} + \mathrm{O}(M), \quad w = w_{\mathrm{conv}} + \mathrm{O}(M), \\
p_{\mathrm{Q-HD}} &= p'_S(z'_S)[1 + M^2\pi_{\mathrm{conv}} + \cdots], \\
T_{\mathrm{Q-HD}} &= T'_S(z'_S)[1 + M\theta_{\mathrm{conv}} + \cdots], \\
\rho_{\mathrm{Q-HD}} &= \rho'_S(z'_S)[1 + M\omega_{\mathrm{conv}} + \cdots],
\end{aligned} \tag{9.17}$$

a set of hydrostatic viscous, non-adiabatic Boussinesq-convective equations, for the functions $\mathbf{v}_{\mathrm{conv}}$, w_{conv}, π_{conv}, θ_{conv}, and ω_{conv}. Namely (dropping the prime $'$), these Boussinesq Hydrostatic-Convection (BH-C) equations (*à la* Zeytounian) are written in the following form:

$$\mathbf{D} \cdot \mathbf{v}_{\mathrm{conv}} + \frac{\partial w_{\mathrm{conv}}}{\partial z} = 0; \tag{9.18a}$$

$$S\frac{\mathrm{d}\mathbf{v}_{\mathrm{conv}}}{\mathrm{d}t} + \left(\frac{1}{\mathrm{Ro}}\right)(\mathbf{k} \wedge \mathbf{v}_{\mathrm{conv}}) + \left(\frac{1}{\gamma}\right)\mathbf{D}\pi_{\mathrm{conv}} = \mathrm{Gr}_{\perp}^{-1/2}\frac{\partial^2 \mathbf{v}_{\mathrm{conv}}}{\partial z^2}; \tag{9.18b}$$

$$\frac{\partial \pi_{\mathrm{conv}}}{\partial z} + \tau^*\theta_{\mathrm{conv}} = 0; \tag{9.18c}$$

$$S\frac{\mathrm{d}\theta_{\mathrm{conv}}}{\mathrm{d}t} + \tau^*\Lambda(0)w_{\mathrm{conv}} = \left(\frac{1}{\mathrm{Pr}}\right)\mathrm{Gr}_{\perp}^{-1/2}\frac{\partial^2\theta_{\mathrm{conv}}}{\partial z^2}; \tag{9.18d}$$

$$\omega_{\mathrm{conv}} = -\theta_{\mathrm{conv}}, \tag{9.18e}$$

where

$$S\frac{\mathrm{d}}{\mathrm{d}t} = S\frac{\partial}{\partial t} + \mathbf{v}_{\mathrm{conv}} \cdot \mathbf{D} + w_{\mathrm{conv}}\frac{\partial}{\partial z}. \tag{9.19a}$$

In equation (9.18d), as stratification parameter, we have

$$\Lambda(0) = \left[\frac{(\gamma - 1)}{\gamma}\right] + \left[\frac{\mathrm{d}T_S(z_S)}{\mathrm{d}z_S}\right]_0, \tag{9.19b}$$

and instead of $1/\mathrm{Re}_\perp$ we have introduced the Grashof number defined by

$$\mathrm{Gr}_\perp(\equiv \mathrm{Re}_\perp^2) \equiv \lambda^2\,\mathrm{Gr} \quad \text{with } \mathrm{Gr} = \gamma R\frac{[h_0(\Delta T)_0/\nu_0]^2}{T_S(0)}. \tag{9.19c}$$

Indeed, the model equations (9.18a–e), with (9.19a–c), for the free circulation-convection breeze phenomena, can be considered as an inner (significant, boundary layer) degeneracy of full atmospheric NSF equations, when two limit processes, (9.9c) and (9.16), with (9.17), are performed. But, because the corresponding outer degeneracy gives the trivial zero solution (as a consequence of the local character of the free circulation-convection phenomenon), we can write the behavior of the free circulation far (when $z \uparrow \infty$, there is no outer solution) from the assumed flat ground surface $z = 0$. Namely, for model equations (9.18a–e), the following boundary conditions (see (9.7) with (9.17) and (9.16)) are obtained:

$$z \uparrow \infty: \quad (\mathbf{v}_{\mathrm{conv}}, w_{\mathrm{conv}}, \pi_{\mathrm{conv}}, \theta_{\mathrm{conv}}) \to 0; \tag{9.20a}$$

$$z = 0: \quad \mathbf{v}_{\mathrm{conv}} = w_{\mathrm{conv}} = 0, \tag{9.20b}$$

$$z = 0: \quad \theta_{\mathrm{conv}} = \tau^*\Xi(t, x, y) \text{ when } (x, y) \in D, \tag{9.20c}$$

$$z = 0: \quad (\mathbf{v}_{\mathrm{conv}}, w_{\mathrm{conv}}, \pi_{\mathrm{conv}}, \theta_{\mathrm{conv}}) \equiv 0, \quad (x, y) \notin D. \tag{9.20d}$$

We observe that the breeze is a localized (regional) phenomenon and, in fact, it is active mainly on the bounded region D at the Earth's surface.

Concerning the initial conditions at $t = 0$, for $\mathbf{v}_{\mathrm{conv}}$ and θ_{conv}, it is necessary to consider an unsteady adjustment problem if we want to study the initial, near $t = 0$, strongly unsteady transition, short-time scale phenomenon – as in [8], particular attention must be given to short-time-scale unsteady solutions near $t = 0$ (see, for instance Section 10.8). But here we have, in fact, two unsteady adjustment, short-time scale, problems!

We observe also that the order of equations (9.18a–e), with respect to z, does not allow us (because we have the reduced equation (9.18c) instead of a full unsteady equation for w) to specify a behavior condition for w_{conv} when $z \to +\infty$! But, in particular, from equation (9.18d) for θ_{conv}, when $\Lambda(0) \neq 0$, since $\theta_{\mathrm{conv}} \to 0$ when $z \to +\infty$, we should also have

$$w_{\mathrm{conv}} = 0 \quad \text{at } z = \infty; \tag{9.21}$$

this constraint (9.21), not being a boundary condition, should be satisfied automatically. On the other hand, from (9.18a) with conditions (9.20a, b) we derive also a second constraint:

$$\mathbf{D} \cdot \int_0^\infty \mathbf{v}_{\text{conv}} = 0. \tag{9.22}$$

Consequently, the presence in equation (9.18d) for θ_{conv} of the term $\tau^* \Lambda(0)$, which in general is not small compared with the other terms in (9.18d), leads in a stably stratified standard atmosphere to the formation of a perceptible compensating flow (anti-breeze) which exists above the main breeze, a phenomenon well-known from observations in nature (see, for example, the book by Khrgian [9]).

From the BH-C system of equations (9.18a–e) we can derive two simplified systems for the breeze phenomena, and the reader can verify the accuracy of both these simplified model equations.

First, in the two-dimensional case ($\mathbf{v}_{\text{conv}} \equiv u\mathbf{i}$, $\partial/\partial y = 0$), when the effect of the rotation is not taken into account and in boundary condition (9.20c) on $z = 0$ for the function Ξ, we have the following simple form (when $t > 0$),

$$\Xi = [\alpha + \beta x] \sin t, \tag{9.23a}$$

where α and β are two specified constants. Quantity β can be interpreted as some characteristic gradient of the underlying-surface temperature, α being a maximum difference between the temperature of land and sea, divided by the characteristic length of the phenomenon. This simplified form (9.23a) is satisfied best in some sufficiently small region (and in this case the Coriolis force is not active) in the vicinity of the shore line, to the order of a few kilometers in both directions.

Using (9.23a), the 2D solution with above assumed simplifications is sought in the form

$$u = u(t, z), \quad \theta_{\text{conv}} = \vartheta(t, z) + x\sigma(t, z), \quad \left(\frac{1}{\gamma}\right) \pi_{\text{conv}} = \varphi(t, z) + x\psi(t, z), \tag{9.23b}$$

and in such a case,

$$\mathbf{D} \cdot \mathbf{v}_{\text{conv}} \equiv 0, \quad w_{\text{conv}} = 0, \tag{9.23c}$$

because $w_{\text{conv}} = 0$ at $z = 0$. Substitution of (9.23b, c) in (9.18b–d) yields a simplified system in which the functions are not a function of x:

$$S\frac{\partial u}{\partial t} + \psi = \left(\frac{1}{\sqrt{\text{Gr}_\perp}}\right) \frac{\partial^2 u}{\partial z^2}; \tag{9.24a}$$

$$S\frac{\partial \vartheta}{\partial t} + u\sigma = \left(\frac{1}{\sqrt{\mathrm{Gr}_\perp}}\right)\frac{\partial^2 \vartheta}{\partial z^2}; \tag{9.24b}$$

$$S\frac{\partial \sigma}{\partial t} = \left(\frac{1}{\sqrt{\mathrm{Gr}_\perp}}\right)\frac{\partial^2 \sigma}{\partial z^2}; \tag{9.24c}$$

$$\frac{\partial \varphi}{\partial z} = \left(\frac{\tau^*}{\gamma}\right)\vartheta; \tag{9.24d}$$

$$\frac{\partial \psi}{\partial z} = \left(\frac{\tau^*}{\gamma}\right)\sigma. \tag{9.24e}$$

The system of equations (9.24a–e) should be solved subject to the following boundary conditions:

$$u = 0, \quad \vartheta = \tau^*\alpha \sin t, \quad \sigma = \tau^*\beta \sin t \quad \text{for } z = 0; \tag{9.25a}$$

$$u = 0, \quad \vartheta = 0, \quad \sigma = 0, \quad \varphi = 0, \quad \psi = 0 \quad \text{for } z = \infty. \tag{9.25b}$$

It is clear that the system of equations (9.24a–e) represents a chain of interactions between various physical effects operating in the breeze mechanism, namely:

$$\begin{array}{ccc} (9.24\text{c}) \Rightarrow (9.24\text{e}) \Rightarrow (9.24\text{a}) & & \Rightarrow (9.24\text{d}) \\ & \Downarrow \quad \Uparrow & \\ & (9.24\text{b}) & \end{array}$$

First, from equation (9.24c), for σ, it follows that the horizontal temperature gradient is produced in the atmosphere due to heating of air by conduction of heat from the underlying surface. Then, equation (9.24e) shows that the appearance of this horizontal temperature gradient leads to the appearance of a horizontal pressure gradient. Later, equation (9.24a) indicates that the pressure gradient induces the onset of (breeze) wind; here an important role is played by eddy diffusion. Further, equation (9.24b) demonstrates the opposite effect, exerted by the wind on the temperature field, where the nonlinear term $u\sigma$ represents a negative heat source in the heat conduction equation (9.24b), and it is precisely this term which describes the wind transport of heat. Finally from (9.24d), a main pressure field φ is induced via the temperature field, in particular, the solution for φ, via (9.24d), gives the daily pressure variation at the underlying-surface level ($z = 0$).

Solutions of the above set of equations show that the structure of this breeze model in vicinity of the shore (the x-axis is directed normal to the shore and the coordinate origin ($x = y = 0$) is taken at the shore line on the asumption of a straight and infinite shore, the y-axis being along the shore) is similar to the wind and temperature progressive wave damping out with altitude. It is important that from this model we established the instant when the wind appears at the ground on the onset of breeze – at this instant, $\partial u/\partial z|_{z=0}$ – and thus it is found that the breeze lags behind the variation in the soil temperature by six hours (this is a rough result, since the observations yield from two to five hours). More important, this model points out the cause of this lag, which is inertia moving air.

A second simplified model is linked with a local wind arising above a slope (as in Prandtl's 1944 example, mentioned above) the steepness of which is not less than several degrees and the deviation of the surface temperature of which, from the temperature of the free atmosphere at the same altitude, exceeds in absolute value several degrees centigrades and changes little along the slope; this is a so-called 'slope wind' on the assumption that it develops in an atmosphere at rest.

Our purpose below is to derive (in a rational way) a specific system of model equations for this slope wind, from the equations (9.18a–e); for this purpose, the first problem is the introduction of the topography in equations (9.18a–e)?.

The slope is assumed to be given by the equations

$$z = \left(\frac{|F|}{h_0}\right) F(x, y) \quad \text{with } |F| = \max(F), \text{ when } (x, y) \in D, \tag{9.26a}$$

where h_0 is given by (9.8a).

From Cartesian coordinates (x, y, z) we pass to curvilinear coordinates (ξ, η, ζ) by

$$\xi = x, \quad \eta = y, \quad \zeta = z - F_0 F(\xi, \eta), \quad F_0 \equiv \left(\frac{|F|}{h_0}\right), \tag{9.26b}$$

and we write

$$\frac{\partial}{\partial z} = \frac{\partial}{\partial \zeta}; \tag{9.27a}$$

$$\frac{\partial}{\partial x} = \frac{\partial}{\partial \xi} - F_0 \left(\frac{\partial F}{\partial \xi}\right) \frac{\partial}{\partial \zeta}; \tag{9.27b}$$

$$\frac{\partial}{\partial y} = \frac{\partial}{\partial \eta} - F_0 \left(\frac{\partial F}{\partial \eta}\right) \frac{\partial}{\partial \zeta}; \tag{9.27c}$$

$$\mathbf{D} \cdot \mathbf{v}_{\text{conv}} + \frac{\partial w_{\text{conv}}}{\partial z} = \frac{\partial u_{\text{sl}}}{\partial \xi} + \frac{\partial v_{\text{sl}}}{\partial \eta} + \frac{\partial \omega_{\text{sl}}}{\partial \zeta}; \tag{9.27d}$$

$$S\frac{\mathrm{d}}{\mathrm{d}t} = S\frac{\partial}{\partial t} + \mathbf{v}_{\mathrm{sl}} \cdot \mathbf{D} + \frac{\partial \omega_{\mathrm{sl}}}{\partial \zeta}, \tag{9.27e}$$

where

$$\omega_{\mathrm{sl}} = w_{\mathrm{sl}} - F_0\left[\left(\frac{\partial F}{\partial \xi}\right)u_{\mathrm{sl}} + \left(\frac{\partial F}{\partial \eta}\right)v_{\mathrm{sl}}\right] = w_{\mathrm{sl}} - F_0(\mathbf{v}_{\mathrm{sl}} \cdot \mathbf{D}F), \tag{9.27f}$$

$$\mathbf{D} = \left(\frac{\partial}{\partial \xi}, \frac{\partial}{\partial \eta}\right) \quad \text{and} \quad \mathbf{v}_{\mathrm{sl}} = (u_{\mathrm{sl}}, v_{\mathrm{sl}}). \tag{9.27g}$$

Then elementary transformations yield, instead of the system of equations (9.18a–d), the following equations for u_{sl}, v_{sl}, ω_{sl}, π_{sl}, and θ_{sl}, which are functions of the time t, and curvilinear coordinates, ξ, η, ζ:

$$\mathbf{D} \cdot \mathbf{v}_{\mathrm{sl}} + \frac{\partial \omega_{\mathrm{sl}}}{\partial \zeta} = 0; \tag{9.28a}$$

$$S\frac{\mathrm{d}\mathbf{v}_{\mathrm{sl}}}{\mathrm{d}t} + \left(\frac{1}{\mathrm{Ro}}\right)(\mathbf{k} \wedge \mathbf{v}_{\mathrm{sl}}) + \left(\frac{1}{\gamma}\right)\mathbf{D}\pi_{\mathrm{sl}} - \left(\frac{F_0}{\gamma}\right)\tau^*\mathbf{D}F\theta_{\mathrm{sl}}$$
$$= \mathrm{Gr}_{\perp}^{-1/2}\frac{\partial^2 \mathbf{v}_{\mathrm{sl}}}{\partial \zeta^2}; \tag{9.28b}$$

$$\frac{\partial \pi_{\mathrm{sl}}}{\partial \zeta} + \tau^*\theta_{\mathrm{sl}} = 0; \tag{9.28c}$$

$$S\frac{\mathrm{d}\theta_{\mathrm{sl}}}{\mathrm{d}t} + \tau^*\Lambda(0)[\omega_{\mathrm{sl}} + F_0(\mathbf{v}_{\mathrm{sl}} \cdot \mathbf{D}F)] = \left(\frac{1}{\mathrm{Pr}}\right)\mathrm{Gr}_{\perp}^{-1/2}\frac{\partial^2 \theta_{\mathrm{sl}}}{\partial z^2}. \tag{9.28d}$$

The boundary conditions associated with (9.28a–d) are

$$\zeta = 0: \quad \mathbf{v}_{\mathrm{sl}} = 0; \quad \omega_{\mathrm{sl}} = 0;$$
$$\theta_{\mathrm{sl}} = \tau^*\Xi(t, \xi, \eta), \quad \text{and for } t > 0,\ (\xi, \eta) \in D, \tag{9.29a}$$

$$\zeta \to \infty: \quad |\xi^2 + \eta^2| \to \infty; \quad \mathbf{v}_{\mathrm{sl}} \to 0; \quad \pi_{\mathrm{sl}} \to 0;$$
$$\theta_{\mathrm{sl}} \to 0 \Rightarrow \omega_{\mathrm{sl}} \to 0. \tag{9.29b}$$

However, in the case of the slope wind we see that

$$\frac{1}{\sqrt{\mathrm{Gr}_{\perp}}} \ll 1, \tag{9.30a}$$

and the following (outer) limit process is considered below:

$$\frac{1}{\sqrt{\mathrm{Gr}_\perp}} \to 0 \quad \text{with } t, \xi, \eta, \zeta \text{ fixed.} \tag{9.30b}$$

With (9.30a, b), we obtain only a trivial zero outer solution

$$\mathbf{v}_{\mathrm{sl}}^{\mathrm{out}} = 0, \quad \pi_{\mathrm{sl}}^{\mathrm{out}} = 0, \quad \theta_{\mathrm{sl}}^{\mathrm{out}} = 0, \quad \omega_{\mathrm{sl}}^{\mathrm{out}} = 0. \tag{9.31}$$

As a consequence, it is necessary to derive associated inner equations, via the introduction of an inner vertical coordinate and an inner vertical velocity

$$\zeta^* = \frac{\zeta}{[\mathrm{Gr}]^{-1/4}} \quad \text{and} \quad \omega_{\mathrm{sl}}^* = \frac{\omega_{\mathrm{sl}}}{[Gr]^{-1/4}}, \tag{9.32a}$$

and consider an inner limit process

$$\frac{1}{\sqrt{\mathrm{Gr}_\perp}} \to 0 \quad \text{with } t, \xi, \eta, \zeta^* \text{ fixed.} \tag{9.32b}$$

As a result of (9.32a, b) we derive for the limit values $\mathbf{v}_{\mathrm{sl}}^*$, ω_{sl}^*, π_{sl}^*, θ_{sl}^*, as function of t, ξ, η and ζ^*, the following system of model boundary layer equations which governs the slope wind local phenomenon:

$$\mathbf{D} \cdot \mathbf{v}_{\mathrm{sl}}^* + \frac{\partial \omega_{\mathrm{sl}}^*}{\partial \zeta^*} = 0; \tag{9.33a}$$

$$S\frac{\mathrm{d}\mathbf{v}_{\mathrm{sl}}^*}{\mathrm{d}t} + \left(\frac{1}{\mathrm{Ro}}\right)(\mathbf{k} \wedge \mathbf{v}_{\mathrm{sl}}^*) + \left(\frac{1}{\gamma}\right)\mathbf{D}\pi_{\mathrm{sl}} - \left(\frac{F_0}{\gamma}\right)\tau^*\mathbf{D}F\theta_{\mathrm{sl}}^*$$
$$= \frac{\partial^2 \mathbf{v}_{sl}^*}{\partial \zeta^{*2}}; \tag{9.33b}$$

$$\frac{\partial \pi_{\mathrm{sl}}^*}{\partial \zeta^*} = 0; \tag{9.33c}$$

$$S\frac{\mathrm{d}\theta_{\mathrm{sl}}^*}{\mathrm{d}t} + \tau^*\Lambda(0)F_0(\mathbf{v}_{\mathrm{sl}} \cdot \mathbf{D}F) = \left(\frac{1}{\mathrm{Pr}}\right)\frac{\partial^2 \theta_{\mathrm{sl}}^*}{\partial \zeta^{*2}}. \tag{9.33d}$$

But from (9.33c), as in classical boundary layer theory,

$$\pi_{\mathrm{sl}}^* \equiv \pi_{\mathrm{sl}}^*(t, \xi, \eta) = \pi_{\mathrm{sl}}^{\mathrm{out}}\big|_{\zeta=0} = 0, \tag{9.34a}$$

and from (9.30a),

$$\frac{1}{\sqrt{\mathrm{Gr}_\perp}} \ll 1 \Rightarrow \left(\frac{h_0}{L_0}\right)^2 \mathrm{Re} = \frac{h_0^2 U_0}{\nu_0 L_0} \gg 1,$$

or, according to (9.8e),

$$L_0 \ll h_0^2 \left[\frac{(\Delta T)_0}{\nu_0}\right]\left[\frac{\gamma R}{T_s(0)}\right]^{1/2}. \tag{9.34b}$$

We observe that, when (9.34b) is realized, it is more judicious to determine the reference time t^0 such that the Strouhal number $S = (L_0/U_0)/t_0 \approx 1$, and in this case the relation (9.8c), for Ro, is not realized, – Ro being a high number relative to 1 (and the Coriolis force does not play a role). Finally, from (9.33a–d) with (9.34a, b) we derive for our slope wind the following three model equations, when the term with Coriolis term is neglected:

$$\mathbf{D}\cdot\mathbf{v}_{\mathrm{sl}}^* + \frac{\partial \omega_{\mathrm{sl}}^*}{\partial \zeta^*} = 0; \tag{9.35a}$$

$$S\frac{\mathrm{d}\mathbf{v}_{\mathrm{sl}}^*}{\mathrm{d}t} - \left(\frac{F_0}{\gamma}\right)\tau^*\mathbf{D}F\theta_{\mathrm{sl}}^* = \frac{\partial^2 \mathbf{v}_{\mathrm{sl}}^*}{\partial \zeta^{*2}}; \tag{9.35b}$$

$$S\frac{\mathrm{d}\theta_{\mathrm{sl}}^*}{\mathrm{d}t} + \tau^*\Lambda(0)F_0(\mathbf{v}_{\mathrm{sl}}^*\cdot\mathbf{D}F) = \left(\frac{1}{\mathrm{Pr}}\right)\frac{\partial^2\theta_{\mathrm{sl}}^*}{\partial\zeta^{*2}}. \tag{9.35c}$$

In this slope wind system (9.35a–c) we have that u_{sl}^*, v_{sl}^* and $\omega_{\mathrm{sl}}^* = w_{\mathrm{sl}}^* - F_0(\mathbf{v}_{\mathrm{sl}}^*\cdot\mathbf{D}F)$, are three components of the slope wind velocity along curvilinear coordinates ξ, η, and z. On the other hand, we have two new terms in equations (9.35b) of the motion for u_{sl}^* and v_{sl}^*,

$$-\left(\frac{F_0}{\gamma}\right)\tau^*\frac{\partial F}{\partial \xi}\theta_{\mathrm{sl}}^*, \quad \text{and} \quad -\left(\frac{F_0}{\gamma}\right)\tau^*\frac{\partial F}{\partial \eta}\theta_{\mathrm{sl}}^*,$$

which take into account the buoyancy effect. The term proportional to $1/\mathrm{Ro}$ in equation (9.35b) does not have an important effect, because of the estimation (9.34b) for L_0 and has been neglected. A third new term in equation (9.35c) for θ_{sl}^* describes the wind transport of the heat flux component associated with the stratification of the standard atmosphere,

$$\tau^*\Lambda(0)F_0(\mathbf{v}_{\mathrm{sl}}^*\cdot\mathbf{D}F) \equiv \tau^*\Lambda(0)F_0\left(\mathbf{u}_{\mathrm{sl}}^*\frac{\partial F}{\partial \xi} + \mathbf{v}_{\mathrm{sl}}^*\frac{\partial F}{\partial \eta}\right).$$

As boundary conditions, for the system (9.35a–c) we have:

$$\mathbf{v}_{\mathrm{sl}}^* = 0 \quad \text{and} \quad \omega_{\mathrm{sl}}^* = 0 \quad \text{for } \zeta^* = 0, \tag{9.36a}$$

$$\theta_{\mathrm{sl}}^* = \tau^*\Xi(t,\xi,\eta) \quad \text{for } \zeta^* = 0 \text{ and when } t > 0, \tag{9.36b}$$

$$\mathbf{v}^*_{\rm sl},\quad \omega^*_{\rm sl},\quad \theta^*_{\rm sl} \to 0 \quad \text{for } |\xi^2+\eta^2| \to \infty, \tag{9.36c}$$

$$\mathbf{v}^*_{\rm sl},\quad \frac{\partial \omega^*_{\rm sl}}{\partial \zeta^*},\quad \theta^*_{\rm sl} \to 0 \quad \text{for } \zeta^* \to \infty. \tag{9.36d}$$

The Prandtl example, (9.1a, b) with (9.1c), is a simplified case of the derived model (9.35a–c) with (9.36b–d).

9.3 Model Problem for the Local Thermal Prediction – A Triple Deck Viewpoint

Below we consider only a two-dimensional steady local thermal problem and rewrite the thermal boundary condition (9.7) in the following form:

$$\frac{T}{T_s(0)} = 1 + \tau\,\Xi\left(\frac{x}{l^0}\right) \quad \text{on } z=0, \tag{9.37}$$

where $\Xi \equiv 0$ when $|x/l^0| \le 1$, where l^0 is the local horizontal length scale. Far upstream, when $x \to -\infty$ and $\Xi \equiv 0$, we assume that we have a basic undisturbed flow which is characterized by an Ekman layer profile:

$$U_{\rm Ek}\left(\frac{x}{L_0}, \frac{(z/l^0)}{\kappa_0}\right) = U_G\left(\frac{x}{L_0}\right)\left\{1-\exp\left[-\frac{(z/l^0)}{\kappa_0}\right]\cos\left[-\frac{(z/l^0)}{\kappa_0}\right]\right\}, \tag{9.38a}$$

where

$$\kappa_0 = \frac{[\Omega_0 \sin\varphi_0/\nu_0]^{-1/2}}{l^0} \equiv \left[\frac{\mathrm{Re}_l}{2\mathrm{Ro}_l}\right]^{-1/2}, \tag{9.38b}$$

with[4]

[4] When the 'global' Rossby number is formed via L_0, $\mathrm{Ro}_L = (U_0/L_0)/2\Omega_0 \sin\varphi_0$, then in basic atmospheric flow, undisturbed by the local thermal condition (9.37), we assume that $\mathrm{Ro}_L \ll 1$ and because the Mach number M is also small in equation (9.11b), we consider the double limit process

$$\mathrm{Ro}_L \to 0 \text{ and } M \to 0, \text{ such that } \mathrm{Ro}_L/\gamma M = \mathrm{Go} = \mathrm{O}(1) \text{ and } \mathrm{Re}_\perp = \mathrm{O}(1).$$

As a result we obtain at the leading order (outer limit), when S, Go, τ^*, t, x, y, z are fixed, the so-called 'geostrophic relation'. For instance, when in equation (9.11a), $\mathrm{Re}_\perp = \mathrm{O}(1)$ and $S = \mathrm{O}(1)$ are both fixed, we have [5]:

$$\mathbf{k} \wedge \mathbf{v}G + \tau^* \,\mathrm{Go}\, \mathbf{D}p_G = 0. \tag{*}$$

The above geostrophic relation (*) is strongly singular near the initial time, $t = 0$, and in the vicinity of the ground, $z = 0$. In the vicinity of $z = 0$, when we assume that in (9.11a) $\mathrm{Re}_\perp \gg 1$ such that

$$\mathrm{Re}_l = \frac{U_0 l^0}{\nu_0} \quad \text{and} \quad \mathrm{Ro}_l = \frac{(U_0/l^0)}{2\Omega_0 \sin\varphi_0}, \tag{9.38c}$$

the local Reynolds and Rossby numbers, based on the local horizontal length scale l^0.

If in the local thermal problem we non-dimensionalize the horizontal and vertical coordinates with l^0: $x' = x/l^0$ and $z' = z/l^0$, then in the dimensionless local problem appears, also, a local Boussinesq number

$$\mathrm{Bo}_l = \frac{g l^0}{R T_s(0)}. \tag{9.38d}$$

If $l^0 \approx 10^3$ m, then $\mathrm{Bo}_l \ll 1$ and in such a case, $\mathrm{Re}_l \gg 2\mathrm{Ro}_l \gg 1$. Therefore, in this case we can assume

$$2\mathrm{Ro}_l = (\mathrm{Re}_l)^{-1/a} \Rightarrow \kappa_0 = (\mathrm{Re}_l)^{-1/m}, \tag{9.39a}$$

with $(0 < a < 1)$ and

$$m = \frac{(2-a)}{(1-a)} > 2. \tag{9.39b}$$

For example, if $U_0 \approx 10$ m/sec, $\nu_0 \approx 5$ m^2/sec and $f^0 \equiv 2\Omega_0 \sin\varphi_0 \approx 10^{-4}$ 1/sec, the considered case, $l^0 \approx 10^3$ m, leads to $m = 5$. For this case, we have the possibility to use as

$$l^0 \approx \left(\frac{U_0}{g}\right)\left[\frac{R T_s(0)}{\gamma}\right]^{1/2} \Rightarrow \frac{\mathrm{Bo}_l}{M} = B^* \equiv 1, \tag{9.39c}$$

and the Boussinesq approximation is correct.

$$\mathrm{Ro}_L/\mathrm{Re}_\perp \equiv \mathrm{Ek}_\perp = E^* \mathrm{Ro}_L^2 \text{ with } E^* = \mathrm{O}(1), \tag{**}$$

then, instead of the geostrophic relation (*), when (inner limit): $\mathrm{Ro}_L \to 0$, with E^*, Go, t, x, y, and inner vertical coordinate, $z/\mathrm{Ro} = z^*$ fixed, we derive the equation

$$\mathbf{k} \wedge \mathbf{v}_G + \tau^* \mathrm{Go}\, \mathbf{D} p_G = E^* \frac{\partial^2 \mathbf{v}_G}{\partial z^{*2}}. \tag{***}$$

Via matching between outer and inner limits, we then have the possibility, with (*) and (***), to derive the so-called *Ackerblom's model problem*, and obtain the Ekman layer profile (9.37). In [5, chapter vii, §28], the reader can find a consistent asymptotic derivation of this *Ackerblom's model problem*, which allows us to derive a boundary condition on the ground, taking into account the influence of the Ekman layer, on the main quasi-geostrophic equation for p_G, in (*).

In geostrophic wind $U_G(x/L_0)$, L_0 is a 'global' horizontal length scale and this 'global' horizontal length scale is also present in the Ekman layer U_{Ek} profile (9.38a).

The value $m = 5$ is the same as the one used by Smith and coworkers [10, 11] for the flow over an isolated two-dimensional short hump in the boundary layer. In [12], the Boussinesq stratified fluid flow is also considered.

When $m = 5$ we have that a typical triple deck case exist:

$$\frac{l^0}{L_0} \approx \mathrm{Re}_L^{-3/8} \quad \text{where} \quad \mathrm{Re}_L = \frac{L_0 U_0}{\nu_0}. \tag{9.39d}$$

(in [13, pp. 211–220], the reader can find a more general approach). Now, according to the Boussinesq approximation, taking into account (9.39b), we have the possibility to formulate the following dimensionless local steady thermal problem (where the main small parameter is κ_0^5 and defined by (9.39a):

$$u\frac{\partial u}{\partial x} + w\frac{\partial u}{\partial z} + \left(\frac{1}{\gamma}\right)\frac{\partial \pi}{\partial x} = \kappa_0^5\left(\frac{\partial^2 u}{\partial x} + \frac{\partial^2 u}{\partial z^2}\right), \tag{9.40a}$$

$$u\frac{\partial w}{\partial x} + w\frac{\partial w}{\partial z} + \left(\frac{1}{\gamma}\right)\frac{\partial \pi}{\partial z} - \left(\frac{1}{\gamma}\right)B^*\theta = \kappa_0^5\left(\frac{\partial^2 w}{\partial x} + \frac{\partial^2 w}{\partial z^2}\right), \tag{9.40b}$$

$$\frac{\partial u}{\partial x} + \frac{\partial w}{\partial z} = 0, \tag{9.40c}$$

$$u\frac{\partial \theta}{\partial x} + w\frac{\partial \theta}{\partial z} + B^*\Lambda(0)w = \left(\frac{1}{\mathrm{Pr}}\right)\kappa_0^5\left(\frac{\partial^2 \theta}{\partial x} + \frac{\partial^2}{\partial z^2}\right), \tag{9.40d}$$

$$\omega = -\theta, \tag{9.40e}$$

when we assume, as in (9.16), that (see also (9.8e))

$$\tau \to 0, \quad M \to 0 \quad \text{and} \quad \frac{\tau}{M} = \tau^* \approx 1. \tag{9.41}$$

As boundary conditions we write:

$$z = 0: \quad u = w = 0, \quad \theta = \tau^*\Xi(x), \quad 0 \le x \le 1, \tag{9.42a}$$

and also, according to our formulation of the considered interaction problem between the local thermal spot and Ekman atmospheric layer,

$$x \to -\infty: \quad u \to 1 - \exp\left[-\frac{z}{\kappa_0}\right]\cos\left[-\frac{z}{\kappa_0}\right] \equiv U^\infty\left(\frac{z}{\kappa_0}\right), \quad w, \pi, \theta \to 0, \tag{9.42b}$$

and we note that:

$$\text{if } \frac{z}{\kappa_0} \to \infty, \text{ then } u \to 1, \text{ for } x \to -\infty \tag{9.42c}$$

$$\text{if } \frac{z}{\kappa_0} \to 0, \text{ then } u \approx \frac{z}{\kappa_0}, \text{ for } x \to -\infty. \tag{9.42d}$$

Now, if we need to take into account the boundary conditions on the ground $z = 0$, then it is necessary to introduce an inner coordinate

$$z^* = \frac{z}{\kappa_0^{\alpha}}, \quad \alpha > 1, \quad \text{and in this case } u \approx \kappa_0^{\alpha-1} z^* \text{ for } x \to -\infty. \tag{9.43a}$$

From equation (9.40a) for u we verify that, if $u \approx \kappa_0^{\alpha-1} u^*(x, z^*)$, then for the consistency $(\kappa_0^{\alpha-1} u^* \partial u^*/\partial z^* + \cdots = \kappa_0^{5-2\alpha} \partial^2 u^*/\partial z^{*2} + \cdots)$, it is necessary to assume that

$$\alpha - 1 = 5 - 2\alpha \Rightarrow \alpha = 2. \tag{9.43b}$$

Finally, it is clear that three vertical coordinates (z) are necessary for the asymptotic triple deck analysis (when $\kappa_0 \to 0$) of the system (9.40a–e). Namely:

(a) z, for an 'upper non-viscous region', where

$$u \approx u_{\text{up}} \to 1 \quad \text{when } x \to -\infty, \tag{9.44a}$$

(b) $\zeta = z/\kappa_0$, for a 'middle, intermediate, region', where

$$u \approx u_m \to 1 - e^{\zeta} \cos\zeta \equiv U^{\infty}(\zeta) \quad \text{when } x \to -\infty, \tag{9.44b}$$

(c) $z^* = z/\kappa_0^2$, for a 'lower, wall, viscous region', where

$$u \approx \kappa_0 u^* \quad \text{and} \quad u^* \to z^* \quad \text{when } x \to -\infty. \tag{9.44c}$$

For the other case, when $l^0/L_0 < \text{Re}_L^{-3/8}$ or $l^0/L_0 > \text{Re}_L^{-3/8}$, it is necessary to apply a different asymptotic analysis (see, for example, the approach by Smith et al. in [14]). But the case $m = 6$ and $m = 4$ can be analyzed from the system (9.40a–e). For the case $m = 3$ it is necessary to start from another problem, where the Boussinesq approximation does not emerge – in this case we have $l^0 \approx 10^4$ m and we may neglect again the Coriolis terms in the local, but non-Boussinesq, equations!

Concerning the analysis of the so-called 'triple-deck' structure, the reader can find in [15, chapter 12], a detailed account of this triple deck approach, when a singular – different from the classical *à la* Prandtl – direct coupling between boundary layer and non-viscous fluid flow, coupling of a three layered structure is necessary!

Below, we give only the main result of this triple deck analysis. Namely, in wall, viscous, lower region, where the significant vertical coordinate is z^*

we derive, instead of equations (9.40a–e), for the function u^*, v^*, and θ^*, the following leading-order system:

$$u^*\frac{\partial u^*}{\partial x}+w^*\frac{\partial w^*}{\partial z^*}+\left(\frac{1}{\gamma}\right)B^*\int_{\infty}^{z^*}\left(\frac{\partial\theta}{\partial x}\right)\mathrm{d}z^* + \left(\frac{1}{\gamma}\right)\frac{\mathrm{d}P(x)}{\mathrm{d}x}=\frac{\partial^2 u^*}{\partial z^{*2}}; \tag{9.45a}$$

$$\frac{\partial u^*}{\partial x}+\frac{\partial w^*}{\partial z^*}=0; \tag{9.45b}$$

$$u^*\frac{\partial\theta^*}{\partial x}+w^*\frac{\partial\theta^*}{\partial z^*}=\left(\frac{1}{\mathrm{Pr}}\right)\frac{\partial^2\theta^*}{\partial z^{*2}}, \tag{9.45c}$$

with

$$z^*=0:\quad u^*=w^*=0,\quad \theta^*=\tau^*\Xi(x),\quad 0\le x\le 1, \tag{9.46a}$$

$$z^*\to\infty:\quad u^*\to z^*,\quad w^*\to 0,\quad \theta^*\to 0,\quad P(x)\to 0,\quad A(x)\to 0,\quad \frac{\mathrm{d}A}{\mathrm{d}x}\to 0, \tag{9.46b}$$

$$x\to-\infty:\quad u^*\to z^*+A(x),\quad w^*\to -z^*\frac{\mathrm{d}A}{\mathrm{d}x},\quad \theta^*\to 0, \tag{9.46c}$$

after matching with the middle deck (region via $\zeta\to 0 \Leftrightarrow z^*\to\infty$). To be precise, in the middle deck as solution we have

$$u_m=A(x)\frac{\mathrm{d}U^{\infty}(\zeta)}{\mathrm{d}\zeta}\quad\text{and}\quad w_m=-\left(\frac{\mathrm{d}A(x)}{\mathrm{d}x}\right)U^{\infty}(\zeta), \tag{9.47}$$

which represents simply a vertical displacement of the streamline through a distance $-\kappa_0 A(x)$, the function $A(x)$ being related to the pressure π_{up} in the upper deck by

$$\left[\frac{\partial^2}{\partial x^2}+\frac{\partial^2}{\partial z^2}+K_0^2\right]\frac{\partial\pi_{\mathrm{up}}}{\partial x}=0,\quad K_0^2=\left(\frac{B^{*2}}{g}\right)\Lambda(0), \tag{9.48a}$$

with at $z=0$:

$$\frac{\partial}{\partial x}\left[\frac{\partial\pi_{\mathrm{up}}}{\partial z}\right]=\gamma\left[K_0^2\frac{\mathrm{d}A(x)}{\mathrm{d}x}+\frac{d^3A(x)}{\mathrm{d}x^3}\right]. \tag{9.48b}$$

On the other hand, the flow in the upper deck is driven by an outflow from the middle deck, and far from the above solution (9.47), for w_m, we have

$$\lim_{\zeta\to\infty} w_m(x,\zeta) = -\frac{dA}{dx}. \tag{9.49}$$

In the middle deck we have also

$$U^\infty(\zeta) \approx \zeta \quad \text{and} \quad \frac{dU^\infty(\zeta)}{d\zeta} \approx 1 \quad \text{when } \zeta \to 0. \tag{9.50}$$

Finally, in the lower deck, associated with the problem (9.45a–c), (9.46a–c), we have the following relation between π^* and θ^*:

$$\frac{\partial \pi^*}{\partial z^*} = B^*\theta^* \Rightarrow \pi^* = B^* \int_\infty^{z^*} \theta^*\, dz^* + P(x). \tag{9.51}$$

The specification of the model problem (9.45a–c), (9.46a–c), with (9.49), (9.50) and (9.51), is completed by the relations (9.48a, b) between $P(x)$ and $A(x)$, since

$$\pi_{\text{up}}(x,0) \equiv P(x). \tag{9.52}$$

The well-known interpretation of (9.52) is that the pressure $P(x)$ driving the flow in the lower deck is itself induced in the main (upper) stream, i.e., the upper deck, by the displacement thickness of the lower deck transmitted through the middle deck by the passive effect of displacement of the streamlines. The strong singular self-induced coupling arises because the problem (9.45a–c), (9.46a–c) to be solved in the lower viscous deck (layer) does not accept $P(x)$ as data known prior to the resolution, as is the case in the framework of the classical Prandtl boundary layer problem. On the contrary, this pressure perturbation $P(x)$ must be calculated at the same time as the velocity components u^* and w^*, as well as the temperature perturbation θ^*. Nevertheless, it must be emphasized that $P(x)$ is related to $A(x)$! This relation is explicit in the linear case, when the parameter τ^*, in condition (9.46a) for θ^*, is considered as a small parameter. We observe also that via a Fourier transform

$$F_k(f(x)) \Rightarrow f^F(k),$$

instead of (9.48b) we obtain the following relation between $A^F(k)$ and $P^F(k)$:

$$A^F(k) = \left[\frac{iN^0}{\gamma}\right]\left[\frac{P^F(k)}{(K_0^2 - k^2)}\right], \tag{9.53a}$$

where

$$N^0 = i[k^2 - K_0^2]^{1/2} \quad \text{if } |k| > K_0 \quad \text{and} \quad [K_0^2 - k^2]^{1/2} \quad \text{if } |k| < K_0. \tag{9.53b}$$

Here we have applied (in the upper region) the standard radiation condition for $z \to \infty$, choosing the sign of N for $|k| < K_0$, so that the wave modes carry energy only upwards.

9.4 Free Convection over a Curved Surface – A Singular Problem

Below, a simplified case considered by Schlichting [16] in 1932 has been generalized to a periodic nonlinear free convection over a curvilinear surface. This study leads to the conclusion that, in the case of small surface slope, the convection flow above the surface exhibits a special structure; steady secondary flows are generated above the surface, throughout the region where the convection takes place. Moreover, it appears that these secondary flows do not depend on viscosity and thermal conductivity. Rather they create a supplementary temperature perturbation on the curved surface and a velocity field far from this surrface.

These results were obtained in 1961 in Moscow Meteo-Center (during my research in I. A. Kibel's Dynamic Meteorology Department) and published only in 1968 [17]. For a two-dimensional case the dimensionless starting equations are a particular case of the model equations (9.33a–d), with

$$\mathrm{Ro} = \infty, \quad S = \mathrm{Pr} = 1 \quad \text{and} \quad \Lambda(0) = 0, \tag{9.54}$$

and in such a case we write for the dimensionless velocity components $u(t,s,n)$ and $w(t,s,n)$, and perturbation of the temperature $\theta(t,s,n)$, the following equations:

$$\frac{\partial u}{\partial t} + \beta\left[u\frac{\partial u}{\partial s} + w\frac{\partial u}{\partial n}\right] = \Gamma\theta + \frac{\partial^2 u}{\partial n^2}, \tag{9.55a}$$

$$\frac{\partial u}{\partial s} + \frac{\partial w}{\partial n} = 0, \tag{9.55b}$$

$$\frac{\partial \theta}{\partial t} + \beta\left[u\frac{\partial \theta}{\partial s} + w\frac{\partial \theta}{\partial n}\right] = \frac{\partial^2 \theta}{\partial n^2}, \tag{9.55c}$$

where β is a small parameter which is a measure of the effect of the nonlinear terms and also the supplementary surface perturbation temperature, θ and velocity component u far from the curved surface. The boundary conditions for the system (9.55a–c) are

$$n = 0\colon \quad u = w = 0, \quad \theta = \cos t + \beta A(s), \tag{9.56a}$$

$$n \to \infty\colon \quad u \to \beta B(s), \quad \theta \to 0. \tag{9.56b}$$

If now we introduce the stream function $\psi(t,s,n)$ and write

$$\psi = \psi_0 + \beta\psi_1 + \cdots, \quad \theta = \theta_0 + \beta\theta_1 + \cdots, \tag{9.57}$$

then as first approximation we obtain

$$\psi_0 = (1/2)\Gamma\left\{\sin\left(\frac{\pi}{4}-t\right)-\left(\frac{1}{\sqrt{2}}\right)e^{-n/\sqrt{2}}\left[\cos\left(\frac{n}{\sqrt{2}}-t\right)\right.\right.$$
$$\left.\left.+\left(1+\frac{n}{\sqrt{2}}\right)\sin\left(\frac{n}{\sqrt{2}}-t\right)\right]\right\}; \tag{9.58a}$$

$$\theta_0 = e^{-n/\sqrt{2}}\left[\cos\left(\frac{n}{\sqrt{2}}-t\right)\right]. \tag{9.58b}$$

Concerning the second approximation ψ_1 and θ_1, we can write

$$\psi_1 = \Gamma\frac{d\Gamma}{ds}[\psi_{1p}e^{2it}+\psi_{1\,\mathrm{st}}], \quad \theta_1 = \frac{d\Gamma}{ds}[\theta_{1p}e^{2it}+\theta_{1\,\mathrm{st}}], \tag{9.59}$$

where the subscript 'p' is relative to the periodic part and subscript 'st' to the steady-state part.

Here we write only the formula for $\psi_{1\,\mathrm{st}}$ and $\theta_{1\,\mathrm{st}}$. Namely:

$$\theta_{1\,\mathrm{st}} = -(1/4)e^{-n/\sqrt{2}}\sin\frac{n}{\sqrt{2}} - (1/8\sqrt{2})(2\sqrt{2}+n)e^{-\sqrt{2n}} + C_1 n + C_2; \tag{9.60a}$$

$$\psi_{1\,\mathrm{st}} = -(1/8\sqrt{2})\left\{(1/2)e^{-\sqrt{2n}}\left[n^2+\left(\frac{9}{\sqrt{2}}\right)n+(25/2)\right]\right.$$
$$+2e^{-n/\sqrt{2}}\left[\left(2+\frac{n}{\sqrt{2}}\right)\cos\frac{n}{\sqrt{2}}-2\sin\frac{n}{\sqrt{2}}\right]$$
$$-C_3n^2+C_4n+C_5; \tag{9.60b}$$

From the boundary conditions we obtain

$$C_1 = C_2 = 0 \quad \text{and} \quad C_3 = 0, \quad C_4 = -7/8, \quad C_5 = \frac{41}{32\sqrt{2}}; \tag{9.60c}$$

$$A(s) = -(1/4)\frac{d\Gamma}{ds}, \quad B(s) = -(7/8)\Gamma\frac{d\Gamma}{ds}. \tag{9.60d}$$

Figure 9.1 is the result of a computation, with the help of the derived formula, of the variation in perturbation of the temperature with altitude n for various time t. The dashed line represent the linear case (first approximation) and we see that nonlinear terms change strongly the vertical structure of the perturbed temperature.

Obviously our above approach is based on a particular 'strategy' which allows us to avoid the necessity of introducing an outer representation of

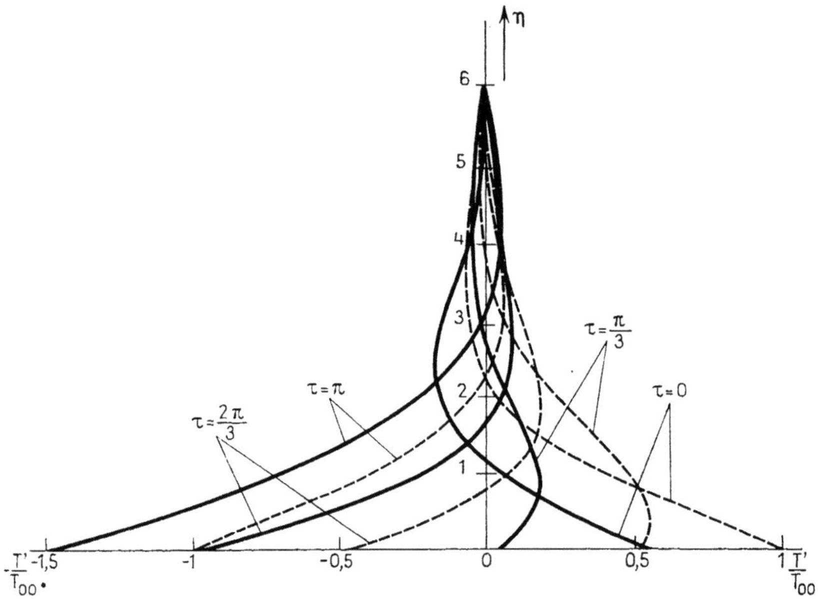

Fig. 9.1 Vertical structure of θ for various time values. Reprinted with kind permission from [17].

the approximate solution far from the heated surface! In Figure 9.1 the position s on the heated curved surface is $s = 0.3$ and the function $\Gamma(s) = [3(3/2)^{1/2}]s(1 - s^2)$. For simplicity of computation, we choose $\beta = 1$! The supplementary perturbations are positives on the heated curved surface when $\Gamma(s)$ decreases and negatives when $\Gamma(s)$ increases. Concerning the secondary current, far from a heated curved surface, it is directed in the direction where the function $\Gamma(s)$ decreases.

In fact, an associated representation near the heated surface is unnecessary. This was shown by Noe in his thesis [18]. Noe considered, for the far region, an outer representation and by matching obtained a uniformly valid approximate solution above the heated surface. In his asymptotic theory (1981) Noe shows the non-existence of a proximate layer near the curved heated surface and as a consequence the singular behavior of θ and ψ is related only, for both fields, to n tending to infinity – far from the heated curved surface.

Obviously the first approximation (9.58a, b) is uniformly valid. Far from the heated curved surface, Noe assumed that the far region, when $n \to \infty$, is characterized by

$$n = \beta^{\gamma} n^* \quad \text{with } \gamma < 0, \tag{9.61a}$$

and wrote

$$\psi - (1/2)\Gamma \left\{ \left[\frac{(1+i)}{2\sqrt{2}} \right] e^{it} + \left[\frac{(1-i)}{2\sqrt{2}} \right] e^{-it} \right\} = \beta^{\varphi} \psi^{*}(t, s, n^{*}; \beta), \tag{9.61b}$$

$$\theta = \beta^{\sigma} \theta^{*}(t, s, n^{*}; \beta). \tag{9.61c}$$

The analysis and matching, between outer (far region) and near (where the classical solution remains valid) regions, give

$$s = 1, \quad 1 + \varphi - \gamma = -2\gamma \quad \text{and} \quad \varphi = 3\gamma \Rightarrow \varphi = -3/4, \quad \gamma = -1/4. \tag{9.61d}$$

We observe that Noe's approach is consistent only when in equation (9.55c), for θ, the stratification term is absent! Curiously, the choice $C_1 = 0$, in our solution (9.60a), according to Noe's analysis, is correct and confirms the analysis performed by Riley in 1965 [19] for the case of 'oscillating viscous flows'. In [13, chapter xi], the reader can find a detailed account of the so-called 'double boundary layer model' of Riley [19] and Stuart [20], relative to the 'high frequency oscillating viscous flow, large Strouhal number, case'.

Finally, for the steady-state dominant problem in the far/outer region, Noe obtained the following coupled problem for ψ^{*}_{st} and θ^{*}_{st}:

$$\left(\frac{\partial \psi^{*}_{\text{st}}}{\partial n^{*}} \right) \frac{\partial^2 \psi^{*}_{\text{st}}}{\partial s \partial n^{*}} - \left(\frac{\partial \psi^{*}_{\text{st}}}{\partial s} \right) \frac{\partial^2 \psi^{*}_{\text{st}}}{\partial n^{*2}} = \Gamma(s) \theta^{*}_{\text{st}} + \frac{\partial^3 \psi^{*}_{\text{st}}}{\partial n^{*3}}, \tag{9.62a}$$

$$\left(\frac{\partial \psi^{*}_{\text{st}}}{\partial n^{*}} \right) \frac{\partial \theta^{*}_{\text{st}}}{\partial s} - \left(\frac{\partial \psi^{*}_{\text{st}}}{\partial s} \right) \frac{\partial \theta^{*}_{\text{st}}}{\partial n^{*}} = \frac{\partial^2 \theta^{*}_{\text{st}}}{\partial n^{*2}}, \tag{9.62b}$$

with, as boundary conditions,

$$n^{*} = 0: \quad \frac{\partial \psi^{*}_{\text{st}}}{\partial n^{*}} = \psi^{*}_{\text{st}} = 0, \quad \theta^{*}_{\text{st}} = (1/4) \frac{\mathrm{d}\Gamma(s)}{\mathrm{d}s}, \tag{9.62c}$$

$$n^{*} \to \infty: \quad \frac{\partial \psi^{*}_{\text{st}}}{\partial n^{*}} \to 0, \quad \theta^{*}_{\text{st}} \to 0. \tag{9.62d}$$

Problem (9.62a–d) was solved by Noe via a method used by Stuart [20], who was inspired by a paper of Fettis [21]. Figure 9.2 is taken from [18] and shows the configuration of the stream lines.

A separating stream line (0) appears which isolates near the slope a rotational flow with a 'slope wind' flow, above, in the outer region which is

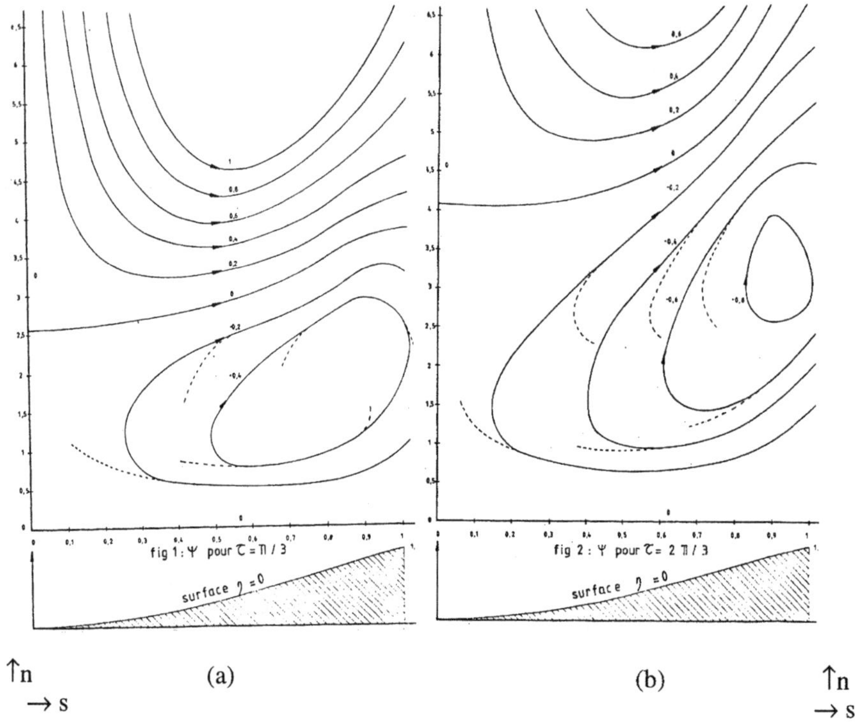

Fig. 9.2 Streamlines over a mountain slope $-n = 0$. Reprinted with kind permission [18].

increasing upwards to the top of the slope. In [18] the reader can also find the asymptotic (outer-inner) solution of a breeze problem over a flat ground; this two-dimensional, without Coriolis force, breeze problem for high Strouhal number,

$$S = \frac{L_0}{U_0 t_0} \quad \text{with} \quad t_0 = \frac{1}{\omega_0} \Rightarrow \varepsilon = \frac{1}{S} = \left(\frac{1}{\omega_0}\right)\frac{U_0}{L_0} \ll 1, \tag{9.63a}$$

is reduced in [18] to the following starting equations, for the stream function ψ and perturbation of the temperature θ:

$$\frac{\partial}{\partial t}\left(\frac{\partial \psi}{\partial n}\right) + \varepsilon\left[\left(\frac{\partial \psi}{\partial n}\right)\frac{\partial}{\partial x}\left(\frac{\partial^2 \psi}{\partial n^2}\right) - \left(\frac{\partial \psi}{\partial x}\right)\frac{\partial}{\partial n}\left(\frac{\partial^2 \psi}{\partial n^2}\right)\right] + \frac{\partial \theta}{\partial x} = \frac{\partial^4 \psi}{\partial n^4}; \tag{9.63b}$$

$$\frac{\partial \theta}{\partial t} + \varepsilon\left[\left(\frac{\partial \psi}{\partial n}\right)\frac{\partial \theta}{\partial x} - \left(\frac{\partial \psi}{\partial x}\right)\frac{\partial \theta}{\partial n}\right] = \lambda \frac{\partial \psi}{\partial x} + \frac{1}{\mathrm{Pr}}\frac{\partial^2 \theta}{\partial n^2}, \tag{9.63c}$$

assuming that the stratification term characterized by λ is proportional to small parameter ε, $\lambda = \varepsilon\mu$, with $\mu = O(1)$. This, according to Noe's rational analysis, leads to (9.63b, c), a more reachable leading-order solution when $\varepsilon \to 0$! For the system of equations (9.63b, c), as boundary conditions, in the case of a periodic free convection, we have

$$n = 0: \quad \frac{\partial \psi}{\partial n} = \psi = 0 \quad \text{and} \quad \theta = (1/2)\Theta(x)[e^{it} + e^{-it}], \tag{9.63d}$$

$$n \to +\infty: \quad \frac{\partial \psi}{\partial n} \to 0, \quad \psi \to 0, \quad \text{and} \quad \theta \to 0. \tag{9.63e}$$

Here, when we write

$$\psi(t, x, n) = (1/2)\left(\frac{\mathrm{d}\Theta(x)}{\mathrm{d}x}\right)[\Psi(n)e^{it} + \Psi^*(n)e^{-it}]$$

we have, at once, at the leading order a singular behavior for $\Psi^*(n)$ when $n \to +\infty$, and when $\varepsilon \to 0$:

$$\Psi(n) \to \Psi_0(n) \quad \text{and} \quad \Psi_0(\infty) = -\frac{1}{2}\cos t.$$

Again it is necessary to consider a far region, near $n = \infty$, with a matching! This is performed in Noe's thesis [18] after a 'laborious' technical analysis. Near $n = 0$ the leading-order solution is correct. We observe that, in reality, in order to obtain a consistent system of outer equations in the far region, it is necessary to assume that $\lambda = \varepsilon^{2/3}\lambda^*$ and in such a case the stratification is active only in the outer far region. Finally, in [18], an outer, in the far region, consistent problem is derived, and in particular, in this far region we have for the stream function ψ and the perturbation of the temperature θ the following representation:

$$\psi = -(1/4)\left(\frac{\mathrm{d}\Theta(x)}{\mathrm{d}x}\right)[e^{it} + e^{-it}] + \varepsilon^{-3/5}\psi_{\mathrm{st}}^{\mathrm{far}}(x, \varepsilon^{2/5}n) + \cdots, \tag{9.64a}$$

$$\theta = \varepsilon\theta_{\mathrm{st}}^{\mathrm{far}}(x, \varepsilon^{2/5}n) + \cdots, \tag{9.64b}$$

the functions $\psi_{\mathrm{st}}^{\mathrm{far}}$ and $\theta_{\mathrm{st}}^{\mathrm{far}}$ being solutions of the following steady-state outer dominant equations ($\varepsilon^{2/5}n = n^*$, is the outer vertical coordinate)

$$\left(\frac{\partial \psi_{\mathrm{st}}^{\mathrm{far}}}{\partial n^*}\right)\frac{\partial}{\partial x}\left(\frac{\partial^2 \psi_{\mathrm{st}}^{\mathrm{far}}}{\partial n^{*2}}\right) - \left(\frac{\partial \psi_{\mathrm{st}}^{\mathrm{far}}}{\partial x}\right)\frac{\partial}{\partial n^*}\left(\frac{\partial^2 \psi_{\mathrm{st}}^{\mathrm{far}}}{\partial n^{*2}}\right) + \frac{\partial \theta_{\mathrm{st}}^{\mathrm{far}}}{\partial x}$$
$$= \frac{\partial^2}{\partial n^{*2}}\left(\frac{\partial^2 \psi_{\mathrm{st}}^{\mathrm{far}}}{\partial n^{*2}}\right), \tag{9.65a}$$

$$\left(\frac{\partial \psi_{\mathrm{st}}^{\mathrm{far}}}{\partial n^*}\right)\frac{\partial \theta_{\mathrm{st}}^{\mathrm{far}}}{\partial x} - \left(\frac{\partial \psi_{\mathrm{st}}^{\mathrm{far}}}{\partial x}\right)\frac{\partial \theta_{\mathrm{st}}^{\mathrm{far}}}{\partial n^*} - \mu\left(\frac{\partial \psi_{\mathrm{st}}^{\mathrm{far}}}{\partial x}\right) = \frac{\partial^2 \theta_{\mathrm{st}}^{\mathrm{far}}}{\partial n^{*2}}, \tag{9.65b}$$

with the boundary conditions

$$n^* = 0: \quad \psi_{\mathrm{st}}^{\mathrm{far}} = 0, \quad \frac{\partial \psi_{\mathrm{st}}^{\mathrm{far}}}{\partial n^*} = 0, \quad \theta_{\mathrm{st}}^{\mathrm{far}} = -(1/8\sqrt{2})\Theta(x)\left(\frac{\mathrm{d}^2\Theta(x)}{\mathrm{d}x^2}\right), \tag{9.65c}$$

$$n^* \to +\infty: \quad \frac{\partial \psi_{\mathrm{st}}^{\mathrm{far}}}{\partial n^*} \to 0 \quad \text{and} \quad \theta_{\mathrm{st}}^{\mathrm{far}} \to 0. \tag{9.65d}$$

9.5 Complements and Remarks

We first comment on some interesting phenomenological features of the 'sea breeze and local winds' from the book by Simpson [22]. In the Foreword, Julian Hunt writes:

> ... but few books have focussed on the special features of the atmosphere caused by the effects of the sea on the climate and weather of land areas near the coasts ...

and further

> This book is an excellent account of them; their history, their different types depending on the coastline or synoptic situation, their connection with local weather such as clouds and rain, their effects on pollution, aircraft and bird flight, ...

The first nine chapters of the Simpson book are aimed at the general reader; they deal with the behavior of the sea breeze and details which can be seen both from the ground and from the air. Other local winds, some of which are closely related, are also dealt with in this section. The last three chapters are slightly more technical and deal with measurements of sea-breeze phenomena. In this book, the theoretician who is mainly concerned with the equations – and often do not even have the opportunity to feel the blast of the sea-breeze close to the seaside – will find much useful information relating to the physical, natural, aspects of the breeze phenomena.

For example, the sea breeze will start to blow when the temperature difference between the land and sea is large enough to overcome any offshore wind. On the coast of southern England (which is also the case, of course, in French Normandy) on a calm day a temperature difference of 1°C is large enough for a sea breeze to form, but to overcome an offshore wind as strong as 8 m/sec a temperature difference of 11°C is needed! Land breezes are quite contrary to sea-breezes. Land breezes blow directly from the shore during the night and rest during the day, while sea breezes blow during the day and rest during the night; thus they alternately succeed each other. The growth and extent of the pressure field at any point is of primary importance as it supplies the driving force for the sea breeze, the thermal wave (produced by the variation of temperature in the lower layers of the atmophere) plays also an important role. The factors which affect land and sea breeze circulation are [22, chapter 12]:

1. Diurnal variation of the ground temperature,
2. Diffusion of heat,
3. Static stability,
4. Coriolis force,
5. Diffusion of momentum,
6. Prevailing wind.

The first three factors are essential, but the fourth, fifth and sixth factors are not necessary to the production of sea breezes, though they do play a role in determining the behavior of such breezes. The fourth factor plays an important part in determining a sea breeze's horizontal dimension and producing a clockwise rotation with time. The fifth factor plays an important part in producing a realistic wind profile near the ground. The sixth factor can play a significant role in that, if it is very strong, a sea breeze cannot be generated, and only if it is moderate can a sea breeze front be formed. Most of the analytical model depends on linear theory, which assumes that the amplitude of the diurnal variation of the ground temperature is much smaller than the vertical temperature difference between the ground and the height affected by the sea-breeze circulation and it neglects several (nonlinear) terms in the governing equations. The sea-breeze circulation as an atmospheric boundary-layer process is strongly influenced (to the lowest 1–2 km of the atmosphere) by viscosity and heat conduction, and its time scale is too long for the Earth's rotation to be ignored, except near the equatorial region. Unfortunately, solutions in a number of analytic (linear) models have been obtained only after simplifying the governing equations to the point where certain of the physical processes listed above were omitted. In Walsh's paper [23] of 1974, all six factors listed above were included and in 1987 Niino [24] used the same model as in Walsh [23], but introduced a different scaling under which the flow fields for various values of external parameters collapse to the same single pattern. The horizontal extent of the sea breeze increases with increasing atmospheric stability and thermal diffusivity. Linear theory does not include the change with time of the vertical stratification of the atmosphere, nor the changes in diffusivity with height. One consequence is that the sea-breeze and land-breeze circulations are symmetric and that the formation of the sea-breeze front cannot be modelled. To overcome these and other limitations, numerical models have been developed. Important steps in the application of a numerical model of the sea breeze to realistic coastlines were taken by Pielke [25], who simulated the sea breeze over Florida. The effect of topography to modify sea-breeze circulation was included in a model by Mahrer and Pielke [26]; depending on the separation of the mountain from the coast, the combined sea breeze and mountain circulations can be more intense during both day and night when they act together. Some more recent numerical studies of the development of the sea-breeze front by Garrat and Physick [27, 28], are of special interest. They simulated gravity current flows in the atmosphere at mesoscale (20–200 km) and examined the effects of turbulent heat transfer from the ground and also that of the Earth's rotation; studying the rates of change of horizontal gradients normal to the front of temperature and velocity. A more recent numerical study has been realized by Xian and Pielke [29], where the effects of width of land masses on the development of sea breezes are investigated. In Simpson's book [22] the reader can find a very good selection of references on sea breezes and local winds. In a book published in 1984 by Pielke [30], the reader can find various useful pieces of information concerning the mesoscale (regional) meteorological modeling. For various theoretical aspects of mesometeorological problems, see [2]. In [31], the reader can find a 'hydrodynamics study of meso-meteo phenomena'. In a recent paper by Robinson et al. [32], devoted to deep convection – which shows that surface heating heterogeneities can indeed control the intensity of deep convection storms, as a linear resonant response of the atmospheric fluid according to dry fluid dynamics, despite the nonlinearity and latent heating occuring in the real system – the reader can find more recent references relative to convection phenomena in the atmosphere.

Here, as a fluid dynamics theoretician we observe only that, it is obvious that the numerical approach to the resolution of the model nonlinear equations is a necessary and fruitful (but also expansive!) job ..., but it is very important to not overlook the 'quality' of the used model and, in particular, to be heedful of the consistency of the analytical method used for the derivation of this model, which must be accurate. Unfortunately, often, in practice, these models (subject to a numerical treatment) are quite *ad hoc* models and their results reflect very poorly the physical reality.

We have already cited the book by Emanuel [3] and now will give some useful results taken from this book. The first part of [3] is devoted to dry convection and, in particular, the Rayleigh–Bénard problem with inclusion of the rotational effects is considered. In such a case a Taylor number appears:

$$\mathrm{Ta} = f^{02}\frac{H^4}{\nu^2}, \tag{9.66a}$$

with $f^0 = 2\Omega^0$, H being a characteristic length scale and ν the kinematic viscosity. The Taylor number is a measure of the relative importance of Coriolis and viscous accelerations. The presence of rotation, as reflected in the Taylor number, increases the critical Rayleigh number and is therefore stabilizing. Note also that for finite f^0 no convection is possible in the limit of vanishing viscosity! Unlike the non-rotating case, viscosity is actually destabilizing for certain ranges of the Taylor number. This result has a general expression in the form of the Taylor–Proudman theorem, which states that sufficiently slow, steady motions in an inviscid rotating fluid cannot vary in the direction of the rotation vector. For sufficiently high rotation rates, convection begins at smaller Rayleigh numbers when no-slip boundaries are used, in contrast to the non-rotating case.

Oscillatory convection can only (in a linear theory) occur when $\mathrm{Pr} < 1$ and $\mathrm{Ta} > 1$ and for large rotation rates and Pr not too small, so that the limit

$$\mathrm{Pr}\,\mathrm{Ta}^2 \to \infty \tag{9.66b}$$

is valid; the oscillatory instability may be expected to dominate for Pr less than about 0.68. At sufficiently high Ta the oscillatory modes have lower critical Rayleigh numbers and, for these modes, the no-slip boundaries have a stabilizing influence.

Concerning the planforms of rotating convection-stationary overturning, as in the non-rotating case, there is a certain indeterminacy in the planform of the convection!

A remarkable aspect of roll convection in rotating fluids was discovered by Veronis (in 1959): in the case of two-dimensional steady overturning, the rotation simply turns the streamlines into planes that are not orthogonal to the convection rolls while preserving the wavelength along the streamlines.

The horizontal planforms of square and hexagonal cell streamlines are presented by Chandrasekhar [33] and reproduced below in Figure 9.3. In each case, fluid spirals into downdrafts and updrafts with cyclonic rotation and out of the drafts with anticyclonic spin.

In many geophysical fluid flows, thermal convection occurs within larger-scale circulations that may vary quite slowly in the horizontal, compared to

(a)

(b)

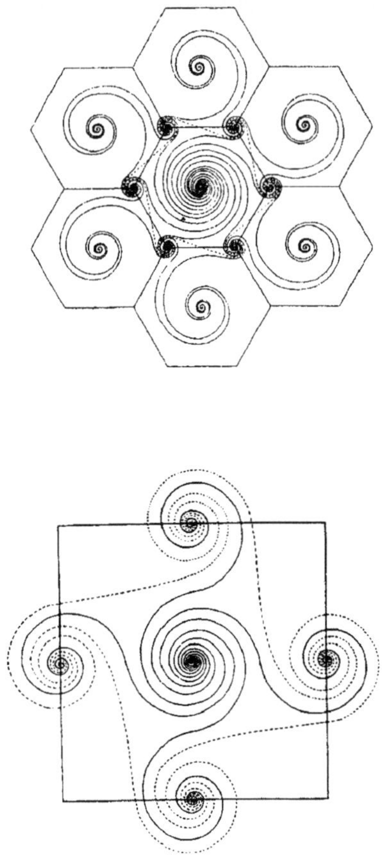

Fig. 9.3 Horizontal streamlines (a) in a square cell and (b) for hexagonal cells, of rotating RB convection.

the scale of the convection, but that may nonetheless exhibit rapid variation in the vertical. In the Earth's atmosphere the large horizontal temperature gradients found in middle latitudes are associated with rather rapid variations of horizontal wind with height, and surface friction gives rise to significant wind shear in the planetary boundary layer. Modelling the effects of these flows on thermal convection can be complicated, since consistency often demands that the limiting processes that lead to the fluid flow be included in the equations governing the dynamics of the convection itself.

For example, the Earth's rotation and horizontal temperature gradients can have significant effects on the character of convection. Moreover, nonlinear aspects of the interaction of buoyant convection with a sheared initial flow can have effects not foreseen by linear dynamics (see [3, chapter 11]).

Part two of [3] provides a fairly rigorous treatment of moist thermodynamics and the stability of moist atmospheres, preparing the student for the subsequent discussion of moist convection that occupies the rest of the book. The physical characteristics and dynamics of individual convective clouds and a general overview of numerical modeling of convective clouds are also considered. Part three of [3] treats the local properties of moist convection, including observed characteristics of precipitating convection and slantwise convection-dynamics of individual convective clouds and cloud systems. Part four treats the global properties of moist convection including most convective boundary layers. The book concludes with an overview of the representation of moist convection in numerical models.

The recent paper by Bois and Kubicki [34] deserves particular attention. It is devoted to some singularities of the instability phenomena related to the double diffusive structure of moist-satured air. The most important conclusion made by the authors concerns the law of molecular diffusion in the chosen medium: following Onsager's assumptions, a generalized expression of Fick's law of diffusion is given. This gives a theoretical model for double diffusive phenomena in cloudy convection, the instability of the cloud being mainly due to moisture, while the instability of the surrounding air is mainly due to heating. In [34] the reader can find a good short introduction to the rheological model (thermodynamics of moist saturated air and diffusion equations), of the Boussinesq, approximate reduced equations and the solution of linear moist Rayleigh–Bénard shallow convection problems, and also the problem of convection in unsaturated air (because a cloud is always confined between layers of clear air). It is clear that the paper by Bois and Kubicki [34] is a very valuable complement to parts two and three of the book [3] by Emanuel, linked with moist cloudy convection.

Another form of convective clouds developing in the atmosphere is related to emergency situations such as explosions and fires as a consequence of a high-power thermal source responsible for the development of a strong convection flow in a local region of the atmosphere and in the formation of clouds having a significant vertical extension. Recently, in [35], this has been considered via a numerical simulation of a set of model equations (approximate, taking into account the Jones–Launder two-parameter k–ε model [36] and using the Favre approach employed in [37]) for the motion, the balance of the total specific water content and content of rain drops, the energy equation and the equations of state outside and inside the cloud. In [35] it is shown that, under the action of energy sources of a different duration, clouds are formed with significantly different dynamic properties. In conclusion, the authors observe that the features of the scientific and technical development of mankind suggest that the probability of emergency situations (fires, explosions) is continuously increasing, and this circumstance makes the development of numerical (but not solely, see for example our recent book [38, section 4.3]) models similar to those described in this study urgent. Among other things, the model problem considered in Section 9.3 can also be used to simulate a localized fire problem!

It seems necessary to observe that, in spite of its paramount importance in the Earth's atmosphere, convection has received comparatively little treatment in textbooks. The book by Emanuel [3] attempts a comprehensive treatment of some facets of the subject. As convection is a broad subject, it has been difficult to define the scope of this book in a sensible way. On the other hand, atmospheric convection is a rapidly evolving subject of research and mainly is now strongly related to environmental aerodynamics, see [39, 40].

In atmospheric convection a particularly simple convection is the so-called 'geostrophic convection', when the Kibel number (instead of the Rossby, Ro, number)

$$\mathrm{Ki} = \frac{U}{f^0 H} \ll 1, \tag{9.66c}$$

the Coriolis force being more important than the nonlinear terms in equations. On the other hand, consideration of the influence of rotation on the development of convection in a fluid layer heated from below, according to Chandrasekhar [33], show that the critical value of the usual Rayleigh number,

$$\mathrm{Ra} = \frac{\alpha g \Delta T H^3}{k\nu}, \tag{9.66d}$$

is increasing, when Ta (given by (9.66a)) is also increasing; rotation makes the development of convection difficult. Concerning heat transfer, we have as a characterizing parameter the Nusselt number

$$\mathrm{Nu} = \frac{F_l H}{\rho C_p k \Delta T}, \tag{9.66e}$$

where F_l is the total heat flux transferable through fluid and, from a similarity argument we have, in a rotating fluid, the dependence

$$\mathrm{Ki} = \mathrm{Ki}(\mathrm{Ra}, \mathrm{Ta}, \mathrm{Pr}) \quad \text{and} \quad \mathrm{Nu} = \mathrm{Nu}(\mathrm{Ra}, \mathrm{Ta}, \mathrm{Pr}), \tag{9.67}$$

where $\mathrm{Pr} = \nu/k$ is the Prandtl number and Ki is assumed to be $\ll 1$. In the atmosphere, convection occurs mainly in boundary layers, e.g., in Ekman layers, and also in thermal boundary layers.

In the framework of the Boussinesq approximation, for $\omega = \text{rot}\,\mathbf{u}$ we can write, from the Navier–Stokes equations, the following equation for ω:

$$\frac{\mathrm{d}\omega}{\mathrm{d}t} - ([\omega + 2\mathbf{\Omega}] \cdot \nabla)\mathbf{u} = \nabla\alpha T \wedge \mathbf{g} + \nu\mathbf{\Delta}\omega + \nabla\nu \wedge \mathbf{\Delta u}. \tag{9.68a}$$

In particular when Ki $\ll 1$, α and ν constant, instead of (9.68a), we have

$$2(\mathbf{\Omega} \cdot \nabla)\mathbf{u} = \alpha\nabla T \wedge \mathbf{g} + \nu\mathbf{\Delta}\omega, \tag{9.68b}$$

and for Ta $\gg 1$, the viscous term is important only in Ekman layers of thickness

$$\delta_{\text{Ek}} \approx \left(\frac{\nu}{2\Omega^0}\right)^{1/2} = H(\text{Ta})^{-1/4}, \tag{9.69}$$

and outside of these layers the viscosity is negligible. If now, $\mathbf{\Omega}$ and $\mathbf{g}$ are only components along the vertical axis z then, outside of the Ekman layers, equation (9.68b) is reduced to an equation for the thermal wind:

$$2\mathbf{\Omega}\frac{\partial \mathbf{u}}{\partial z} = \alpha\nabla T \wedge \mathbf{g}, \tag{9.70a}$$

and

$$\frac{\partial w}{\partial z} = 0. \tag{9.70b}$$

Conseqently, the vertical component of the velocity is generated in an Ekman boundary layer. In a paper by Golitsyn [41], for a static steady convection when the fluid is heated from below, the dissipation of the specific kinetic energy is given by

$$\varepsilon = \frac{\alpha g F_l}{[1 - \text{Nu}^{-1}]\rho C_p}, \tag{9.71a}$$

but, on the other hand,

$$\varepsilon = \nu\left(\frac{\partial u_i}{\partial x_k}\right)\left[\left(\frac{\partial u_i}{\partial x_k}\right) + \left(\frac{\partial u_k}{\partial x_i}\right)\right]. \tag{9.71b}$$

If in (9.71b) we assume that all derivatives are of one order and recall that variation of the velocity takes place on scale δ_{Ek}, we obtain $\varepsilon \approx \nu U^2/\delta_{\text{Ek}}^2$. Hence, with (9.71a, b) we obtain

$$U^2 \approx \left(\frac{\alpha g F_l}{2\Omega^0[1 - \text{Nu}^{-1}]\rho C_p}\right). \tag{9.72}$$

The problem is now to determine F_l and for this it is necessary to consider the Fourier equation for the temperature T,

$$\frac{\partial T}{\partial t} + u_i \frac{\partial T}{\partial x_i} = k \frac{\partial^2 T}{\partial x_i^2}. \tag{9.73}$$

We assume that the rate ΔT is given and in dimensionless form, in front of the non-dimensional term $\partial^2 T/\partial x_i^2$, we have the Péclet number

$$\mathrm{Pe} = \frac{UH}{k}. \tag{9.74a}$$

If $\mathrm{Pe} \gg 1$, in the case of a developed convection, then the main variations of the temperature take place inside of a thermal boundary layer of thickness

$$\delta \approx \frac{H}{\mathrm{Pe}^{1/2}}, \tag{9.74b}$$

and heat flux across this layer with a sharp temperature gradient is estimated as

$$F_l \approx \frac{\rho C_p k \Delta T}{\delta}. \tag{9.74c}$$

As a consequence, with (9.66e), we write

$$\mathrm{Nu} \approx \frac{H}{\delta} \approx \mathrm{Pe}^{1/2}. \tag{9.75}$$

Now we introduce a Rayleigh number relative to heat flux F_l such that

$$R_{Fl} = \frac{\alpha g F_l H^4}{\rho C_p k^2 \nu} = \mathrm{Ra}\,\mathrm{Nu}. \tag{9.76}$$

From (9.72), for U, and taking into account (9.74a), (9.76) and (9.66a), we obtain

$$\mathrm{Nu} \approx [R_{Fl}(\mathrm{Nu} - 1)]^{1/4}\,\mathrm{Ta}^{-1/8}. \tag{9.77}$$

When $(4\mathrm{Nu})^{-1} \ll 1$, then instead of $(\mathrm{Nu} - 1)^{1/4}$ we write $\mathrm{Nu}^{1/4}$ and we obtain

$$\mathrm{Nu} \approx R_{Fl}^{1/3}\,\mathrm{Ta}^{-1/6}. \tag{9.78}$$

From (9.78), the heat transfer must fall when the angular velocity increase as $(\Omega^0)^{-1/3}$.

Finally, the condition (9.66c) is written [41] as

$$\mathrm{Ki} = R_{Fl}^{2/3}\,\mathrm{Ta}^{-5/6}\,\mathrm{Pe}^{-1} \ll 1, \tag{9.79}$$

and from $\delta_{\mathrm{Ek}} > \delta$ we obtain

$$\mathrm{Ra} < \mathrm{Ta}^{5/4}. \tag{9.80}$$

The experimental investigations of Rossby in 1969 [42], concerning heat transfer in rotating fluids heated from below, allows us to verify some of the above derived criteria.

In a recent short paper [43], thermocapillary forces are shown to induce wave motions in rotating fluids (around the vertical axis) and subject to a (horizontal) temperature gradient directed along the surface (due to the thermocapillary effect); these motions are similar to geophysical ones. In order to construct a model, the authors consider the theory of a shallow incompressible fluid in the quasi-geostrophic approximation (when $\mathrm{Ki} \ll 1$). The influence of the temperature gradient is supposed to manifest itself only in the appearance of surface forces due to the surface tension dependency on temperature (thermocapillary effect).

A simple linear model of air flow distortion by the effect of breeze is considered in the paper [44]. As heating or cooling of the land leads to breeze circulation only in the region of the temperature gradient and, if for instance, the land is heated during the day, large temperature gradients are observed at small distances; the temperature remains practically constant inland, its gradient tends to zero and therefore a breeze will decay. To model such behavior of a breeze, the authors consider a linear approach describing the influence of local temperature inhomogeneities of the underlying surface on perturbations in temperature and wind velocity fields in the atmosphere. Provided the temperature drops slowly downstream, such a simplified model allows one to evaluate perturbations in real breeze circulations. The authors consider three cases: in the first case, temperature drops at the same rate as it grows; in the second case, there is a strong drop of temperature of the underlying surface; in the third case, there is a very slow drop of the underlying surface temperature perturbation, which is more closed to real breeze circulation.

As a complement of the modelling, performed in Section 9.2, we start from the full NSF equations for a heavy, compressible viscous fluid in rotation, and the corresponding initital and boundary conditions (see for instance [45]). With a proper choice of non-dimensional quantities, the NSF equations depend, in particular, on the Grashof (Gr), Strouhal (S) and Rossby (Ro) numbers. The boundary condition for the temperature on a flat wall includes the parameter τ (similar to an Eckert number). It is assumed that $\mathrm{Gr} \gg 1$, in the framework of an asymptotic modelling of the atmospheric free convection problem, and that $1/\sqrt{\mathrm{Gr}}$ is our main small parameter, such that

$$\tau \text{ and } 1/\sqrt{\mathrm{Gr}}, \text{ are simultaneously very small,} \tag{9.81a}$$

as this indeed is the case for atmospheric convection problems, and satisfies the similarity relation

$$\tau = G\left(\frac{1}{\sqrt{\mathrm{Gr}}}\right)^{\alpha}, \quad G = \mathrm{O}(1), \tag{9.81b}$$

with $\alpha > 0$ a real number to be determined. The analysis shows [45] that two types of inner degeneracies occur corresponding to the values

$$\alpha = 1 \quad \text{and} \quad \alpha = 2/5. \tag{9.81c}$$

The former yields linear model equations and the latter, under the complementary assumptions (large Rossby number but small Strouhal number)

$$S = \sigma\left(\frac{1}{\sqrt{\mathrm{Gr}}}\right)^{1/5} \quad \text{and} \quad \mathrm{Ro} = \mu\left(\frac{1}{\sqrt{\mathrm{Gr}}}\right)^{-1/5} \quad \sigma \text{ and } \mu = \mathrm{O}(1), \tag{9.81d}$$

yields nonlinear model equations.

The two outer degeneracies give the trivial zero solution which determines the behavior of the inner asymptotic representation far from the flat heated wall.

It is of interest to note that our approach allows one to determine the exact form of the inner and outer asymptotic representations and leaves the opportunity, if this is necessary, of going beyond the derived limiting leading-order model equations.

We observe also that via our approach we define the validity of the derived model equations thanks to similarity relations (9.81b) and (9.81d) between τ, S, Ro and $(1/\sqrt{\mathrm{Gr}})$. With the non-dimensional quantities, in the nonlinear convection case, the coordinates are

$$x = x_1, \quad y = x_2 \quad \text{and} \quad z = \frac{x_3}{(1/\mathrm{Gr}^{1/5})}, \tag{9.82a}$$

the vertical coordinate z being an inner coordinate. For the horizontal velocity vector $\mathbf{v}$, vertical component w of the velocity vector, perturbation π of the pressure, and perturbation θ of the temperature, we have the relations

$$\mathbf{v} = \left[\frac{u_1}{(1/\mathrm{Gr}^{1/10})}, \frac{u_2}{(1/\mathrm{Gr}^{1/10})}\right], \quad w = \frac{u_3}{(1/\mathrm{Gr}^{1/5})}, \tag{9.82b}$$

$$\pi = \frac{(1 - p/p_s)}{(1/\mathrm{Gr}^{2/5})}, \quad = (1 - T/Ts)/(1/Gr1/5), \tag{9.82c}$$

where $p_s(x_3)$ and $T_s(x_3)$ characterize the standard atmosphere.

For $\mathbf{v}$, w, π and θ dependent on inner coordinates t, x, y and z, we have as model convection equations, when $1/\sqrt{\mathrm{Gr}} \to 0$,

$$\sigma \frac{\partial \mathbf{v}}{\partial t} + (\mathbf{v} \cdot \mathbf{D})\mathbf{v} + w \frac{\partial \mathbf{v}}{\partial z} + \left(\frac{1}{\mu}\right)(\mathbf{k} \wedge \mathbf{v}) = -\left(\frac{1}{\mathrm{Bo}}\right)\mathbf{D}\pi + \frac{\partial^2 \mathbf{v}}{\partial z^2}, \quad (9.83\mathrm{a})$$

$$\mathbf{D} \cdot \mathbf{v} + \frac{\partial w}{\partial z} = 0, \quad (9.83\mathrm{b})$$

$$\frac{\partial \pi}{\partial z} = \mathrm{Bo}\,\theta, \quad (9.83\mathrm{c})$$

$$\sigma \frac{\partial \theta}{\partial t} + (\mathbf{v} \cdot \mathbf{D})\theta + w \frac{\partial \theta}{\partial z} + \mathrm{Bo}\,\Lambda_0 w = \left(\frac{1}{\mathrm{Pr}}\right)\frac{\partial^2 \theta}{\partial z^2}, \quad (9.83\mathrm{d})$$

where $\mathbf{D} = (\partial/\partial x, \partial/\partial y)$, Bo is the Boussinesq number and Λ_0 the parameter of the stratification. For the above convection equations (9.83a–d), the boundary convection conditions, at $z = 0$ are

$$\theta = \Theta(t, x, y), \quad t > 0, \quad \mathbf{v} = 0 \quad \text{and} \quad w = 0, \quad (9.84\mathrm{a})$$

and at $t = 0$:

$$\mathbf{v} = 0, \quad \theta = 0 \Rightarrow w = 0 \quad \text{and} \quad \pi = 0, \quad (9.84\mathrm{b})$$

for $x \to \infty$ and $y \to \infty$:

$$\mathbf{v} \to 0, \quad w \to 0, \quad \pi \to 0, \theta \to 0. \quad (9.84\mathrm{c})$$

In an asymptotic approach, the conditions far from the flat heated wall, are derived via a matching with an outer degeneracy where the dimensionless vertical coordinate is simply x_3. The outer equations for the outer functions (dependent on t, $x_1 = x$, $x_2 = y$ and x_3), associated with the inner functions (9.82b, c), with (9.81c, d), lead when $1/\sqrt{\mathrm{Gr}} \to 0$, to a very degenerate model system of non-viscous equations for an unsteady two-dimensional incompressible fluid flow for the horizontal outer velocity vector and the outer perturbation of the pressure, dependent both on the time t, and on horizontal coordinates, $x_1 = x$ and $x_2 = y$. In the case considered here, a free convection problem, the solution of this two-dimensional unsteady Euler problem is zero! As a consequence, for the above free convection problem, (9.83a–d), (9.84a–c), we have the following behavior far from the flat heated wall $z = 0$:

$$\mathbf{v} \to 0, \quad \pi \to 0, \theta \to 0 \quad \text{when } z \to \infty, \quad (9.85\mathrm{a})$$

and also, when $\Lambda_0 = O(1)$ and fixed for $1/\sqrt{\mathrm{Gr}} \to 0$, we have obviously

$$\theta \to 0 \quad \text{when } z \to +\infty \Rightarrow w = 0. \tag{9.85b}$$

In reality, in free convection problems the account of the variation of dynamic (μ) and thermal (k) exchange coefficients with the altitude is important and in [46], a simple case has been considered. From the hydrostatic equations (*à la* Boussinesq as above), governing a free convection phenomenon in the vicinity of the thermally non-homogeneous flat ground, in [46] we have exhibited an asymptotic model with three layers linked at the variation of exchange coefficient with the altitude. We give the formulation of the corresponding second approximation boundary layer problem which takes into account the influence of the dissipation sublayer appearing in the vicinity of the heated flat ground. A simple case is considered (as an explanatory example) and it allows us to obtain an explicit solution for the perturbation of the temperature in a periodic convection with time. In [46], the starting dimensionless equations are slightly different from the above equations (9.83a–c). Namely:

$$\sigma \frac{D\mathbf{v}}{Dt} + \left(\frac{1}{\mathrm{Ro}}\right)(\mathbf{k} \wedge \mathbf{v}) = -\mathbf{D}\pi + \left(\frac{1}{\mathrm{Re}_\perp}\right)\frac{\partial}{\partial z}\left(\mu \frac{\partial \mathbf{v}}{\partial z}\right), \tag{9.86a}$$

$$\mathbf{D} \cdot \mathbf{v} + \frac{\partial w}{\partial z} = 0, \tag{9.86b}$$

$$\frac{\partial \pi}{\partial z} = \left(\frac{B^*}{\gamma}\right)\theta, \tag{9.86c}$$

$$\sigma \frac{D\theta}{Dt} + \Lambda_{00} w = \left(\frac{1}{\mathrm{Re}_\perp \mathrm{Pr}}\right)\frac{\partial}{\partial z}\left(k \frac{\partial \theta}{\partial z}\right), \tag{9.86d}$$

with $\sigma D/Dt = \sigma \partial/\partial t + (\mathbf{v} \cdot \mathbf{D}) + w\partial/\partial z$, and conditions (as a free local problem),

$$z = 0\colon \quad \mathbf{v} = 0, \quad w = 0, \quad \theta = \Theta(t, x, y), \tag{9.87a}$$

$$\text{at infinity:} \quad r^2 \to \infty, \quad \mathbf{v} = 0, \quad w = 0, \quad \theta = 0, \quad \pi = 0, \tag{9.87b}$$

$$\text{at } t = 0\colon \quad \text{at rest.} \tag{9.87c}$$

In equations (9.86a) and (9.86d) we assume:

$$\mu = 1 + \mu^*\left(\frac{z}{\delta}\right) \quad \text{and} \quad k = 1 + k^*\left(\frac{z}{\delta}\right), \tag{9.88a}$$

with $\delta \equiv \delta(\varepsilon)$, and:

$$\frac{\delta(\varepsilon)}{\varepsilon} \equiv \Delta(\varepsilon) \to 0 \quad \text{with } \varepsilon \to 0, \tag{9.88b}$$

and the parameter

$$\varepsilon = \mathrm{Re}_{\perp}^{-1/2} \tag{9.89a}$$

being our main small parameter.

As a consequence of (9.88a) we must consider three limiting processes:

(a) $\varepsilon \to 0$, with z fixed – *non-viscous, adiabatic layer* case,
(b) $\varepsilon \to 0$, with $z^* = z/\varepsilon$ fixed – *classical boundary layer* case,
(c) $\varepsilon \to 0$, with $\zeta = z/\delta(\varepsilon)$ fixed – *dissipative sublayer* case.

It is shown that

$$\delta(\varepsilon) = \varepsilon^2 \quad \text{and} \quad \Delta(\varepsilon) \equiv \varepsilon. \tag{9.89b}$$

The equations and boundary conditions, with matching, for the three above-mentioned layers has been derived in [46]. A set of second-order boundary layer equations has also been derived, where in conditions at $z^* = 0$, we have the influence of the variability of $\mu^*(\zeta)$ and $k^*(\zeta)$. When

$$\sigma = 1, \quad B^* = 1, \quad \mathrm{Pr} = 1, \quad \mathrm{Ro} = \infty, \quad \mathbf{D} = \left(\frac{\partial}{\partial x}\right)\mathbf{i}, \quad \mathbf{v} = u\mathbf{i}, \tag{9.90a}$$

and, at $z^* = 0$,

$$\Theta(t, x) = (a^0 + b^0 x)\sin t, \tag{9.90b}$$

we have a very simple solution of the classical boundary equations, for

$$\lim_{\varepsilon \to 0}(v, w/\varepsilon, \pi, \theta) = (u_0^*, w_1^*, \theta_0^*) \tag{9.90c}$$

$$\Rightarrow u_0^* = 0, \quad w_1^* = 0, \quad \theta_0^* = (a^0 + b^0 x)\exp\left(-\frac{z^*}{\sqrt{2}}\right)\sin\left[t - \frac{z^*}{\sqrt{2}}\right], \tag{9.90d}$$

and for θ_1^*, in the framework of the second-order boundary layer equations, we derive the solution

$$\theta_1^* = -\Phi(k^*)\left\{(a^0 + b^0 x)\exp\left(-\frac{z^*}{\sqrt{2}}\right)\sin\left[t + \left(\frac{\pi}{4}\right) - \frac{z^*}{\sqrt{2}}\right]\right\}, \tag{9.90e}$$

where the function

$$\Phi(k^*) = \int_0^\infty \left\{\left[\frac{1}{[1 + k^*(v)]}\right] - 1\right\} \mathrm{d}v, \tag{9.91}$$

takes into account the effect of the dissipative sublayer emerging as a consequence of the variability with altitude of the coefficient $k^*(z/\varepsilon^2)$, of the thermal exchange.

Concerning the linear convective instability in atmosphere specifically, when we consider for the unknown atmospheric functions U a solution of the form

$$U = U_B(z)\exp[\sigma t - i\mathbf{k}\cdot\mathbf{x}], \tag{9.92}$$

we observe that, in the framework of an uniformly Boussinesq approximation for a half-space, Bois [47] applied the Boussinesq atmospheric equations to the study of the atmospheric motions which are generated from the state of rest by convective instability.

It was first shown in [47] that *if the Brunt–Väisälä frequency of the medium is always real, the atmosphere is stable. Then, if this frequency is imaginary in a certain zone, an instability threshold occurs only through a stationary state.*

The existence of cellular atmospheric flows is also studied in Bois [47] – an instability threshold is determinated, and a critical Rayleigh number is defined, for which the cellular flows occur – such flows are discernible only in the zone where the Brunt–Väisälä frequency is imaginary. In [47], the reader can find also a proof of the principle of exchange of stabilities, which establishes that the threshold between a stable state and an unstable state is a stationary state.

More precisely, we note that the natural instability of a medium, which is related to 'inverse' temperature profiles, namely, temperature profiles for which ($\Xi(\zeta)$ is the potential temperature of the medium and $\zeta = Mz$, for small Mach number M) the function

$$h(\zeta) = \left(\frac{1}{\Xi}\right)\frac{\mathrm{d}\Xi}{\mathrm{d}\zeta} \tag{9.93}$$

is negative.

Common measurements show that this situation can effectively arise in the troposphere, where $h(\zeta)$ can considerably vary from one day to another because of radiation from the ground. When $h(\zeta)$ is negative, it is possible that there appear instability effects due to the wave propagation for which the velocity is of the form (9.92), with a real σ. The corresponding motions are unstable of the Rayleigh–Bénard type, and the question of their existence is a Bénard problem (heated from below but for an infinite medium). The theoretical justifications of this existence [48] were proposed, in general, by assuming that the medium is confined between two levels – the ground and a 'free surface' – and a Rayleigh number can be defined as in the case of the classical Bénard problem. Cellular flows appear from a critical value of this Rayleigh number. In fact, in [47] a slowly varying $\mathrm{Ra}(\zeta)$ is introduced and

Bois draws a stability curve in the $(K, \mathrm{Ra}(0))$ plane for numerical values which correspond to those of air at the ground level (see Figure 9.4). The results show that there exists a critical Rayleigh number, from which the cellular flows appear.

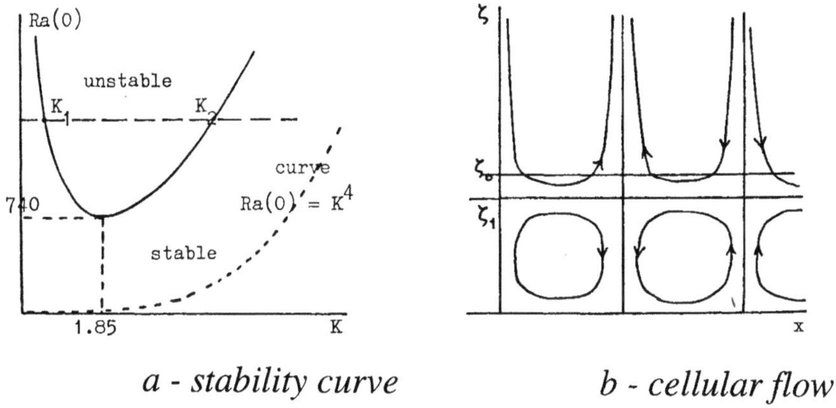

Fig. 9.4 Bénard problem in unbounded atmosphere. Reprinted with kind permission from [52].

Finally, I finish Chapter 9, which is devoted to thermal convection in the atmosphere, with some results obtained in 1961 in our 'Kandidat dissertation' (a Russian PhD, defended in Moscow University, and published in 1964 [49]), devoted to 'Hydrodynamic Study of Local Winds over a Thermally Inhomogeneous Mountain Slope'. In [49], the temporal evolution, from at rest, of a local wind (mountain breeze) along a mountain slope, has been investigated as a consequence of a thermal convection generated by the thermal inhomogeneities of the slope ground. The slope ground temperature was assumed known as a power series expansion of

$$\tau = \sqrt{t}, \tag{9.94}$$

and in such a case, correspondingly, the solution of the unsteady-state atmospheric boundary layer equations (with associated initial and boundary conditions) is determined also in a series of τ. As regards the choice of method to be applied to expansion in power series of τ, it should be pointed out that Oleinik [50] succeeded in substantiating its merit by furnishing conclusive proof that series so constructed are true representations of the solution, the incident error being of the order of magnitude of the neglected term.

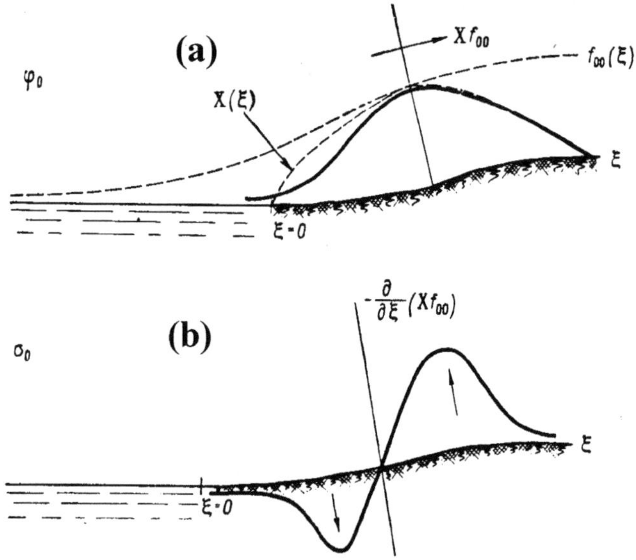

Fig. 9.5a Initial evolution of the 'mountain breeze'. Reprinted with kind permission from [49].

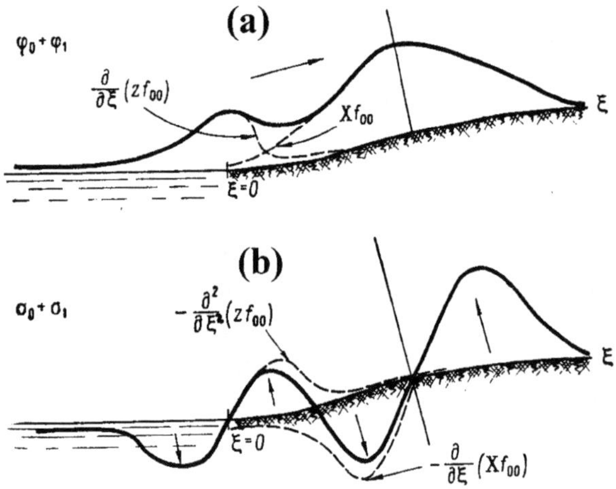

Fig. 9.5b Intermediate evolution of the mountain slope wind. Reprinted with kind permission from [49].

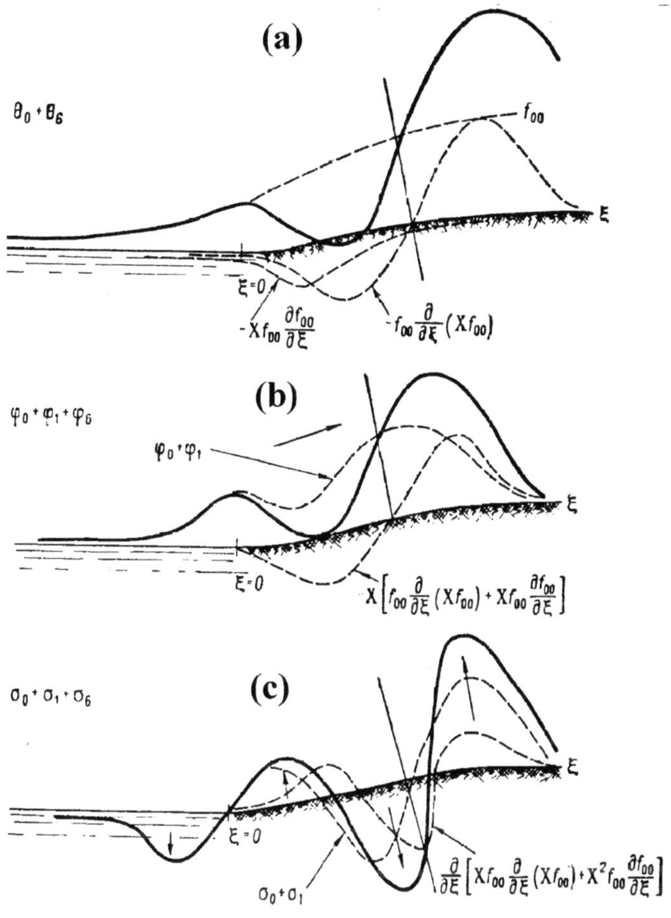

Fig. 9.5c Global effect of the mountain slope wind. Reprinted with kind permission from [49].

Obviously, it would be futile to expect the method to do something beyond its scope, using it to obtain the steady-state limit solution! For instance, for perturbation of the temperature θ, horizontal component φ and vertical component σ of the velocity vector, we derive from the starting equations the following set of equations for the terms in power series of τ: $\theta_0, \theta_6, \varphi_0, \varphi_1, \varphi_6, \sigma_0$):

$$\frac{\partial^2 \theta_0}{\partial s^2} + 2s\frac{\partial \theta_0}{\partial s} - 4\theta_0 = 0, \quad \theta_{0|s=0} = f_{00}(\xi), \quad \theta_{0|s=\infty} = 0; \qquad (9.95a)$$

$$\frac{\partial^2 \theta_6}{\partial s^2} + 2s\frac{\partial \theta_6}{\partial s} - 16\theta_6 = 4\left(\varphi_0\frac{\partial \theta_0}{\partial \xi} + \sigma_0\frac{\partial \theta_0}{\partial s}\right); \tag{9.95b}$$

$$\frac{\partial^2 \varphi_0}{\partial s^2} + 2s\frac{\partial \varphi_0}{\partial s} - 8\varphi_{00} = -4X\theta_6, \quad \varphi_{0|s=0} = 0, \quad \varphi_{0|s=\infty} = 0; \tag{9.95c}$$

$$\frac{\partial^2 \varphi_1}{\partial s^2} + 2s\frac{\partial \varphi_1}{\partial s} - 10\varphi_1 = 8A_0\frac{\partial}{\partial \xi}\left[Z\int_{\infty}^{s}\theta_0\,\mathrm{d}s\right]; \tag{9.95d}$$

$$\frac{\partial^2 \varphi_6}{\partial s^2} + 2s\frac{\partial \varphi_6}{\partial s} - 20\varphi_6 = -4X\theta_6 + 4\left(\varphi_0\frac{\partial \theta_0}{\partial \xi} + \sigma_0\frac{\partial \varphi_0}{\partial s}\right); \tag{9.95e}$$

$$\frac{\partial \sigma_n}{\partial s} = -\frac{\partial \varphi_n}{\partial \xi}, \quad n = 0, 1, 6. \tag{9.95f}$$

In these equations the vertical coordinate $s = \zeta/2\tau$ where ζ is the vertical curvilinear coordinate directed along the outer normal to the slope and the curvilinear coordinate ξ is the abcissa along the slope surface ($\zeta = s = 0$). The coefficients A_0, X and Z are dimensionless scalars that are linked with the geometry of the mountain slope. The function $f_{00}(\xi)$ simulates the temperature distribution (for θ) along the slope. In various parts of Figure 9.5 above, the evolution of a local wind, along the mountan slope, under the double action of the slope and the thermal field at the slope is represented.

In Figures 9.5a and b, the continuous, boldface lines are relative to horizontal (a) and vertical (b) components of the wind. In Figure 9.5c the bold lines (a) are relative to temperature and (b) and (c) are relative to horizontal and vertical components of the wind.

We conclude this chapter with a special mention of the book by Monin [51], where the reader can find, in chapter 8, a pertinent account of the various problems relative to atmospheric general circulation. We also mention the book by Boubnov and Glolitsyn [52], where the 'Convection in Rotating Fluids' is investigated. Finally, in [53] an asymptotic theory of Boussinesq waves in the atmosphere is presented.

References

1. M.G. Velarde and Ch. Normand, Convection. *Scientific American* **243**(1), 92–108, 1980.
2. L.N. Gutman, *Introduction to the Nonlinear Theory of Mesoscale Meteorological Processes*. Israel Program for Scientific Translations, Jerusalem, 1972 [translated from Russian].
3. K.A. Emanuel, *Amospheric Convection*. Oxford University Press, New York, 1994.
4. R.Kh. Zeytounian, *Topics in Hyposonic Flow Theory*. Lecture Notes in Physics, Vol. 672, Springer-Verlag, Berlin/Heidelberg, 2006.

5. R.Kh. Zeytounian, *Meteorological Fluid Mechanics*. Lecture Notes in Physics, Vol. m5, Springer-Verlag, Heidelberg, 1991.
6. R.Kh. Zeytounian, On the foundations of the Boussinesq approximation applicable to atmospheric motions. *Izvestiya 'Atmospheric and Oceanic Physics'* **39**, Suppl. 1, S1–S14, 2003.
7. I.A. Kibel, *An Introduction to the Hydrodynamical Methods of Short Period Weather Forecasting*. Macmillan, London, 1963 [translated from Russian].
8. B.S. Dandapat and P.C. Ray, *Int. J. Non-Linear Mech.* **28**(5), 489–501, 1993.
9. A.Kh. Khrgian, *Physics of the Atmosphere*. Gos-Tekh-Izdat, Moscow, 1953.
10. F.T. Smith, *J. Fluid Mech.* **57**(4), 803–824, 1973.
11. F.T. Smith, R.I. Sykes and P.W.M. Brighton, *J. Fluid Mech.* **83**(1), 163–176, 1977.
12. R.I. Sykes, *Proc. Roy. Soc. London* **A361**, 225–243, 1978.
13. R.Kh. Zeytounian, *Les modèles asymptotiques de la mécanique des fluides, II*. Lecture Notes in Physics, Vol. 276, Springer-Verlag, Berlin/Heidelberg, 1987.
14. F.T. Smith, P.W.M. Brighton, P.S. Jackson and J.C.R. Hunt, *J. Fluid Mech.* **113**, 123, 1981.
15. R.Kh. Zeytounian, *Asymptotic Modelling of Fluid Flow Phenomena*. Kluwer Academic Publishers, Dordrecht, 2002.
16. H. Schlichting, Berechnung ebener periodisher Grenzschichtströmungen. *Phys. Z.*, **33**(8), 337, 1932.
17. R.Kh. Zeytounian, Convection naturelle périodique au-dessus d'une surface courbe. *J. Méc. (France)* **7**(2), 231–247, 1968.
18. J.M. Noe, Sur une theorie asymptotique de la convection naturelle. Thèse de Doctorat de 3ème Cycle, Université des Sciences et Techniques de Lille I, No. d'ordre 884, March 1981.
19. N. Riley, Oscillating viscous flows. *Mathematika* **12**, 161–175, 1965.
20. J.T. Stuart, Double boundary layers in oscillatory viscous flow. *J. Fluid Mech.* **24**(4), 673–687, 1966.
21. H.E. Fettis, On the integration of a class of differential equations occuring in boundary-layer and other hydrodynamic problems. In *Proc. 4th Mid West Conference on Fluid Mechanichs*, Purdue University, pp. 93–114, 1955.
22. J.E. Simpson, *Sea Breeze and Local Winds*. Cambridge University Press, 1994.
23. J.E. Walsh, Sea breeze theory and applications. *J. Atmos. Sci.* **31**, 2012–2026, 1974.
24. H. Niino, On the linear theory of land and sea breeze circulation. *J. Meteorol. Soc. Japan* **65**, 901-921, 1987.
25. R.A. Pielke, A comparaison of three-dimensional and two-dimensional numerical prediction of sea breezes. *J. Atmos. Sci.* **31**, 1577–1585, 1974.
26. Y. Maher and R.A. Pielke, The effects of topography on sea and land breezes in a two-dimensional numerical model. *Mon. Weather Rev.* **105**, 1151–1162, 1977.
27. J.R. Garrat and W.L. Physick, *Beitr. Phys. Atmos.* **59**, 282–300, 1986.
28. J.R. Garrat and W.L. Physick, *Beitr. Phys. Atmos.* **60**, 88–102, 1987.
29. Z. Xian and R.A. Pielke, *J. Appl. Meteorol.* **30**, 1280–1304, 1991.
30. R.A. Pielke, *Mesoscale Meteorological Modeling*. Academic Press, Orlando, 1984.
31. R.Kh. Zeytounian, *Étude Hydrodynamique des Phénomènes Mésométéorologiques*. L'École de la Météorologie, Direction de la Météorologie Nationale, Paris, 1968.
32. F.J. Robinson, S.C. Sherwood and Y. Li, *J. Atmos. Sci.* **65**, 276–286, January 2008.
33. S. Chandrasekhar, *Hydrodynamic and Hydromagnetic Stability*. Clarendon Press, Oxford, 1961.
34. P.-A. Bois and A. Kubicki, A theoretical model for double diffusive phenomena in cloudy convection. *Ann. Geophys.* **21**, 2201–2218, 2003.

35. N.E. Veremei, Yu. Dovgalyuk and E.N. Stankova, *Izvestiya, Atmospheric and Oceanic Physics* **43**(6), 731–744, 2007.
36. W.P. Jones and B.E. Launder, *Int. J. Heat Transfer* **16**, 1119–1130, 1973.
37. E.N. Stankova, In *Proc. Conference of Young Scientists and Specialists of the Voeikov Main Geophys. Observ.*, Leningrad, pp. 47–53, 1990 [in Russian].
38. R.Kh. Zeytounian, *Theory and Applications of Viscous Fluid Flows*. Springer-Verlag, Berlin/Heidelberg, 2004.
39. R.S. Scorer, *Environmental Aerodynamics*. Wiley, 1978.
40. J.R.C. Hunt, *Environmental Fluid Mechanics*. IUTAM Conference, pp. 13–31, 1980.
41. G.S. Golitsyn, Simple theoretical and experimental study of convection with some geophysical applications and analogies. *J. Fluid Mech.* **95**(3), 567–608, 1979.
42. H.T. Rossby, *J. Fluid Mech.* **36**(2), 1970.
43. G.S. Kirichenko and P. Poritskiy, Thermocapillary analogue of Rossby waves. *Atmospheric and Oceanic Physis* **31**(5), 608–610, April 1996 [English translation].
44. Zhang Meigen, Linear model of air flow distortion by the effect of breeze. *Atmospheric and Oceanic Physis* **31**(6), 787–791, June 1996 [English translation].
45. R.Kh. Zeytounian, Sur une formulation rigoureuse du problème de la convection libre atmosphérique. *J. Engng. Math.* **11**(3), 241–247, July 1977.
46. R.Kh. Zeytounian and A. Mahdjoub, Prise en compte d'une sous-couche de dissipation dans les phénomènes de convection libre. *ZAMP* **40**, 931–939, November 1989.
47. P.A. Bois, Effets dissipatifs dans les écoulements atmosphériques. Deuxième partie: Instabilité linéaire convective dans l'atmosphère. *J. Méc. (France)* **18**(4), 633–660, 1979.
48. M.J. Manton, Convection in the lower atmosphere. *Austr. J. Phys.* **27**, 495–509, 1975.
49. R.Kh. Zeytounian, Hydrodynamical study of the initial stage of the development of local winds. Doctoral Thesis, Trudy of the World Meteo Centre, Vol. 3, pp. 19–74, 1964 [in Russian].
50. O.A. Oleinik, *PMM* **33**(3), 441, 1969 [in Russian].
51. A.S. Monin, *Fundamentals of Geophysical Fluid Dynamics*. GidrometeoIzdat, Leningrad, 1988 [in Russian, but an English translation is available].
52. B.M. Boubnov and G.S. Golitsyn, *Convection in Rotating Fluids*. Fluid Mechanics and Its Applications, Vol. 29, Kluwer Academic Publishers, Dordrecht, 1995.
53. P.A. Bois, Asymptotic theory of Boussinesq waves in the atmosphere. In: Lecture in CISM Course, Udine (Italy), October 1983. Publ. IRMA, Université de Lille-I, Vol. VI, Fasc. 4, No. 2, pp. II.1 to II.89, 1984

Chapter 10
Miscellaneous: Various Convection Model Problems

10.1 Introduction

It is obvious that the reader, after having digested the preceding nine chapters of this book, will know that many convection phenomena in fluids have not been considered in our discussion! But this is 'inevitable' because convection is a very 'broad' subject with many and various facets. It is true that, for convection in liquids, it has been relatively easy to define the scope of Chapters 3 to 7 in a sensible way, restricting ourselves to three main facets (as thoroughly discussed in Chapter 8) of the classical Bénard convection problem, heated from below. However, in the cases when we consider gases, liquid mixtures, the effect of surfactants, dipolar fluid, or other complementary effects (Dufour and Soret, evaporation, surface active agents, etc.), a clear definition of a precise method for a rational treatment of these various cases is shown to be a difficult affair; obtaining corresponding model equations has often, unfortunately, had to be carried out in an *ad hoc* manner! Despite the inconsistency of these models, the valuable rigourous mathematical results (see, for instance, the book [1] by Straughan, which presents rigorous convection studies, mainly by the 'energy method', in a variety of fluid and porous media contexts) obtained for such models are of little practical value, mainly because we have no assurance that the derived (or proposed) approximate models are really significant and or consistent?

In fact, we are faced with an essential problem of 'confidence' concerning this modeling approach that is intended to be an aid to numerical simulation!

My aim in writing this book has been to present the fundamentals, and to attain this goal it has been necessary to limit the material covered by selecting problems that illustrate unity in convection phenomena and at the same time put forward its essentials. Consequently, a number of important aspects of convection in fluids have ben omitted. In spite of this, the reader can find here in Sections 10.2 to 10.9 some interesting convection model problems, giving a more complete idea of the wide range of phenomena relative to convection in fluids and also the vast possibilities to apply rational analysis and asymptotic modelling.

In Section 10.2, the convection problem in the Earth's outer core is considered and, in particular, we analyze the paper of Jöhnk and Svendsen [2] and give some information concerning this convection problem. Section 10.3 is devoted to various comments related to the magneto-convection, electro-thermo-convection, ferro-hydrodynamic convection, chemical convection, solar convection, oceanic circulation and penetrative convection which have mainly been inspired by Straughan's book [1], and also by various survey papers relative to the convection problems mentioned above, which have been published in *Annual Review of Fluid Mechanics* (Palo Alto, California, USA). Section 10.4 is devoted to a brief discussion of the 'averaged, integral boundary layer (IBL)', technique for the non-isothermal case, first considered by Zeytounian [3] and subsequently improved by Kalliadasis et al. in [4], and in two recent papers [5, 6]; here I give my personal opinion and a short 'history' of events. In Section 10.5, the results of Golovin et al. [7] and also Kazhdan et al. [8], where the existence of two monotonic modes (short-scale mode and long-scale mode) of surface tension driven convective instability is obtained, is discussed. Section 10.6 concerns thermosolutal convection, when the density varies both with temperature and concentration/salinity, and the corresponding diffusivities are very different (double-diffusive convection [9]), Pr being quite different from Sc; this is, for instance, essential in solidifying alloys (inferior metal mixed with gold or silver), where the interface rejects a solute into the liquid phase. The main theoretical sources here are the papers by Coullet and Spiegel [10], Knobloch et al. [11], and Proctor and Weiss [12]. In Section 10.7, as a complement to Chapter 9, we consider the so-called 'anelastic (deep) approximation for the atmospheric thermal convection' – the derivation of these anelastic dissipative deep equations adapted for an atmospheric (deep, viscous, non-adiabatic case) convection problem, is inspired by our monograph [13, chapter 10, section 10.2]. In Section 10.8 an interesting convection, initial-boundary value problem is linked with a thin liquid film over a cold/hot rotating disk, according to Dandapat and Ray [14], where the matching between inner (for short time, near $t = 0$) and outer (for evolution time, far from $t = 0$) is accurately performed; this latter paper has proved to be a valuable prototype problem for many other unsteady film problems. In Section 10.9, solitary wave phenomena in a convection regime are discussed in view of the results obtained, in particular by Christov and Velarde [15], and also by Rednikov et al. [16]. Finally, in Section 10.10, some complementary references and comments concerning some other convection phenomena are given.

This chapter, with various complementary convection phenomena, might be itself the subject of a full book. Nevertheless, limiting my investigations to Sections 10.2 to 10.9 mentioned above, I have considered Chapter 10 as a 'useful informative extension' of the main part of this book, where the reader is invited to survey the vast panorama of the 'convection world'.

In the following sections, my approach is the same, but at the same time in some sections I do not have the possibility to develop a detailed rational analysis and an asymptotic modelling. Readers should expect to find there many unresolved questions which might be subject to careful theoretical research, but also to various modes of personal reasoning!

10.2 Convection Problem in the Earth's Outer Core

The Earth's magnetic field is generally thought to be generated by convection in its outer core, a process influenced among other things by the stratification present in the outer core. The paper [17] by Fearn and Loper give interesting information concerning the compositional convection and stratification in the Earth's fluid core and the book [18] by Melchior is devoted to the 'Physics of the Earth's Core'. Concerning the 'Dynamics of the Earth's Inner and Outer Cores', see the paper by Smylie and Szeto [19]. In Stacey's book [20], the reader can find 'Applications of Thermodynamics to Fundamental Earth Physics'.

In a review article by Jöhnk and Swendsen [2], a thermodynamic formulation of the equations of motion and buoyancy frequency for Earth's fluid outer core is given. These authors present a precise formulation of the balance and constitutive relations appropriate to the modeling of the motion of a fluid outer core, including a full thermodynamic analysis and derivation of buoyancy frequency and its role in the equations of motion for the fluid outer core, and the dependence of this equation on the thermodynamic state of the outer core. By appropriate scaling, a variety of different approximations arise from this formulation, using the thermodynamic definition of the buoyancy frequency. Special interest is put on scaling for the outer core eigenmodes yielding consistent formulations for the Boussinesq, and the subseismic, approximation. The starting equations of motion take into account the Coriolis and centrifugal acceleration terms (the fluid is rotating) and also the gravitational acceleration, $\mathbf{g} = \nabla G$, where G is the gravitational potential for a stratified barotropic ($p = P(p)$) Newtonian fluid with constant bulk and shear viscosity. But, in fact, in the framework of a thermodynamics for a two-component fluid, the authors consider the specific internal energy as

$$e = E(S, \rho, C_1, C_2), \tag{10.1a}$$

where S and ρ represent the specific entropy and mass density of the mixture (C_α, $\alpha = 1, 2$, denote the mass fraction of component α in the mixture). We have

$$C_1 + C_2 = 1 \Rightarrow C := C_1 = 1 - C_2 \tag{10.1b}$$

and

$$\rho \frac{\mathrm{d}C}{\mathrm{d}t} = -(\nabla \cdot \mathbf{i}), \tag{10.1c}$$

which is an evolution equation for mass fraction C of component 1 depending on the divergence of the corresponding mass flux $\mathbf{i}$. We observe that ρC represents the partial density of component 1, and equation of state (10.1a) is written as

$$e = E^*(S, \rho, C). \tag{10.1d}$$

In particular, using the standard thermodynamic definitions

$$K_s := \rho \left[\frac{\partial P^*}{\partial \rho}\right]_{S,C}, \quad \text{where} \quad p := \rho^2 \left[\frac{\partial E^*}{\partial \rho}\right]_{S,C} = P^*(S, \rho, C), \tag{10.2a}$$

$$\gamma := \left(\frac{\rho}{\theta}\right) \left[\frac{\partial \Theta^*}{\partial \rho}\right]_{S,C}, \quad \text{where} \quad \theta := \left[\frac{\partial E^*}{\partial S}\right]_{\rho,C} = \Theta^*(S, \rho, C), \tag{10.2b}$$

$$C_v = \frac{\theta}{[\partial \Theta^*/\partial S]_{\rho,C}} \tag{10.2c}$$

of the mixture of Grüneisen parameter γ, specific heat at constant volume (and concentration) C_v, and isentropic bulk modulus K_s, respectively, one obtains the relation (see also [20])

$$\left(\frac{1}{\rho^2}\right) \left[\frac{\partial P^*}{\partial \rho}\right]_{S,C} = \frac{\partial^2 E^*}{\partial \rho \partial S} = \left[\frac{\partial \Theta^*}{\partial \rho}\right]_{S,C} = \frac{\gamma \theta}{\rho}. \tag{10.3}$$

As a result, a complete system of equations of motion for a rotating, gravitating, stratified, heat conducting, 2-component Newtonian fluid is derived.

As a further result, we have nine differential equations for the nine quantities $(G, \rho, p, \theta, S, C, u)$; the complexity of the differential equations has increased considerably via the incorporation of entropy and concentration into the model.

Under realistic physical assumptions for the problem of normal modes, the thermodynamic processes, represented explicitly in a complete system of equations, can be reduced in essence (according to the authors!) to a single parameter, i.e., the buoyancy frequency. This complete system, although very complex, does not completely represent an exact thermodynamic description of the Earth's outer core! On the other hand, the stratification of a (fluid) system is intimately related to its so-called buoyancy frequency. This buoyancy frequency may be interpreted as the oscillation frequency or eigenfrequency in the system considered below, (In a complicated system the oscillation and buoyancy frequencies may no longer coincide, as is the case in our simple model.) In this model, the expressions 'stratification' and 'buoyancy frequency' will, for simplicity, always refer to N^2 (the square of the buoyancy or Brunt–Väisälä frequency). Usually, estimates of N^2 in the literature are based on

$$N^2 = -\left(\frac{g}{\rho}\right)\left[\left(\frac{g\rho^2}{K_s}\right) + \rho'_E\right], \tag{10.4}$$

where in a simple model we can interpret ρ'_E as the density stratification of the fluid. In [2, pp. 83–89], the reader can find a pertinent discussion concerning the 'stratification and the buoyancy frequency'. The resulting non-dimensional form of the system of nine equations for $(G, \rho, p, \theta, S, C, u)$, derived via a dimensional analysis and approximations for core undertones (subdued tones), contain certain non-dimensional numbers, reflecting the order of the terms with which they are associated.

The list of characteristic values (Table 10.1) for the outer core (see [18, 20]) has been taken from [2, p. 90]. A particularly important simplification arising out of this non-dimensional analysis is that temperature and concentration (to lowest order) play a role in core oscillations only through the dependence of the buoyancy frequency on these variables. In addition, the non-dimensional analysis validates the Boussinesq approximation of the balance relations, in which the general buoyancy frequency discussed above enters as the sole stratification parameter.

This dimensionless analysis is performed by Jöhnk and Svendsen [2, pp. 90–97] and deserves (at least, from my point of view) a further careful examination (and maybe a complementary investigation). The main feature is (because the characteristic velocity U is very small $\approx 3 \times 10^{-4}$ m/s) re-

Table 10.1 Some characteristic values for the Earth's outer core. Reprinted with kind permission from [2].

Gravitational constant	$\Gamma = 6.67 \times 10^{-11}$	$m^3\ kg^{-1}\ s^{-2}$
Angular velocity	$\Omega = 7.3 \times 10^{-5}$	s^{-1}
Thickness of the OC	$L = 2.3 \times 10^{6}$	m
Velocity (westward drift)	$U = 3 \times 10^{-4}$	$m\ s^{-1}$
Mean gravity	$[g] = 7.5$	$m\ s^{-2}$
Mean density	$[\rho] = 1.1 \times 10^{4}$	$kg\ m^{-3}$
Density variation	$\Delta\rho = 2.3 \times 10^{3}$	$kg\ m^{-3}$
Mean temperature	$[\theta] = 4.4 \times 10^{3}$	K
Temperature variation	$\Delta\theta = 1.2 \times 10^{3}$	K
Mean concentration (LEs)	$[c] = 2 \times 10^{-1}$	
Concentration variation	$\Delta c = 1 \times 10^{-1}$	
Isentropic bulk modulus	$[K] = 8.5 \times 10^{11}$	Pa
Specific heat at constant volume	$[c_v] = 6.8 \times 10^{2}$	$m^2\ s^{-2}\ K^{-1}$
Grüneisen parameter	$[\gamma] = 1.3$	
Thermal expansion	$[\alpha_\theta] = 1.1 \times 10^{-5}$	K^{-1}
Thermal conductivity	$[k] = 3.3 \times 10^{1}$	$kg\ m\ s^{-3}\ K^{-1}$
Mass diffusivity	$[D] = 6 \times 10^{-9}$	$m^2\ s^{-1}$
Kinematic viscosity	$[\nu] = 1 \times 10^{-6}$	$m^2\ s^{-1}$

lated to a small Rossby number (see the value of Ω^0 and L in Table 10.1), $\mathrm{Ro} = U/\Omega^0 L \ll 1$, leading to an important simplification of the equations of motion. However, it is now well known [21] that the limiting process, $\mathrm{Ro} \to 0$ with time-space coordinates fixed, is very singular in time t (near $t = 0$ and an adjustment problem must be considered to geostrophy) and also on the boundary (Ekman layer problem). With Ro we have also the following dimensionless parameters:

$$\mathrm{Ek} = \frac{[\nu]}{\Omega^0 L^2}, \quad \frac{\Delta\theta}{[\theta]}, \quad \frac{\Delta\rho}{[\rho]}, \quad \frac{\mathrm{Ek}}{\mathrm{Pr}} = \frac{[k]}{\Omega^0 L^2 [C_v][\rho]}, \tag{10.5a}$$

$$\frac{g}{4\pi} = \frac{L\Gamma[\rho]}{[g]}, \quad C = \frac{[K_s]}{[g][\rho]L}, \quad \frac{\mathrm{Fr}}{\mathrm{Ro}} = \frac{\Omega^{02} L}{[g]}, \tag{10.5b}$$

which are respectively the Ekman number, relative temperature and density change, Ekman–Prandtl number, self-gravitational number, compressibility number and Froude–Rossby number.

In Table 10.2, the magnitude for the above dimensionless numbers are given.

In [2, sections 4.2–4.6], the authors consider respectively, 'Scaling of the basic fields', 'Buckingham Π theorem', 'Von Zeipel's (1924) Result', 'Dimensionless, Perturbed Balance' and 'Constitutive Relations and Approxi-

Table 10.2 Definition and some magnitude for four dimensionless parameters. Reprinted with kind permission from [2].

$[\rho]$, $\Delta\rho$	density stratification
$[\theta]$, $\Delta\theta$	temperature stratification
$[K_s]$	compressibility
$[g]$	gravity
Γ	self gravitation
$[c_v]$	heat capacity
$[k]$	heat conduction
$[\nu]$	viscosity
$[D]$	diffusion
L	length
Ω	frequency
U	velocity

$\mathcal{E} = \dfrac{[\nu]}{\Omega L^2}$	Ekman number
$\mathcal{R} = \dfrac{U}{\Omega L^2}$	Rossby number
$\mathcal{T} = \dfrac{\Delta\theta}{[\theta]}$	relative temperature change
$\mathcal{D} = \dfrac{\Delta\rho}{[\rho]}$	relative density change

mation'. It seems to me that their approach is not very clear and in any case does not lead to a rational derivation of approximate model equations; again, a consistent asymptotic approach is necessary!

But, assuredly, 'the use of the thermodynamic approximation significantly simplifies the equations of motion for a heat conducting, two-component fluid outer core'.

A rational derivation of a set of approximate model equations, inspired from our above discussion, remains a challenging open problem! As a complement to [2], the reader can read also the two (now classical) papers by Roberts and Soward [22] and Wood [23].

10.3 Magneto-Hydrodynamic, Electro, Ferro, Chemical, Solar, Oceanic, Rotating, Penetrative Convections

The Boussinesq approximation, which allows one to consider for various convection problems a *Boussinesquian* (*à la* Boussinesq) fluid, is perhaps the most widely used simplification of convection in fluid modelling. We have already, in preceding chapters of this present book, dealt with this Boussinesq approximation. In Chapters 3 and 5 it appears in relation to the 'Rayleigh' simple thermal convection problem and the thermal, *à la* Rayleigh–Bénard, shallow convection problem. In Chapter 9, we have used this Boussinesq approximation in various parts to formulate atmospheric convection models. Finally, in Section 10.2, the Boussinesq approximation plays a significant role in the derivation of simplified, approximate, model equations for convection in the Earth's outer core.

A very good illustration of the plurality of the 'Boussinesquian thermal convection' is the numerous survey papers (in various volumes of *Annual Review of Fluid Mechanics*) where this Boussinesq approximation is the basis for mathematical formulation of the model problems. Examples are: convection in mushy layers [24]; solar convection and magnetic buoyancy [25]; magneto-convection [22]; mantle convection [26]; oceanic general circulation [27]; convection in rotating systems [28]; convection involving thermal and salt fields [9]; dynamic of Jovian atmospheres [29]; etc. These examples are applied to a variety of gases and liquids or to more complicated fluids with various complementary effects.

It is interesting to observe that Oberbeck, before Boussinesq (see [30]), uses a Boussinesq type approximation in meteorological studies of the Hadley thermal regime for the trade-winds arising from the deflecting effect of the Earth's rotation. Concerning this rotating convection, the reader can find recent (1998!) developments in the well-documented paper [31] by Knobloch; this paper is, in fact, a pertinent complement to chapter III of the famous book by Chandrasekhar [32], where the effect of rotation on Rayleigh–Bénard convection is taken into account (formulated as a stability problem for an infinite layer with rotation parallel to gravity, which involves several assumptions; see for instance the brief discussion in [31, p. 1422]). In [32, chapter IV] the reader can find the effect of a magnetic field on Rayleigh–Bénard convection and some details about the effect of Alfvén waves. In [32, chapter V] the coupled effect of rotation and a magnetic field is considered, where the propagation of hydromagnetic waves in a rotating fluid and onset of thermal instability in the presence of rotation and magnetic field are both investigated. On the other hand, in Straughan's recent book [1], the reader can find a presentation of nonlinear energy stability results obtained in convection problems by means of an integral inequality technique (energy method). In [1, chapter 7], this energy method is systematically developed for geophysical problems (for example, for patterned ground formation) and in [1, chapter 9] for the Bénard problem in the case of a micropolar fluid. In [1, chapters 11–13], a different line of attack is adopted; these chapters are concentrated on technologically relevant and relatively new theories: dielectric fluids and electro-hydrodynamic/electrothermal-convection, ferro-hydrodynamics convection and chemical convection (for reacting viscous fluids far from equilibrium).

More precisely, we consider first the Magneto-Hydrodynamic, MHD, Convection (see [22, 23], and [1, chapter 11]), which is the study of the interaction between magnetic fields and fluid conductors of electricity. The body force acting on the fluid is the Lorentz force that arises when electric current flows at an angle to the direction of an impressed magnetic field. The MHD convection problem is a very important one because it has intrinsic applications to the behaviour of planetary and stellar interiors (see, for instance, the paper by Hughes and Proctor [25]), and in particular, to the behaviour inside of the Earth (as has been discussed here in Section 10.2).

In MHD convection, the relevant equations may be written (Eulerian case, a perfect plasma and adiabatic motion) as [33]:

$$\rho \frac{\mathrm{d}U}{\mathrm{d}t} + 2\Omega \wedge \mathbf{U} = \nabla p + g\rho \mathbf{k} = \left(\frac{1}{\mu}\right)(\nabla \wedge \mathbf{B}) \wedge \mathbf{B}, \tag{10.6a}$$

$$\frac{\partial \mathbf{B}}{\partial t} + \nabla \wedge (\mathbf{B} \wedge \mathbf{U}) = 0, \tag{10.6b}$$

$$\nabla \cdot \mathbf{B} = 0, \tag{10.6c}$$

$$\frac{\mathrm{d}\rho}{\mathrm{d}t} + \nabla \cdot \mathbf{U} = 0, \tag{10.6d}$$

$$\frac{\mathrm{d}T}{\mathrm{d}t} - \left[\frac{(\gamma-1)}{/\gamma}\right]\left[\frac{T}{p}\right]\frac{\mathrm{d}p}{\mathrm{d}t} = 0, \tag{10.6e}$$

$$p = R\rho T, \tag{10.6f}$$

which is a closed system for p, ρ, T, $\mathbf{U}$ and $\mathbf{B}$, the pressure, density, temperature, velocity and magnetic induction with constant magnetic permeability μ. In dimensionless form in these equations we have five non-dimensional parameters:

$$\mathrm{St} = \frac{L_0}{t_0 U_0}, \quad \mathrm{Ro} = \frac{U_0}{2\Omega_0 L_0}, \quad M = \frac{U_0}{(\gamma R T_0)^{1/2}}, \quad \mathrm{Fr} = \frac{U_0}{(g L_0)^{1/2}}, \tag{10.7a}$$

$$A = \frac{U_0}{[B_0/(\mu\rho_0)^{1/2}]}, \tag{10.7b}$$

the Strouhal, Rossby, Mach, Froude and Alfvén numbers. Below it is assumed that A and M are simultaneously very small and satisfy the similarity relation [34]

$$A^2 = a_0 M \quad \text{with } a_0 = \mathrm{O}(1) \text{ is a constant.} \tag{10.8a}$$

In a quasi-steady case, $\mathrm{St} = 0$, the case when

$$\mathrm{Fr} \to 0, \text{ such that } A^2 = b^0\,\mathrm{Fr}^2, \text{ with } b_0 = \mathrm{O}(1), \text{ a second constant,} \tag{10.8b}$$

leads to limit (superscript '0'), $M = (b_0/a_0)\mathrm{Fr}^2 \to 0$, a static equilibrium approximation (Ro $= \infty$; see the paper by Grad [35]) written with dimensionless quantities:

$$(\nabla \wedge \mathbf{B}^0) \wedge \mathbf{B}^0 = \left(\frac{a_0}{\gamma}\right) \nabla p^1 + b_0 \mathbf{k} \tag{10.9a}$$

and

$$\nabla \cdot \mathbf{B}^0 = 0, \tag{10.9b}$$

where $p^1 = \lim[(p-1)/M]$ for $M \to 0$. If from $\nabla \cdot \mathbf{B}^0 = 0$, we write $\mathbf{B}^0 = \nabla\varphi \wedge \nabla\psi$, then equation (10.9a) gives $\mathbf{B}^0 \cdot \nabla P = 0$, or $P \equiv (a_0/\gamma)p^1 + b_0 z = \Pi(\varphi, \psi)$, and [36]

$$(\nabla \wedge \mathbf{B}^0) \cdot \nabla\varphi = \frac{\partial \Pi}{\partial \psi} \tag{10.9c}$$

and

$$(\nabla \wedge \mathbf{B}^0) \cdot \nabla\psi = -\frac{\partial \Pi}{\partial \varphi}, \tag{10.9d}$$

both these 'first integrals' (10.9c, d) being equivalent to (10.9a, b). When Ro $= c_0 A^2 = c_0 a_0 M \to 0$, instead of the two equations (10.9c, d), we derive the following limit equation (from the dimensionless form of (10.6a)):

$$\left(\frac{1}{c_0}\right)(\Omega \wedge \mathbf{U}^0) + \left(\frac{a_0}{\gamma}\right) \nabla p^1 + b_0 \mathbf{k} = (\nabla \wedge \mathbf{B}^0) \wedge \mathbf{B}^0, \tag{10.10a}$$

and if $T^1 = \lim[(T-1)/M]$ for $M \to 0$, then to the above limit equation (10.10a) we can associate the following set of quasi-steady limit equations [34]:

$$\nabla \cdot \mathbf{U}^0 = 0 \quad \text{and} \quad \nabla \cdot \mathbf{B}^0 = 0, \tag{10.10b}$$

$$\nabla \wedge (\mathbf{B}^0 \wedge \mathbf{U}^0) = 0, \tag{10.10c}$$

$$\mathbf{U}^0 \cdot \nabla T^1 = \left[\frac{(\gamma - 1)}{\gamma}\right] \mathbf{U}^0 \cdot \nabla p^1, \tag{10.10d}$$

$$\rho^1 = p^1 - T^1, \tag{10.10e}$$

which are consistent only when (because $\beta = \gamma(M/\mathrm{Fr})^2 \equiv (b_0/a_0)M \to 0$)

$$\frac{RT_0}{g} \gg L_0 \gg \frac{\mathbf{U}_0^2}{g}. \tag{10.10f}$$

If Fr $\to 0$ such that $\beta = \gamma(\mathrm{M}/\mathrm{Fr})^2 = \mathrm{O}(1)$ then, with (10.8a), a strongly coupled limit system is derived in [34], similar to a deep convection system, which is adapted to the study of atmospheres in various planets of the solar system (and, concerning this problem, see section 23 in Monin's book [37]) for which the characteristic angular velocity of rotation is not very high ($A^2/\mathrm{Ro} \ll 1$). This model system is well balanced for a description of magneto-convective motions within a relatively thick layer (when the gravity is low and reference temperature at the ground is high, then the layer is more thick). Another interesting case is linked with the limit

$$\beta = \beta^* M \quad \text{with } \beta^* = \mathrm{O}(1), \tag{10.10g}$$

and in this case, as limit steady-state system, we derive a set of model equations which is similar to a Boussinesq system for low Rossby number (see, for instance, [38]). This limit system seems pertinent for a rational analysis of the development of the sun-spot for which the magnetic and convection effects are strongly coupled.

Finally, we note that the theory of heavy-magneto-fluids at low Alfvén number is very similar to the theory of heavy-rotating-fluids at low Rossby number, and on the other hand we observe an analogy between the static equilibrium approximation (10.9a,b), in MHD, and the classical quasi-geostrophic approximation in dynamic meteorology [37].

Many papers deal also with the effect of a uniform magnetic field on the onset of Bénard–Marangoni convection in a layer of conducting fluid; in Wilson's paper [39] the reader can find various references. In particular, a vertical magnetic field always has a stabilizing effect, but when the free surface is deformable and includes a sufficiently large Marangoni number, it will always have unstable modes no matter how strong the applied magnetic field is! In a recent survey paper by Zhang and Schubert [40], the MHD phenomena, in rapidly rotating spherical systems, is discussed. (Although this is not the place for a full dicussion, I would like to mention a particularly interesting and singular effect: in the presence of a non-uniform magnetic field, most classical fluids – for instance, water and aquous solutions – can exhibit some diamagnetism or paramagnetism and their behavior may be strongly affected (convection suppressed, or convection present when the fluid is heated from below).

On the other hand, the domain of MHD effects in liquid metals is also quite important, namely when thermo-electric effects (for instance, the *Seebeck effect*) generate an electric current and, as soon as a magnetic field is present, results in an electrically driven flow.

Finally, I think that our above investigations can be generalized when in starting equations (10.6a–f) the viscosity and heat conduction are taken into account. In this case the equations are more complicated but the characteristic Ekman number being small, it is possible to consider a boundary layer approximation near the wall.

In a paper by Roberts and Soward [22], a complete dimensionless system of MHD equations is given (see [22, p. 126]). The first use of the energy method in MHD was by Rionero in 1967 (establishing existence of a maximizing solution in the energy variational problem) and in 1971 he also included the Hall effect.

The linear theory of Chandrasekhar (1981) shows that as the field strength is increased for the MHD convection problem with the magnetic field perpendicular to the layer, the magnetic field has a strongly inhibiting effect on the onset of convection motion. The first analyses to confirm this stabilizing effect from a nonlinear energy point of view are due to Galdi and Straughan (1985) and Galdi (1985) introduced a highly non-trivial generalized energy that contains gradients as well as the fields themselves.This energy has some resemblance to the one needed to obtain stabilization in the rotating Bénard problem (for references of the above papers by Rionero, Chandrasekhar, Galdi and Straughan, see [1]).

Now, concerning Electro-Hydrodynamic, EHD, Convection, the branch of fluid mechanics concerned with electric force effects, this topic is relatively new and has been attracting increasing attention in the theoretical and engineerng literature. Some aspects of this EHD convection are well presented in Straughan's book [1, chapter 11], who gives Rosensweig's [41, pp. 1, 2] explanation:

> the force interaction arising in EHD is often due to the free electric charge acted upon by an electric force field.

In [1], most of the analysis described is via linear instability theory, since energy theory has been so far successful only in certain cases. This is indeed a rich area for future research. Roberts [42] first allows the dielectric constant of the fluid to vary with temperature T, and assumes the homogeneous insulating fluid at rest in a layer (with d as depth) with vertical, parallel applied gradients of temperature and electrostatic potential V. The electric displacement is denoted by $\mathbf{D}$, and the body force per unit volume $\mathbf{f}$ on an isotropic dielectric fluid is given (see in [43, eq. (15.15)] or in [1, p. 163]) as a function of the pressure p, electrical field $\mathbf{E}$ and density ρ, by

$$\mathbf{f} = -\nabla p + \left(\frac{\rho}{8\pi}\right)\nabla\left\{E^2\left[\frac{\partial\varepsilon}{\partial\rho}\right]_T\right\} - \left(\frac{E^2}{8\pi}\right)\left[\frac{\partial\varepsilon}{\partial T}\right]_\rho \nabla T. \tag{10.11a}$$

The appropriate Maxwell equations are

$$\nabla \cdot \mathbf{D} = 0, \quad \operatorname{curl} \mathbf{E} = 0 \quad \text{with } \mathbf{D} = \varepsilon(\rho, T)\mathbf{E} \text{ and } \mathbf{E} = -\nabla V. \tag{10.11b}$$

For a Newtonian fluid, from a similar to a Boussinesq approximation, Roberts considers the following equations (u_i are the components of the velocity vector $\mathbf{u}$):

$$\frac{\partial u_i}{\partial t} + u_j \frac{\partial u_i}{\partial x_j} = g_i + \nu \mathbf{\Delta} u_i, \tag{10.11c}$$

with

$$g_i = -\frac{\partial \omega}{\partial x_i} - \left(\frac{1}{\rho_0}\right)\left(\frac{E^2}{8\pi}\right)\left[\frac{\partial \varepsilon}{\partial T}\right]_\rho \frac{\partial T}{\partial x_i} - g[1 - \alpha(T - T_0)]\mathbf{k}_i, \tag{10.11d}$$

g being gravity, α the thermal expansion coefficient, and $\mathbf{k} = (0, 0, 1)$, with

$$\omega = \left(\frac{p}{\rho_0}\right) - \left(\frac{E^2}{8\pi}\right)\left[\frac{\partial \varepsilon}{\partial \rho}\right]_T. \tag{10.11e}$$

For the given T we have

$$\frac{\partial T}{\partial t} + u_j \frac{\partial T}{\partial x_j} = \kappa \mathbf{\Delta} T, \tag{10.11f}$$

κ being the thermal diffusivity.

It seems to me that the first problem in the framework of a rational analysis and asymptotic modelling is to derive, with (10.11a), for $\mathbf{f}$ – as a phenomenological relation – a dimensionless system in the compressible, viscous and heat conducting with viscous dissipation case. Then one can verify whether the above Roberts, approximate (*ad hoc*) system is really derived. Unless such a procedure is implemented, it seems to me that it is difficult to trust the further results of Roberts in [1, chapter 11].

Another approximate model was also considered by Turnbull [44], which, curiously, considers that:

Ohm's law (J is current and Q the volume space charge),

$$\mathbf{J} = \sigma \mathbf{E} + Q\mathbf{u}, \quad \text{with charge conservation, } \partial Q/\partial t + \nabla \cdot \mathbf{J} = 0, \tag{10.11g}$$

is only an approximation for poorly conducting liquids, and, therefore, it makes no sense to solve the equations exactly since the model is only an approximation! A third contribution was made by Deo and Richardson [45] which considers a generalized energy method in EHD stability theory. According to Straughan:

> ... the paper [45] represents a substantial contribution to energy stability theory, especially since, as far as I can determine, it is the first to address the very interesting EHD problem.

The authors of [45] work with dimensionless quantities and introduce various non-dimensional parameters, in this case the governing equations of the considered problem ('charge injection induced instability and the nonlinear energy stability analysis'; see [1, section 11.4]) admit a steady one-dimensional hydrostatic equilibrium solution with the equilibrium electric potential. As a consequence, a system of equations for perturbations in liquid velocity, electric field and space-charge density is derived in [45, pp. 173, 174]. From my point of view, precisely the consistency of this system (making sure that all the terms in equations of this system are the same order of magnitude) is an interesting problem in a framework of a rational analysis and asymptotic modelling. Only in this case do the results of Deo and Richardson really represent a 'substantial contribution'!

Now, Ferro-Hydrodynamic, FHD, Convection, which is the subject of chapter 12 in [1], is of great interest because the fluids of concern possess (as is written in [1]) a giant magnetic response, as a consequence, in particular, of a spontaneous formation of a labyrinthine pattern in thin layers and, also, the enhanced convective cooling in a ferrofluid that has a temperature-dependent magnetic moment. The book by Rosensweig [41] is a perfect introduction to this fascinating subject, and in [1, chapter 12] the reader can find the 'relevant basic equations of FHD'. It seems that the papers by Cowley and Rosensweig [46] and Gailitis [47] are interesting for the problem of interfacial stability of a ferromagnetic fluid. On the other hand the temperature dependence of magnetization is important in thermo-convective instability in FHD.

Chapter 13 of [1] is devoted to 'Chemical Convection'. In particular, the phenomenon of double-diffusive convection in a fluid layer, where two scalar fields (see Section 10.5 here, where these scalar fields are heat and salinity concentration) affect the density in fluid, is closely related to this CC; the behaviour in the double-diffusive case is often more diverse than for the Bénard classical convection problem. But, when temperature and one or more species are present and interactions between species are allowed, then the system becomes increasingly richer (see, for instance, the comments in [48]); as a consequence, reaction-diffusion equations for mixtures of viscous fluids play an important role in everyday life.

Unfortunately, the relevant equations for various cases are usually written down in an *ad hoc* manner and often vary considerably. A rational derivation of a consistent system of model equations would seem appropriate, and in

[1, chapter 13], the reader can find some results in this direction, i.e., basic equations for a chemical reacting mixture (according to Morro and Straughan [49]), a model for reactions far from equilibrium and convection in a layer.

Finally, we note that recently in [50] the authors are interested in the influence of Marangoni effects on the propagation of chemical fronts in thin solution layers, open to the air, in the absence of any buoyancy- driven effects.

Many fascinating solar phenomena (in particular, Solar Convection) can be attributed to the influence of the Sun's magnetic field, well described in the review paper of Hughes and Proctor [25]. According to this paper:

> The interior of the Sun (which has a radius of 7×10^5 km) is divided into three main sections – an inner core (occupying 25% by radius) in which the (thermo)-nuclear reactions take place, an intermediate region (extending to approximately 70% of the solar radius) where heat is transported by radiation, and an outer region where heat is carried by convective motions.

In [25] the authors are concerned with the magnetic field in, and just below, the convection zone and they concentrate their investigations only on two particular facets of solar magneto-hydrodynamics: first, on the behaviour of a vertical field in the surface layers of the Sun and, second, on the evolution of a horizontal field in somewhat deeper regions.

The complex dynamo mechanism by which the field is regenerated and the effects due to solar rotation are not considered in [25]. The interesting feature is the convection in the surface layers of the Sun which is dominated by cells of two discrete scales: granules with average size 1500 km, and supergranules that are about 20 times as large. It is postulated that beneath these cells the heat will be transported by giant cells spanning almost the entire depth of the convection zone! The use of Boussinesq approximation does not permit discussion of several important phenomena. In fact, the convection zone of the Sun is highly turbulent and the fields within it highly intermittent; the time scale for changes in the granulation pattern is only 15 minutes or so, and that for the supergranulation pattern a few hours. In any case, it is hard to justify a theory in which viscosity (presumably turbulent) is retained while magnetic diffusivity (typically of comparable size) is ignored!

More recent work has attempted to set the problem of flux-sheet concentration and evacuation in the context of a proper treatment of nonlinear compressible convection in the framework of the full equations – for instance in [25, eqs. (2.1)–(2.5), pp. 189–190], describing convection in the outer regions of the Sun. However, it seems that a full theory of radiative transfer is required to give a correct model of convection (with or without a magnetic field) in the surface layers of the Sun! In fact, equations (2.1)–(2.5) in [25] are extremely complicated, even in the absence of a magnetic field. We observe also that the curvature forces exerted by the magnetic field can lead to the propagation of waves along field lines (giving an oscillatory convection). In [25] the reader can find various pertinent references up to 1987.

Again in [1, chapter 18], the reader can find a review of the application of nonlinear stability in Ocean Circulation Models. In [1], first the stability of wind-driven convective motions in the upper layers of the ocean – known as *Langmuir circulations* – is considered in the framework of the Lebovich and Paolucci paper [51]. Then, the Stommel–Veronis (quasi-geostrophic) model is considered in [1, chapter 18, according to a series of papers by Crisciani and his co-workers relative to nonlinear energy stability (see, for instance [52]). This quasi-geostrophic model is relative to a streamfunction of two-dimensional motion and is developed and discussed in detail in chapter 2 of Pedlosky's book [53].

On the other hand, the review paper by McWilliams [27] is devoted to history, formulation and solution of numerical models for oceanic general circulation under equilibrium surface wind stress and buoyancy flux forcing. In the framework of the theory of convection in fluids, obviously the fundamental fluid dynamics equations of oceanic circulation are the Navier–Stokes(–Fourier) equations on the rotating Earth for a compressible liquid, seawater, which is comprised of water plus a suite of dissolved salts that occur in nearly constant ratio but variable amount (the salinity **s**), with an empirically determined equation of state. In fact, the model equations used (see, for instance, the very pertinent paper [54] by Veronis, who considers the 'large amplitude Bénard convection in a rotating fluid') are based on several substantial simplifications to this above-mentioned fundamental set! An interesting convection aspect is the 'buoyancy-driven circulation', as the geographical patterns of buoyancy forcing differ for heat and fresh water. Tropical heating and polar cooling tend to force a circulation with sinking at high latitudes, whereas tropical excess evaporation (largest somewhat away from the equator) and polar excess precipitation tend to force sinking at low latitudes. This creates the possibility of multiple equilibria with alternative overturning circulation patterns.

It is necessary to observe that quasi-geostrophic approximation, large-scale (planetary) geostrophic equations, balance equations, and primitive equations are usually used for atmospheric motions; see, for instance, the books [13, 55] and also [21], where these model equations are derived from a careful rational analysis coupled with an asymptotic modelling. I think that such a consistent approach is also possible in the problem of the derivation of model approximate equations for oceanic circulation – a remarkable (and rather unusual; see, for instance, our recent paper [56]) aspect of such an approach is the possibility (by asymptotic matching) to associate to the main non-adiabatic, dissipative hydrostatic (N-A DH) model evolution equations, two local systems of model equations – on the one hand near the initial time (for an unsteady adjustment (UA) and, on the other hand for a meso- scale prediction (which takes into account the influence of the vertically propagating gravity waves (M-SP)).

The N-ADH main equations are derived via the limit process:

$$\varepsilon = \frac{H}{L} \to 0 \quad \text{and} \quad \mathrm{Re} = \frac{UL}{\nu} \to \infty, \quad \text{such that } \varepsilon^2\,\mathrm{Re} = \mathrm{O}(1), \tag{10.12a}$$

or

$$L \approx \left(\frac{U}{\nu}\right) H^2, \tag{10.12b}$$

for the velocities (u_M, v_M, w_M), pressure p_M, density ρ_M and temperature T_M, with the space-time dimensionless coordinates (t, x, y, z) fixed and $\mathrm{O}(1)$. More precisely,

$$x = a\cos\varphi_0\lambda, \quad y = a(\varphi - \varphi_0), \quad z = r - a, \tag{10.12c}$$

where φ_0 is a reference latitude – the origin of this right-handed curvilinear coordinateless system lies on the Earth's surface (for a flat ground we have $r = a$) at latitude φ_0 and longitude $\lambda = 0$.

The UA local equations, valid near initial time, are derived also from (10.12a), for the velocities ($u_l = u_M$, $v_l = v_M$, and $w_l = \varepsilon w_M$), pressure $p_l = p_M$, density $\rho_l = \rho_M$ and temperature $T_l = T_M$, but with a short dimensionless time

$$\tau = \frac{t}{\varepsilon^2} \quad \text{and space coordinates } x, y, \zeta = z/\varepsilon \text{ fixed and } \mathrm{O}(1). \tag{10.12d}$$

The M-SP local equations, valid in a meso-scale region situated around the point $(x^0, y^0, z = 0)$, are also derived from (10.12a), for (u_l, v_l, w_l), p_l, ρ_l and T_l, but, as functions of τ^* and x^*, y^* and z fixed and $\mathrm{O}(1)$, where

$$\tau^* = \frac{t}{\varepsilon}, \quad x^* = \frac{(x - x^0)}{\varepsilon}, \quad y^* = \frac{(y - y^0)}{\varepsilon} \quad \text{and } z \text{ are fixed and } \mathrm{O}(1). \tag{10.12e}$$

In particular, both of the above local models, with (10.12d) and (10.12e), give by matching the possibility to obtain, on the one hand, consistent initial conditions for the main N-ADH model with (10.12a, b) for a 'large' weather prediction and, on the other hand, the required lateral (horizontal) conditions at infinity for the meso-scale-regional prediction, once the large weather prediction by N-ADH model is known at the position $(x_0, y_0, z = 0)$. This allows us to take into account the meso-scale topographic roughness which also influences large-scale circulation (neglected in the large weather prediction by the N-ADH model).

Finally, concerning the problems related to the atmospheric and oceanic general circulations, see chapter 8 in Monin's book [57]; on the other hand, the book by Marchuk and Sarkisyan [58] gives a very pertinent and complete account of various computational algorithms for numerical simulations of ocean circulation.

Concerning Rotating Convection, see the two references in the beginning of Section 10.3. The article by Knobloch [31], cited there for work as recent as 1998, is a pertinent complement to chapter III of Chandrasekhar's famous book [32]. Knobloch summarizes some of the developments (up to 1996) in bifurcation theory and their relevance to the study of rotating convection. From the point of view of these authors:

> ... undoubtedly, the most fundamental property of rotating systems is the absence of steady-state bifurcations. As a result patterns in rotating systems invariably precess.

In [31] the author noted that this fact is related to the absence of reflection symmetry in vertical planes in such systems, and pointed out that the imposition of translation invariance in addition to rotation invariance changes the above picture in a dramatic way.

In [31, section 2], relative to convection in a rotating cylinder, the author writes for the case of small Froude numbers the equations of motion, nondimensionalized with respect to the thermal diffusion time in the vertical, in the following form:

$$\left[\left(\frac{1}{\mathrm{Pr}}\right)\frac{D}{Dt} - \nabla^2\right]\mathbf{u} = -\nabla p + \mathrm{Ra}\Theta\mathbf{k} + 2\Omega^0\mathbf{u}\wedge\mathbf{k}, \tag{10.13a}$$

$$\left[\frac{D}{Dt} - \nabla^2\right]\Theta = w, \tag{10.13b}$$

$$\nabla\cdot\mathbf{u} = 0, \tag{10.13c}$$

where $\mathbf{u} = u\mathbf{e}_r + v\mathbf{e}_\varphi + w\mathbf{k}$ is the velocity field, Θ and p are the departures of the temperature and pressure from their conduction profiles, and $\mathbf{k}$ is the unit vector in the vertical direction. In equation (10.13a) we have

$$\Omega^2 \equiv \frac{d^2 \Omega_{\text{phys}}}{\nu}, \quad \text{Ra} \equiv \frac{g\alpha \Delta T d^3}{\kappa \nu}, \quad \frac{1}{\text{Pr}} \equiv \frac{\kappa}{\nu}, \tag{10.13d}$$

Ω^0 being the dimensionless angular velocity. We observe that, only when the Froude number $d(\Omega_{\text{phys}})^2/g$ is small, one can suppose that gravity is essentially vertical everywhere within a domain of characteristic size d. In [31], Knobloch considered an interesting nonlinear regime, perhaps the most interesting phenomenon in a rotating layer, the Küppers–Lortz (KL) instability (see [59]; this KL instability is, in fact, related to the absence of reflection symmetry in vertical planes in horizontally unbounded layers). When instability in a pattern of parallel rolls becomes unstable to another set of rolls oriented at an angle with respect to the first, once the rotation rate exceeds a critical value, this new set is itself unstable in the same fashion, etc., resulting in complex dynamics right at the onset. In [31] Knobloch discusses the KL instability for traveling waves and shows that structurally stable heteroclinic cycles involving traveling roll states are possible, and result in quite unusual dynamical behavior.

Concerning the theory of dynamical systemm and chaos, the reader can consult the book by Wiggins [60]. In [61], the reader can find numerous references and comments concerning various aspects of the rotating Rayleigh–Bénard convection, for instance, rotating about a vertical axis [62, 63], pattern formation [64], direct transition to turbulence [65], pattern dynamics [66], and chaotic domain structure [67].

Furthermore, concerning stability results, I mention the paper by Mulone and Rionero [68] where new stability results for any Prandtl and Taylor numbers are obtained.

Finally, concerning Penetrative Convection, we mention the book of Straughan [69] devoted to 'mathematical aspects' and also chapter 17 of his more recent book [1] where, at the beginning of this particular chapter, the author pertinently writes

> A pioneering piece of work on penetrative convection is the beautiful paper (1963) of Veronis [70].

From the physics' point of view, there are many geophysical and astrophysical settings which involve a layer of gravitationally stable fluid, bounded above or below by a layer which is in convective motion. A typical example is the case of the Earth's atmosphere bounded below by the ground (or ocean). This bounding surface is heated by solar radiation, and the air close to the surface then becomes warmer than the upper air, and so a gravitationally unstable system results; when convection occurs, the warm air rises and penetrates into regions that are stably stratified.

There are various ways to obtain the penetrative effect of convection. First, if one assumes (as Veronis does in [70]) that the equation of state is no longer linear in the temperature, one introduces a 'non-Boussinesq' effect. A second way is to assume, as in a classical Boussinesq approximation case, that the density is linear in the temperature field, but introduce an internal heat source in the layer via the equation for the temperature. A third way is, in fact, the case similar to the atmospheric case when the lower bounding surface is heated by a radiating heating. For instance in a usual Rayleigh–Bénard shallow convection system of equations, as a boundary condition for the temperature T, at $z = 0$, we write a heat flux condition (with $q^0 = \text{const}$)

$$\kappa \frac{\partial T}{\partial z}\bigg|_{z=0} = \kappa q^0, \tag{10.14}$$

and when $q^0 > 0$, this means heat is being taken out of the fluid layer, whereas $q^0 < 0$ means that heat is being put into the fluid layer through the boundary $z = 0$.

Obviously, in a more general case, the three cases discussed above can be combined; in particular, I invite the reader to read [1, pp. 347, 348], where a beautiful experiment (performed by Krishnamurti [71]) is described.

A fourth way to study penetrative convection is via a multi-layer theory, by assuming there are two (or more) layers present with different temperature gradients, at least one being destabilizing, with suitable boundary conditions at the interface(s).

Penetrative convection in fluid layers with internal heat sources is considered by Ames and Straughan in [72]. Penetrative convection in the upper ocean due to surface cooling is the subject of a paper by Cushman-Roisin [73]. Denton and Wood [74] consider penetrative convection at low Péclet number. Numerical studies of penetrative convective instabilities are considered by Faller and Kaylor in [75]. For the penetrative convection in porous media, see the paper by George et al. [76]. Penetrative convection in rotating fluids, a model for the base of stellar convection zones, is considered by Jennings in [77]. Penetrative convective flows induced by internal heating and mantle compressibility is investigated by Machetel and Yuen in [78]. For a model for the onset of penetrative convection, see Matthews [79]. Nonlinear penetrative convection is studied by Moore and Weiss in [80]. In [81], Straus considers penetrative convection in a layer of fluid heated from within, and Zahn et al. in [82], perform a nonlinear modal analysis of penetrative convection.

10.4 Averaged Integral Boundary Layer Approach: Non-Isothermal Case

We have already explained, in Section 7.5, the main point of the 'averaged, integral boundary layer (IBL)' technique, for isothermal (Shkadov in 1967 [83]) and non-isothermal cases (first considered by Zeytounian in 1998 [3], see the system (7.94a–c) in Section 7.5). In fact, the non-isothermal system (7.94a–c) was derived in 1995 (during our employment at the LML of the University ST of Lille) in the framework of our research related to obtaining finite-dimensional dynamical-system models for numerical simulation of the destabilizing Marangoni effect on route to chaos and in appearance of strange attractors (see, for instance [3, section 4.4]). Thanks to an invitation from Professor Manuel G. Velarde, Director de la Unidad de Fluidos del Instituto Pluridisciplinar de la Universidad de Madrid (Spain), in November 1999, I had the opportunity to expound a part of this research, including obtaining of this system of non-isothermal averaged equations, similar to (7.94a–c), which, from my point of view, is of unquestionable interest for investigation of the destabilizing Marangoni effect. Curiously, in 2003, Kalliadasis et al. [84] also considered the non-isothermal case but did not mention my 1998 approach; these authors included heat transport effects and obtaining of an averaged energy equation for temperature distribution on the free surface of a film heated from below by a local heat source,

$$T_w = f(x). \tag{10.15a}$$

In sections 4 and 5 of the paper by Trevelyan and Kalliadasis [85] (where also my 1998 approach is not mentioned) the reader can find, as a complement to [84], a simple explanation of the weighted-residuals approach. For the isothermal falling-film problem, Ruyer-Quil and Manneville [86, 87], showed that a Galerkin projection for the velocity field with just one test function (the self-similar profile assumed by Shkadov in 1967), and with the weight function as the test function itself, fully corrects the critical Reynolds number obtained from the Shkadov IBL approximation; in fact, the Kármán–Pohlhausen averaging method employed by Shkadov in 1967 can be viewed as a special weighted-residual method with as weight function $w_u \equiv 1$, and the reader is referred to [85] for details.

On the other hand, in [4], Kiyashko is 'replaced' by Ruyer-Quil and Velarde, both of whom are familiar with my 1998 review paper and with both of whom I had fruitful discussions at the Instituto Pluridisciplinar (Madrid), during my several visits in the years 2000–2003. Pages 308 and 309 in [4] are devoted to a treatment of the energy equation, but assuming that, in equation (10.15a), $f(x) \equiv 0$. It is shown that the IBL treatment of the energy equation adopted by Kalliadasis, Kiyashko and Demekhin in [84] is effectively a 'tau' method [88] in which the trial function does not satisfy the equation or all boundary conditions. Like the momentum equations, the first step is the assumption of a self-similar temperature profile

$$T(t, x, y, z) = b(t, x, z)g(\eta) \quad \text{with } \eta = y/h(t, x), \tag{10.15b}$$

where the amplitude b and the test function g have to be specified. Such a functional form was originally proposed by Zeytounian [3] for the case $\mathrm{Bi} = 0$. Zeytounian used for the amplitude b the averaged temperature across the film $\int_0^h T\,\mathrm{d}y$. Then Kalliadasis et al. [4] chose to put the emphasis on the temperature at the interface $T_{y=h}$ since it appears directly in Newton's law of cooling and the assumed velocity profile; therefore,

$$b \equiv T_{y=h} \quad \text{and} \quad g(1) = 1 \Rightarrow T = T_{y=h}\eta. \tag{10.15c}$$

In fact, like the velocity profile, the temperature profile is assumed corresponding to the flat film solution, and the assumption is that the linear temperature profile obtained for a flat film $T_S = T_S(y/d)$ persists even when the interface is no longer flat. It is clear that this temperature distribution does not satisfy Newton's law of cooling; in fact this mixed Dirichlet–Neumann boundary condition cannot be satisfied simply by choosing function g in (10.15b), the approximation in (10.15c) for the temperature T being a variant of the Galerkin method (called the 'tau' method). Because the averaged diffusive term across the film requires knowledge of the temperature gradient on the wall $y = 0$, it is necessary to consider an inner product with a weight function $W(\eta)$ applying to the energy equation. When as $W(\eta)$ we choose a parabolic profile $2\eta - \eta^2$, then we obtain

$$\int_0^h (2\eta - \eta^2)\frac{\partial^2 T}{\partial y^2}\,\mathrm{d}y = \left.\frac{\partial T}{\partial y}\right|_{y=h} - \left(\frac{2}{h^2}\right)\int_0^h T\,\mathrm{d}y, \tag{10.15d}$$

which leads to the choice made by Zeytounian [3] of an amplitude corresponding to the averaged temperature across the flow. In [4, 84, 85], as $W(\eta)$, we have simply η and

$$\int_0^h \eta\frac{\partial^2 T}{\partial y^2}\,\mathrm{d}y = \left.\frac{\partial T}{\partial y}\right|_{y=h} - \left(\frac{1}{h}\right)[T_{y=h} - T_w] \tag{10.15e}$$

evaluates the term $\partial T/\partial y|_{y=h}$ from Newton's law of cooling and not (10.15c), applying all boundary conditions prior to substituting the linear approximation for T in (10.15c). As a result, although (10.15c) does not satisfy the free-surface boundary condition, the averaged energy equation (*à la* Kalliadasis, the two-dimensional case))

$$\frac{\partial T_{y=h}}{\partial t} + (7/40)\left[\frac{T_{y=h}}{h}\right]\frac{\partial q}{\partial x} + (27/20)\left[\frac{q}{h}\right]\frac{\partial T_{y=h}}{\partial x}$$
$$+ \left(\frac{3}{\text{Pé}}\right)\text{Bi}\left(\frac{1}{h}\right)(T_{y=h} - T_A) + \left(\frac{1}{h^2}\right)T_{y=h} = 0, \quad (10.15\text{f})$$

does!

In 2005, Ruyer-Quil et al. [5], and Scheid et al. [6] attacked the problem of the modellization of a thermocapillary flow (a thin liquid film) falling down a uniformly heated wall. Their approach (developed by Ruyer-Quil and Manneville [86, 87, 89]) was to use a gradient expansion combined with a Galerkin projection with polynomial test functions-weighted-residuals for both velocity and temperature fields. In particular (see [5, p. 208], the set of equations (4.18a–c)), for $q = \int_0^h u\,\mathrm{d}y$, h and $\theta \equiv T_{y=h}$, a model consistent at $\mathrm{O}(d/\lambda)$ can be formulated in a long-wave approximation. Specifically, we derive the following three evolution equations:

$$\frac{\partial h}{\partial t} = -\frac{\partial q}{\partial x}, \quad (10.15\text{g})$$

$$\frac{\partial q}{\partial t} = (5/6)h - (5/2)\left[\frac{q}{h}\right]\frac{\partial q}{\partial x} + \left\{(9/7)\left[\frac{q}{h}\right]^2 - (5/6)\cot\beta h\right\}\frac{\partial h}{\partial x}$$
$$- (5/4)\text{Ma}\,\frac{\partial \theta}{\partial x} + (5/6)\Gamma h\frac{\partial^3 h}{\partial x^3}, \quad (10.15\text{h})$$

$$\text{Pr}\,\frac{\partial \theta}{\partial t} = \left(\frac{3}{h^2}\right)[1 - (1 + \text{Bi}\,h)\theta]$$
$$+ \text{Pr}\left\{(7/40)\left[\frac{(1-\theta)}{h^2}\right]\frac{\partial q}{\partial x} - (27/20)\left[\frac{q}{h}\right]\frac{\partial \theta}{\partial x}\right\}, \quad (10.15\text{i})$$

where

$$\Gamma = \frac{\sigma(T_A)}{\rho_0 \nu^{4/3}(g\sin\beta)^{1/3}} \quad (10.15\text{j})$$

is the Kapitza number.

This set of equations (10.15g–i) can be contrasted with the model derived by Kalliadasis et al. [4] and Kalliadasis et al. [84]. The functional form of the first-order averaged heat equation (10.15i) is very similar to equation (10.15f), since for the derivation of both equations, the temperature across the film is essentially a self-similar profile with weight function for the energy equations chosen within the Galerkin framework, e.g., the first-order polynomial y/h. The two averaged heat equations (10.15i) and equation (10.15f) differ in [4] only in the choice of scaling for the temperature and the presence of a Prandtl number in (10.15i) instead of a Péclet (Pe) number in (10.15f). In fact, the two models really differ in the treatment of the momentum equation; equation (10.15h) contains the same terms with the corresponding averaged momentum equation in [4, 84] but with different coefficients. In these equations we have *seven terms*, while in our system of averaged equations (7.94a–c) *eight terms* are present in equation (7.94a) for q and the coupling is stronger. In Ruyer-Quil et al.'s 2005 paper [5], a so-called 'reduced regularized model' of three equations is also derived for q, h and θ:

$$\frac{\partial h}{\partial t} = -\frac{\partial q}{\partial x}, \tag{10.15k}$$

$$\begin{aligned}\frac{\partial q}{\partial t} &= (9/7)\left[\frac{q}{h}\right]^2\frac{\partial h}{\partial x} - (17/7)\left[\frac{q}{h}\right]\frac{\partial q}{\partial x} \\ &\quad + \left\{1 - (1/70)q\frac{\partial h}{\partial x} + (5/56)\mathrm{Ma}\left(\frac{1}{h}\right)\frac{\partial \theta}{\partial x}\right\}^{-1}\Bigg[(5/6)h \\ &\quad - (5/2)\left[\frac{q}{h^2}\right] + 4\left[\frac{q}{h^2}\right]\left(\frac{\partial h}{\partial x}\right)^2 \\ &\quad - (9/2h)\left(\frac{\partial q}{\partial x}\right)\frac{\partial h}{\partial x} - 6\left[\frac{q}{h}\right]\frac{\partial^2 h}{\partial x^2} + (9/2)\frac{\partial^2 q}{\partial x^2} \\ &\quad - (5/6)\cot\beta h\frac{\partial h}{\partial x} + (5/6)\Gamma h\frac{\partial^3 h}{\partial x^3} \\ &\quad - \mathrm{Ma}\left[(5/4)\frac{\partial \theta}{\partial x} - (1/224)hq\frac{\partial^2\theta}{\partial x^2}\right]\Bigg],\end{aligned} \tag{10.15l}$$

$$\begin{aligned}\mathrm{Pr}\frac{\partial \theta}{\partial t} &= \left(\frac{3}{h^2}\right)[1 - (1 + \mathrm{Bi}\,h)\theta] + \mathrm{Pr}\Bigg[(7/40)\left[\frac{(1-\theta)}{h}\right]\frac{\partial q}{\partial x} \\ &\quad - (27/20)\left[\frac{q}{h}\right]\frac{\partial \theta}{\partial x}\Bigg] + [1 - \theta - (3/2)\mathrm{Bi}\,h\theta]\left[\frac{(\partial h/\partial x)}{h}\right]^2\end{aligned}$$

$$+\left(\frac{1}{h}\right)\left(\frac{\partial\theta}{\partial x}\right)\frac{\partial h}{\partial x}+(1-\theta)\left(\frac{1}{h}\right)\frac{\partial^2 h}{\partial x^2}+\frac{\partial^2\theta}{\partial x^2}. \qquad (10.15\text{m})$$

Despite the limitations of the above reduced regularized model (10.15k–m) for large Péclet numbers, the model has substantially extended the region of validity of the Benney long-wave expansion which exhibits a turning point with branch multiplicity at an O(1) value of Re, and for all Péclet numbers (see [4]), while in these regions the above model has no turning point and predicts the continuing existence of solitary waves for all Reynolds numbers. But the reduced regularized model (10.15k–m) should give results in reasonable agreement with experiments for waves of smaller amplitude for which no recirculation zones are observed.

10.5 Interaction between Short-Scale Marangoni Convection and Long-Scale Deformational Instability

It is important to note that in the case of the Bénard–Marangoni convection in a liquid layer with a deformable interface, as was previously shown by Takashima [90] through a linear stability analysis (see Section 7.6), there exist two monotonous modes of surface tension driven instability. One, a short-scale mode, is caused by surface tension gradients alone, without surface deformation, and it leads to formation of a stationary convection with a characteristic scale of the order of the liquid layer depth. The other, a long-scale mode, is influenced also by gravity and capillary (Laplace) forces, and surface deformation plays a crucial role in its development. This instability mode results in a large-scale convection and in the growth of long surface deformations in which the characteristic scale is large in comparison with the thickness of the liquid layer. As shown by Golovin et al. [7], these two types of Marangoni convection, having different scales, can interact with each other in the course of their nonlinear evolution. There are two mechanisms of coupling between them. On the one hand, surface deformation changes locally the Marangoni number, which depends on the depth of the liquid layer; this leads to a space-dependent growth rate of the short-scale convection and, hence, its intensity becomes also space dependent. On the other hand, the short-scale convection generates an additional mean mass/heat flux from the bottom to the free surface, which is proportional to its intensity (square of the amplitude). When the intensity is not uniform, this leads to additional long-scale surface tension gradients affecting the evolution of the long-scale mode. Indeed, the coupling effects are most pronounced in the case when the long-and the short-scale modes have instability thresholds close to each other. In the above cited paper [7], Golovin et al. studied these effects analytically (via a weakly nonlinear analysis) in the vicinity of the instability thresholds. Close to the bifurcation point, the mean long-scale flow generated by the short-scale convection is very weak, to the order of ε^3, and it will considerably affect the long-scale surface deformations only if the latter are also small, to the order of ε^2, where ε is the amplitude of the deformationless convective mode ; this happens when the surface tension is sufficiently large. According to Golovin et al. [7], near the instability threshold, the nonlinear evolution and interaction between the two modes can be described by a system of two coupled nonlinear equations (derived by a multiscale asymptotic technique with the elimination of singular terms, as in Section 5.4 and also in Section 7.6). Namely:

$$\frac{\partial A}{\partial T} = (\pm A) + \frac{\partial^2 A}{\partial x^2} + A|A|^2 + \eta A, \tag{10.16a}$$

$$\frac{\partial \eta}{\partial T} = -(\pm m)\frac{\partial^2 \eta}{\partial x^2} - w\frac{\partial^4 \eta}{\partial x^4} + s\frac{\partial^2 |A|^2}{\partial x^2}, \tag{10.16b}$$

where the parameters m, w, s are all positive. The parameter m characterizes the effect of surface tension gradients and gravity, the parameter w corresponds to the Laplace pressure and s is the interaction parameter characterizing the coupling between the two modes of Marangoni convection. The complex amplitude of the short-scale convection A undergoes a long-scale evolution, described by the Ginsburg–Landau equation (10.16a), but the latter, however, contains an additional nonlinear (quadratic) term ηA, connected with the surface deformation η. This term adds to the linear growth rate for the amplitude A and describes, in fact, a non-uniform space-dependent supercriticality. It plays a stabilizing role when the surface is elevated ($\eta > 0$), and suppresses the short-scale convection under the surface deflections. The surfave deformation η is governed by a nonlinear evolution equation of the fourth order (10.16b). However, the only nonlinear term in this equation is the coupling term, $s\partial^2|A|^2/\partial x^2$, describing the effect of the mean flow generated by the short-scale convection; this term always plays a stabilizing role.

In two equations (10.16a, b) the various signs of the terms correspond to the four cases described by Golovin et al. in [7]. In fact:

(i) when in (10.16a) we have $+A$ and in (10.16b) $+m$, both the short-scale deformations mode and the long-scale deformational one are unstable;
(ii) if in (10.16a) we have $+A$ and in (10.16b) $-m$, only the short-scale mode is unstable;

(iii) when in (10.16a) we have $-A$ and in (10.16b) $+m$, the deformational mode is unstable;
(iv) if in (10.16a) we have $-A$ and in (10.16b) $-m$, both modes linearly stable, but their nonlinear interaction may lead to an instability.

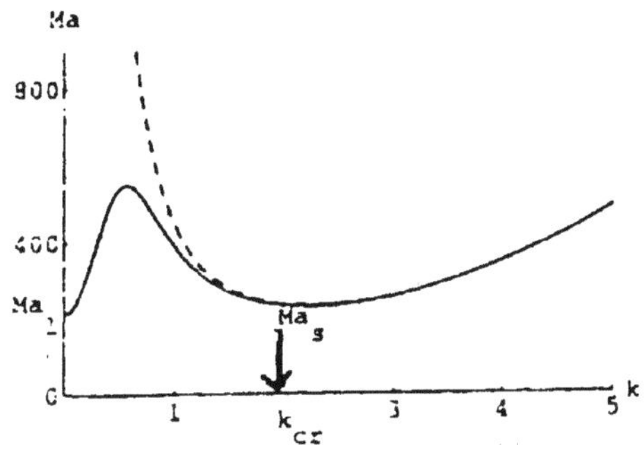

Fig. 10.1 Neutral stability curves for the Marangoni convection. Reprinted with kind permission from [7].

The typical neutral stability curve Ma(k), represented in Figure 10.1, has two minima: first Ma_l corresponding to $k = 0$, describes the long-wave instability, and then Ma_s, related to $k_c \neq 0$, indicates the threshold of short-scale convection. In Figure 10.1, the dashed line corresponds to the layer with undeformable interface and in this case only one minimum exists, Ma_s.

The surface deformation η is a real quantity, whereas the amplitude A of the small-scale convection is complex. If we assume, for the sake of simplicity, that A is also real, thus considering the evolution of the short-scale convective structure with a fixed wave number $k = k_c$, then we write for the derivation of a three-mode truncated model

$$A(T) = A_0(T) + A_1(T)\cos(k_m x) + \cdots, \qquad \eta = B_1(T)\cos(k_m x) + \cdots, \tag{10.16c}$$

with $k_m = (m/2w)^{1/2}$, corresponding to the maximum linear growth rate of the first harmonic. In this case, substituting (10.16c) into (10.16a, b) and considering the third case ($-A$, $+m$), after appropriate rescaling, we obtain the following dynamical system for $A_0(T)$, $A_1(T)$, and $B_1(T)$:

$$\frac{\mathrm{d}A_0}{\mathrm{d}T} = -A_0[1 + A_0^2 + (3/2)A_1^2] + (1/2)A_1B_1, \tag{10.16d}$$

$$\frac{\mathrm{d}A_1}{\mathrm{d}T} = -A_1\left[1 + \left(\frac{m}{2w}\right) + 3A_0^2 + (3/4)A_1^2\right] + A_0B_1, \tag{10.16e}$$

$$\frac{dB_1}{dT} = -\mu B_1 - \sigma A_0 A_1, \tag{10.16f}$$

where

$$\mu(k_m) = mk_m^2 - wk_m^4, \quad \sigma = 2k_m^2 s.$$

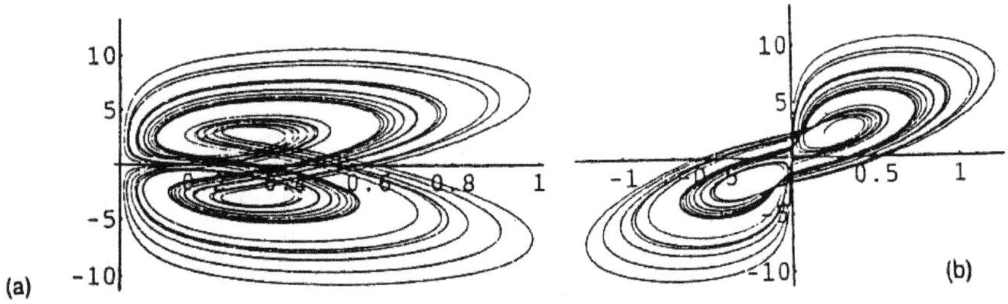

Fig. 10.2 Chaotic attractor of system (10.16d–f); $k_m = 1$, $\mu = 1$, $\sigma = 30$; and time interval $T \in [0, 150]$. Reprinted with kind permission from [8].

The system (10.16d–f) describes the time evolution of a periodic surface deformation (mode B_1), which can generate not only a periodic mode of the short-scale convection (A_1) following the surface deformation, but also a uniform zero mode (A_0). Figure 10.2 above shows projection of one of the (strange) chaotic attractors on the planes: (a) (A_0, B_1) and (b) (A_1, B_1). Thus, the coupling between the short-scale convection and large-scale deformations of the interface can lead to stochastization of the system and be one of the causes of interfacial turbulence of the thin film.

The reader can find also in [8] the numerical analysis of the system of two nonlinear coupled equations (10.16a, b), which confirms the predictions of weakly nonlinear analysis and shows the existence of either standing or travelling waves in the proper parametric regions, at low supercriticality. With increasing supercriticality, the waves undergo various transformations leading to the formation of pulsating travelling waves, non-harmonic standing waves as well as irregular wavy behavior resembling 'interfacial turbulence'.

Figure 10.3, according to a numerical study of Kazhdan et al. [8], presents a long-time series for the surface deformation η and the amplitude A (dashed line) at a fixed location, normalized by their maximum absolute value. It can be seen that oscillations of η and A are highly correlated: the amplitude of the short-scale convection follows the surface oscillations, being large beneath surface elevations and small under surface depressions.

Figure 10.4 shows a projection of the phase portrait of the system (10.16a, b) on the plane (η, $\partial\eta/\partial T$). The motion is apparently chaotic. In the Fourier

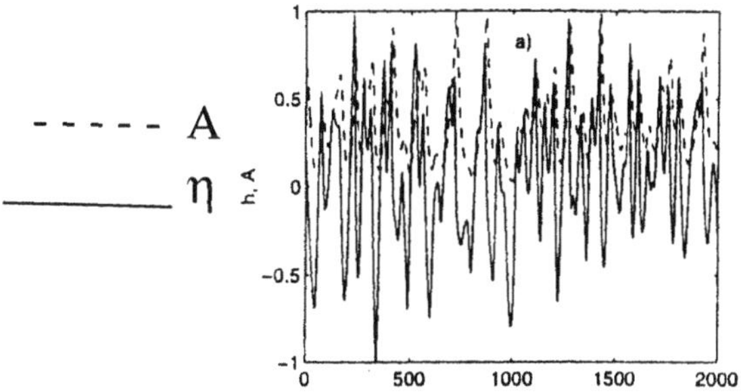

Fig. 10.3 Chaotic oscillations of surface deformation h and of the amplitude of the short-scale convection A at a fixed space point, in case $s = 70$, $m = 4.5$, $w = 1$; $Ma_l < Ma < Ma_s$. Reprinted with kind permission from [8].

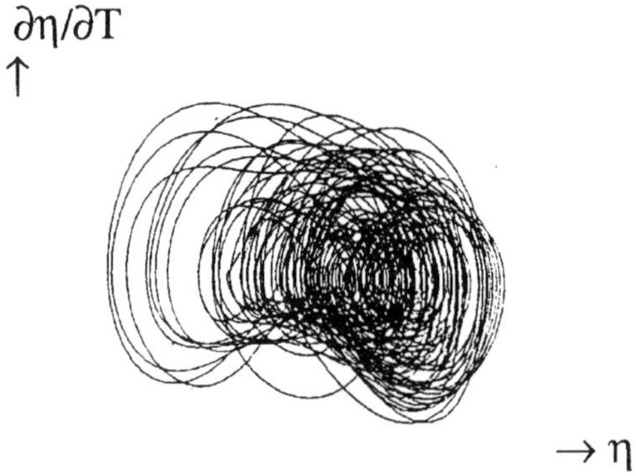

Fig. 10.4 Phase portrait of the chaotic oscillations of the surface deformation at a fixed point shown in Figure 10.3 ($T < 6000$). Reprinted with kind permission from [8].

space spectrum, due to strong damping of the higher harmonics caused by the fourth derivative in the evolution equation (10.16b) for η, only a small number of modes are excited (less than eight), and the number of modes does not change significantly with an increase of the coupling parameter. Hence, the observed irregular pattern can be characterized as a low-mode chaotic system. With increasing s (the coupling parameter in equations (10.16a, b)), the growth of surface deformations is suppressed by short-scale convection and coupling between the two modes gives rise to long-scale standing waves modulating the short-scale roll convection pattern.

With increasing s (the coupling parameter in equations (10.16a, b)), the growth of surface deformations is suppressed by short-scale convection and the coupling between the two modes gives rise to long-scale standing waves modulating the short-scale roll convection pattern. As the coupling becomes

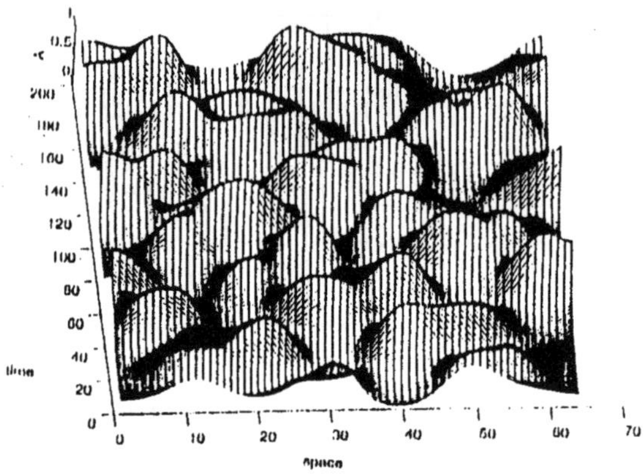

Fig. 10.5 Long-scale modulation of the amplitude of the short-scale convection generated by the surface deformation. Reprinted with kind permission from [8].

stronger, these oscillations decay and a stationary large-scale structure appears instead. This structure consists of narrow well-separated depressions, with the rest of the interface being almost flat; under surface depressions the fluid is almost quiescent, while in flat regions convection has an almost constant amplitude.

With further increase of the coupling parameter s, the stationary pattern becomes unstable and both long-scale surface deformations and the amplitude of the short-scale roll convection undergo irregular oscillations as shown in Figure 10.5 in the case $s = 70$, $m = 4.5$, $w = 1$ and $\mathrm{Ma}_l < \mathrm{Ma} < \mathrm{Ma}_s$.

10.6 Some Aspects of Thermosolutal Convection

Themosolutal convection is one facet of multicomponent convection; the reader can find various valuable information on this topic in the review paper by Turner [9]. In general, this (also sometimes called) ‘double-diffusive convection’ is a generalization of the process of thermal convection in a thin fluid layer, which arises when spatial variations of a second component with a different molecular diffusivity are added to the thermal gradients; thermohaline convection was (often) introduced to describe the heat-salt system. Below we give some conclusions directly inspired from [9], where the reader can find a pertinent review of the relevant laboratory experiments, stability analysis and various extensions and applications (oceanography, chemical studies, etc.).

Where one layer of fluid is placed above another (denser) layer having different diffusive properties, two basic types of convective instabilities arise, in the 'diffusive' and 'finger' configurations. In both cases, the double-diffusive fluxes can be much larger than the vertical transport in a single-component fluid because of the coupling between diffusive and convective processes. A 'diffusive' interface results when component T (temperature in the heat-salt, T–S, system) is heavy at the top and S is heavy at the bottom, with the lower layer being denser (we assume that T is the component with larger molecular diffusivity

$$D_T > D_S \quad \text{such that } D_S/D_T = \tau < 1, \tag{10.17a}$$

which means that cold, dilute solution lies above hot, concentrated solution, and the vertical transport of T and S across the hydrostatically stable central core of the interface occur in this case entirely by molecular diffusion. But because $D_T > D_S$, the edges of the interface can become marginally unstable. The resulting unstable buoyancy flux into the layers above and below drive large-scale convection that keeps the two layers well stirred and the interface sharpened and in this case the downward flux of T (expressed in density terms as $\alpha \mathbf{F}_T$ through the relation

$$\mathrm{d}\rho(T, S) = \rho_0[\alpha \Delta T + \beta \Delta S] \tag{10.17b}$$

is greater than the upward flux of salt $\beta \mathbf{F}_S$; as a consequence the potential energy of the system as a whole is continually decreasing and density difference between the two layers increases in time. Over a wide range of conditions, $\beta \mathbf{F}_S/\alpha \mathbf{F}_T$ is nearly constant and close to $\sqrt{\tau}$ for a heat-solute system. The structure shown in figure 2,in [9], is an opposite configuration, with a layer of S above a layer of T. In this case small disturbances can grow rapidly, and long, narrow, vertical convection cells called 'salt fingers' are formed and extend through the interface. It is now the more rapid horizontal diffusion of T relative to S over the width of the fingers that allows the release of the potential energy stored in the S field; when τ is small (e.g., for heat-salt fingers), even a tiny fraction of salt in the warmer top layer will lead to the formation of persistent fingers. The finger mechanism of transport, in which gravity plays a vital role, is described in terms of this familiar system. Each downward-moving finger is surrounded by upward-moving fingers, and vice versa. The downgoing fingers continually lose heat (by horizontal conduction) to the neighboring upgoing fingers, and therefore the dowgoing fingers become more dense and the upgoing fingers less dense. There is a slower transfer of salt sideways, which results in a small vertical salinity gradient. Thus the small-scale (finger) convective motions are driven by the local density anomalies between fingers and lead to $\beta \mathbf{F}_S > \alpha \mathbf{F}_T$. However, the horizontally averaged vertical density gradient through the interface remains stable and dominated by the T gradient; there is an unstable boundary layer at the edge of the salt-finger interface that drives convection in the layers. The potential energy of the whole system is again decreasing and the density difference between the layers is increasing, but now, since $\beta \mathbf{F}_S > \alpha \mathbf{F}_T$, this energy can be regarded as derived from the salt field. Now, layers can form when a smooth statically stable gradient of one property has an opposing gradient of a second property superimposed on it, or when there is a vertical flux of the destabilizing component, the formation of the double-diffusive layers being most readily demonstrated in the second case. More precisely, when a linear, stable salinity gradient is heated from below, for example, the fluid does not immediately convect from top to bottom. First, the heated layer immediately above the boundary breaks down to form a thin convecting layer. The depth d of this layer, and also the temperature and salinity steps ΔT and ΔS at its top, grow in time t like $\sqrt{t}$, in such a way that

$$\alpha \Delta T = \beta \Delta S, \tag{10.17c}$$

there is no net density step (as the two gradients become nearly equal but opposite), and the layer properties depend less and less on the boundary flux, which then just acts as a trigger for an internal instability. The thickness δ of the diffusive thermal boundary layer growing ahead of the convecting layer is also proportional to $\sqrt{t}$ and therefore to d, the multiplying constant being

$$Q = \frac{\delta}{d} = D_T \left[\frac{S^*}{H^*} \right],$$

where $H^* = -g\alpha \mathbf{F}_T/\rho C$ is the imposed buoyancy flux corresponding to the heat flux $\mathbf{F}_T$ into a fluid of specific heat C, and $S^* = -g\beta \, \mathrm{d}S/\mathrm{d}z$ is a measure of the initial salinity gradient. When Q and τ are small, the criterion for instability of the thermal boundary layer is a critical Rayleigh number based on δ and ΔT; after this criterion is achieved at a certain thickness, the depth d remains constant, and a new layer grows on top of the first. Finally, convecting layers can also be produced in the 'finger' situation, with a flux of sugar (S) imposed at the top of a salt (T) gradient. Fingers form, grow, and break up because of a collective instability having the form of an internal wave; this process produces a convecting layer that deepens, bounded below by an interface containing shorter, stable fingers, but these fingers in turn grow and become unstable, thus producing a second convecting layer.

In a paper by Kaloni and Qiao [91] a 'nonlinear convection, induced by inclined thermal and solutal gradients with mass flow' is investigated.

In the book by Joseph [92, chap. VIII, pp. 4–5], where it is observed that the crux of the Boussinesq approximation for a non-homogeneous (stratified), heat conducting viscous and compressible (dilatable) fluid (liquid) motion in the gravity field, is that: (i) the variation of the density perturbation is neglected in the mass continuity equation and in the equation for the horizontal motion; (ii) but this density perturbation is taken into account in the equation for the vertical motion through its influence as a buoyancy term; (iii) the influence of pressure perturbation on the buoyancy and in the equation of energy (written for the temperature perturbation) can be neglected; and (iv) the influence of perturbation of pressure in the equation of state can be also neglected and the rate of viscous dissipation is neglected in the equation for the temperature perturbation. When all of the simplifications are present, the Navier–Stokes–Fourier (NSF) exact equations for compressible heat conducting and diffusive flow of a viscous, non-homogeneous fluid can be approximated by the following set of (so-called Oberbeck–Boussinesq, OB) equations:

$$\rho = \rho_0[1 - \alpha_0(T - T_0) + \beta_0(S - S_0)], \tag{10.18a}$$

$$\nabla \cdot \mathbf{U} = 0, \tag{10.18b}$$

$$\rho_0 \left[\frac{\partial \mathbf{U}}{\partial t} + (\mathbf{U} \cdot \nabla)\mathbf{U} \right] + \nabla P - \rho_0 g[1 - \alpha_0(T - T_0) + \beta_0(S - S_0)] = \nabla \cdot \mathbf{S}, \tag{10.18c}$$

$$\frac{\partial T}{\partial t} + \mathbf{U} \cdot \nabla T = \kappa_T \nabla^2 T + Q_T(t, \mathbf{x}), \tag{10.18d}$$

$$\frac{\partial S}{\partial t} + \mathbf{U} \cdot \nabla S = \kappa_S \nabla^2 S + Q_S(t, x). \tag{10.18e}$$

In (10.18c), $\mathbf{T} = -P\mathbf{I} + \mathbf{S}$ is the stress, $\mathbf{S} = 2\mu\mathbf{D}[\mathbf{U}]$ is the extra stress, $\mathbf{U}$ is the (solenoidal) velocity and $\mathbf{g}$ is a body-force field (typically gravity). In equation (10.18d) for the temperature $T(t, \mathbf{x})$, κ_T is the thermal diffusivity and $Q_T(t, \mathbf{x})$ is a prescribed heat source field. Finally, in equation (10.18e) for the solute concentration $S(t, \mathbf{x})$, κ_S is the solute diffusivity and $Q_S(t, \mathbf{x})$ is a prescribed field specifying the distribution of solute sources. However, in various papers, as a starting (approximate) system, the authors consider a dimensionless system of equations for a two-dimensional thermosolutal convection in a Boussinesq fluid (see, for instance, Hupper and Moore [93]). Specifically,

$$\left(\frac{1}{\mathrm{Pr}}\right) \frac{D\omega}{Dt} = \mathrm{Ra}\frac{\partial \theta}{\partial x} + R_s \frac{\partial S}{\partial x} + \nabla^2 \omega \tag{10.19a}$$

$$\frac{D\theta}{Dt} = \frac{\partial \psi}{\partial x} + \nabla^2 \theta, \tag{10.19b}$$

$$\frac{DS}{Dt} = \frac{\partial \psi}{\partial x} + \nabla^2 S, \tag{10.19c}$$

$$\nabla^2 \psi = -\omega, \tag{10.19d}$$

where $D/Dt = \partial/\partial t + \partial(\psi, \ )/\partial(x, z)$, ψ the stream function, ω the vorticity, and θ and S perturbations from the static profiles of temperature and solute concentration. The simple boundary conditions are

$$\theta = S = \psi = \omega = 0, \quad z = 0, H\pi, \tag{10.19e}$$

where H is a measure of the dimensionless height of the layer, and

$$\psi = \omega = \frac{\partial \theta}{\partial x} = \frac{\partial S}{\partial x} = 0, \quad x = 0, \pi. \tag{10.19f}$$

The solution of the problem (10.19a–f) is controlled by the four dimensionless parameters

$$\mathrm{Ra} = \frac{g\alpha \Delta T d^3}{\kappa \nu}; \quad R_s = \frac{g\beta \Delta S d^3}{\kappa_s \nu}; \quad \mathrm{Pr} = \frac{\nu}{\kappa}; \quad \tau = \frac{\kappa_s}{\kappa}. \tag{10.19g}$$

We note that a similar problem (10.19a–g) is also considered in [11] and again in [10] as an application of a general approach for the derivation of amplitude equations for a system with competing instabilities. Proctor and Weiss [12] considered normal forms and chaos in thermosolutal convection. The competition between stabilizing and destabilizing effects can lead to oscillations which become chaotic in the nonlinear regime, the competing effects being produced by gradients in temperature and in concentration of a solute (e.g., salt) with a lower diffusivity. Thermosolutal convection has received much attention over the years both for its oceanographic interest and as a paradigm of double convection, relevant to convection in binary fluids or in the presence of rotation or a magnetic field (with industrial, geophysical and astrophysical applications). When cold fresh water lies above hot salty water, convection can occur even if the mean density increases downwards. Motion typically sets in at a Hopf bifurcation which gives rise to standing wave solutions if the system is laterally constrained. There has been particular interest in determining whether oscillatory convection can become chaotic before the original cellular structure is destroyed. Numerical integration of the full system (10.19a–d) has revealed transition to chaos following cascades of period-doubling bifurcations (see, for instance, [11]). But it is obvious that it is important to seek tractable low-order models by a rational approximation to the system of partial differential equations (10.19a–d). This is performed in [12] by an asymptotic method similar to one expounded in Section 5.4, assuming that Ra and R_s are close to their critical value, and that H is large. The authors derive an amplitude equation for $A = a(T)\sin\zeta$, where $T = \delta^2 t$ and $\zeta = \delta z$, with $\delta = (1/H) \ll 1$ (the linearized stability problem for (10.19a–d) being considered in [12] in the limit $H \uparrow \infty$), the solution for (10.19a–d), (10.19a–f), at leading order, when $\theta = \theta_0 + \delta\theta_1 + \cdots$ and $\psi = \psi_0 + \delta\psi_1 + \cdots$ being

$$\psi_0 = A\sin x \quad \text{and} \quad \theta_0 = A\cos x. \tag{10.19h}$$

In [10], the reader can find a description of a more abstract and formal method for extracting amplitude equations from systems governed by partial differential equations when such systems are near to points of bifurcation. Again, the above double convection problem (10.19a–f) is considered in [10] as the application, and the following amplitude equation is derived:

$$\frac{\mathrm{d}^2 A}{\mathrm{d}t^2} + (\mu + \lambda A^2)\frac{\mathrm{d}A}{\mathrm{d}t} + \nu A - \chi A^3 = 0. \tag{10.20}$$

10.7 Anelastic (Deep) Non-Adiabatic and Viscous Equations for the Atmospheric Thermal Convection (*à la* Zeytounian)

Ogura and Phillips [94] derived in 1962 a set of approximate equations which are valid for a Boussinesq number Bo ≈ 1, such that for the vertical scale of the atmospheric (adiabatic and non-viscous) motions, H, we have the estimation (see Section 9.2)

$$H \approx \frac{RT_S(0)}{g} \approx 10^4\text{m}. \tag{10.21a}$$

However, in this case it is necessary that the characteristic value of the Väisälä frequency (with dimension, $N_S(0)$) satisfy the constraint

$$N_S(0) \approx 10^{-3}\ 1/\text{s} \approx \frac{U_0}{H}, \tag{10.21b}$$

where U_0 is a characteristic scale for the velocity ($U_0 \approx 10$ m/s).

However, in [94] the starting equations were the Euler equations for an adiabatic non-viscous fluid. Gough [95] considered non-adiabatic, viscous equations and performed a scale analysis to derive approximate equations governing the motion of anelastic convection. The Gough analysis is valid if the relative density and temperature fluctuations produced by the motion are small. The approach of Gough in [95], unfortunately, is rather confused and the derivation is carried out, in fact, in an *ad hoc* manner, the resulting 'anelastic' equations having a complicated (rather awkward) form with unusual notation!

To begin our derivation of 'dissipative anelastic' equations, we first refer the reader to equations (8-5-8) at the beginning of chapter 8 in our book [13], with the thermal ground condition $z = 0$, and equation (8-21) for θ. We start our derivation from the NSF atmospheric equations written in dimensionless form for the velocity vector $\mathbf{u}$, and the thermodynamic perturbations, π, ω, θ, relative to the standard atmosphere (a function only of $z_s = \text{Bo}\, z$).

In a dimensionless thermal condition for $\theta = [(T - T_S(z_s))]/T_S(z_s)]$ on the ground,

$$\theta = \tau_0 \Theta(t, x, y) \quad \text{on } z = 0, \text{ with } \tau_0 = \Delta T_0 / T_S(0), \tag{10.21c}$$

where ΔT_0 is a characteristic temperature associated with the given distribution $\Theta(t, x, y)$ of temperature perturbation on the ground. In fact, the parameter τ_0 is linked with the Boussinesq number Bo $[= H/(RT_S(0)/g)]$ by the relation

$$\tau_0 = \frac{\text{Bo}}{\nu_0} \quad \text{with} \quad \nu_0 = \frac{H}{h_0}, \text{ where } h_0 = \frac{R\Delta T_0}{g}. \tag{10.21d}$$

We observe that in NSF atmospheric equations our Froude number Fr_L is such that

$$\text{Fr}_L^2 = \frac{U_0^2}{Lg}, \tag{10.21e}$$

where L is a horizontal scale for the atmospheric motion under consideration and we assume below that (local motion)

$$\varepsilon = \frac{H}{L} \approx \text{O}(1) \Rightarrow \text{Ro} \gg 1, \tag{10.21f}$$

because $\text{Ro} \approx \text{O}(1)$, only if $L \approx 10^5$ m. One the other hand, in dimensionless NSF (in the atmospheric equation (8-5) in [13, chap. 8]), for $\mathbf{u}$ we have a term proportional to θ:

$$-(1+\omega)\left(\frac{\text{Bo}}{\varepsilon}\right)\left[\frac{1}{\gamma M_S^2}\right]\theta\mathbf{k} \equiv -(1+\omega)\left(\frac{1}{\text{Fr}_L^2}\right)\theta\mathbf{k},$$

and when, $\mathrm{Fr}_L^2 \downarrow 0$ it is necessary that $\theta \approx \mathrm{Fr}_L^2$ in equation (8.7) for θ, and this implies that the stratification term is

$$(1+\pi)\left[\frac{\alpha_S^0}{\mathrm{Fr}_L^2}\right] N_S^2(z_s)(\mathbf{u}\cdot\mathbf{k}), \tag{10.21g}$$

where

$$\alpha_S^0 N_S^2(z_s) \approx \left[\frac{\mathrm{Bo}}{T_S(0)}\right]\left\{\frac{[\gamma-1]}{\gamma}+\frac{\mathrm{d}T_S}{\mathrm{d}z_S}\right\}. \tag{10.21h}$$

As a consequence it is necessary, in the anelastic (deep-viscous) convection case, to consider the double limiting process

$$\mathrm{Fr}_L^2 \to 0 \quad \text{and} \quad \alpha_S^0 \to 0 \quad \text{with} \quad \frac{\alpha_S^0}{\mathrm{Fr}_L^2} = \mathrm{O}(1). \tag{10.22a}$$

However, if we write for the vertical displacement $\delta\ (= h_0\Delta)$ of a fluid particle (at a fixed dimensionless time t) with respect to its position z_s in the standard atmosphere, we obtain the following dimensionless relation between z_s and z:

$$z_s = \mathrm{Bo}\left[z - \left(\frac{\Delta}{\nu_0}\right)\right], \tag{10.22b}$$

and in the limit $\nu_0 \to \infty$ (because $H \gg R\Delta T_0/g$) we recover the relation

$$z_s = \mathrm{Bo}\, z. \tag{10.22c}$$

For this, it seems more judicious to consider the limiting process

$$\mathrm{Fr}_L^2 \to 0 \quad \text{and} \quad \nu_0 \to \infty \tag{10.22d}$$

with Bo fixed and O(1) and the similarity rule (with $\varepsilon \equiv 1$):

$$\frac{\mathrm{Fr}_L^2}{(1/\nu_0)} \equiv \frac{U_0^2}{R\Delta T_0} = \gamma m_0^2 = \mathrm{O}(1), \tag{10.22e}$$

where $m_0^2 = U_0^2/\gamma R\Delta T_0$ is a squared Mach number related to the celerity $(\Delta c_0)^2 = \gamma R\Delta T_0$ linked with ΔT_0. With (10.22d) and (10.22e) the solution of NSF atmospheric dimensionless equations can be sought in the form ($\mathrm{Fr}_H^2 = U_0^2/gH \approx \mathrm{Fr}_L^2$)

$$\mathbf{u} = \mathbf{u}_d + \cdots, \quad \pi = \mathrm{Fr}_H^2 \pi_d + \cdots, \quad \omega = \mathrm{Fr}_H^2 \omega_d + \cdots, \quad \theta = \mathrm{Fr}_H^2 \theta_d + \cdots. \tag{10.23}$$

With the expansion (10.23) we first obtain, for $\mathbf{u}_d$, π_d, θ_d, the following two approximate leading-order equations:

$$S\frac{D\mathbf{u}_d}{Dt} + \left(\frac{1}{\text{Bo}}\right) T_{Sd}\nabla\pi_d - \theta_d\mathbf{k}$$

$$= \left(\frac{1}{\rho_{Sd}}\right)\left(\frac{1}{\text{Re}}\right)\left\{\frac{\partial^2 \mathbf{u}d}{\partial z^2} + \Delta_2\mathbf{u}_d + \left(\frac{\text{Bo}}{3\gamma}\right)\left[\frac{(\mathbf{u}_d\cdot\mathbf{k})}{T_{Sd}}\right]\right\}, \quad (10.24\text{a})$$

and

$$\nabla\cdot\mathbf{u}_d = \left(\frac{\text{Bo}}{\gamma}\right)\left[\frac{(\mathbf{u}_d\cdot\mathbf{k})}{T_{Sd}}\right]. \quad (10.24\text{b})$$

where $\Delta_2 = \partial^2/\partial x^2 + \partial^2/\partial y^2$.

We ensure now that, from the similarity rule $\alpha_S^0/\text{Fr}_L^2 = \text{O}(1)$, in (10.22a), we can write for $\text{d}T_S/\text{d}z_S$ the following relation (see (10.21h)):

$$\frac{\text{d}T_S}{\text{d}z_S} = \lambda^0\Gamma_{Sd}(\text{Bo}\, z)\text{Fr}_H^2 - \frac{[\gamma-1]}{\gamma}, \quad (10.24\text{c})$$

where $\lambda^0 = \text{const}$ and $\Gamma_S(z_s)$ is an arbitrary function of the order unity which takes into account a 'weak stratification', with the standard altitude z_s, of the standard atmosphere. The relation (10.24c) is a necessary condition on $\text{d}T_S/\text{d}z_S$ if we want to derive a consistent limiting approximate (deep-non-adiabatic) equation for θ_d from the full energy equation (equation (8-7) in [13]).

First, from (10.24c) we obtain for the density $\rho_S(z_S)$ in the standard atmosphere (as a function of z_s),

$$\frac{\text{d}\rho_S}{\text{d}z_S} = \left[\lambda^0\Gamma_S(z_s)\text{Fr}_H^2 - \left(\frac{1}{\gamma}\right)\right]\left[\frac{1}{T_S(z_S)}\right]. \quad (10.24\text{d})$$

However, in the framework of the limiting process (10.22d), associated with (10.22e) and (10.23), when $\nu_0 \to \infty$, we have the relation (10.22c).

Finally, for the limit values of T_{Sd} and ρ_{Sd}, in approximate equations (10.24a) and (10.24b) we have the following expressions:

$$T_{Sd} = \left[1 - \frac{[\gamma-1]}{\gamma}\right]\text{Bo}\, z \quad \text{and} \quad \rho_{Sd} = [T_{Sd}]^{1/(\gamma-1)}, \quad (10.24\text{e})$$

and

$$\frac{\text{d}\log\rho_{Sd}}{\text{d}z} = -\left(\frac{\text{Bo}}{\gamma}\right)\left[\frac{1}{T_{Sd}}\right] \Rightarrow \nabla\cdot[\rho_{Sd}(z)\mathbf{u}_d] = 0. \quad (10.24\text{f})$$

Our approximate anelastic-deep equation for θ_d is then written in the following dimensionless form:

$$S\frac{D\theta_d}{Dt} - \frac{[\gamma - 1]}{\gamma} S\frac{D\pi_d}{Dt} + \lambda^0 \,\mathrm{Bo}\left[\frac{\Gamma_{Sd}(\mathrm{Bo}\, z)}{T_{Sd}(\mathrm{Bo}\, z)}\right](\mathbf{u}_d \cdot \mathbf{k})$$
$$= \left[\frac{1}{\rho_{Sd}(\mathrm{Bo}\, z)}\right]\left[\frac{1}{\mathrm{Pr}\,\mathrm{Re}}\right]\Bigg\{\Delta_2\theta_d$$
$$+ \left[\frac{\partial^2\theta_d}{\partial z^2} - \frac{2[\gamma - 1]}{\gamma}\right]\left[\frac{\mathrm{Bo}}{T_{Sd}(\mathrm{Bo}\, z)]}\frac{\partial\theta_d}{\partial z}\right]\Bigg\}$$
$$+ \frac{[\gamma - 1]/\gamma}{[\rho_{Sd}(\mathrm{Bo}\, z) T_{Sd}(\mathrm{Bo}\, z)]}\left(\frac{\mathrm{Bo}}{\mathrm{Re}}\right)\Phi_d, \qquad (10.24\mathrm{g})$$

where Φ_d is the limiting value of the viscous dissipation.

In anelastic-deep adiabatic and viscous equations (10.24a) and (10.24g), $SD/Dt = S\partial/\partial t + \mathbf{u}_d \cdot \nabla$ and as thermal boundary condition on $z = 0$ we have

$$\theta_d = \left[\frac{\mathrm{Bo}}{\gamma m_0^2}\right]\Theta(t, x, y). \qquad (10.24\mathrm{h})$$

For the perturbation of the density ω_d, we derive the relation (the atmospheric, dry, air being a thermally perfect gas)

$$\omega_d = \pi_d - \theta_d. \qquad (10.24\mathrm{i})$$

When instead of (10.21f), we assume

$$\varepsilon = \frac{H}{L} \ll 1 \quad \text{such that } \mathrm{Ro} = \mathrm{O}(1), \qquad (10.24\mathrm{j})$$

we obtain the hydro-static deep viscous and non-adiabatic convection equations, where instead of Re we have

$$\mathrm{Re}_\perp = \varepsilon^2 \,\mathrm{Re} = \mathrm{O}(1), \qquad (10.24\mathrm{k})$$

and in this case the Coriolis force is active in the equation for $\mathbf{u}_{hd}$, but instead of the full equation for the vertical velocity component $(\mathbf{u}_{hd} \cdot \mathbf{k}) = w_{hd}$, we have the hydro-static balance

$$T_{Sd}\frac{\partial\pi_{hd}}{\partial z} = \mathrm{Bo}\,\theta_{hd}. \qquad (10.24\mathrm{l})$$

10.8 Flow of a Thin Liquid Film over a Rotating Disk

The unsteady flow of a thin liquid film on a cold/hot rotating disk is analysed by means of matched asymptotic expansion under the assumption of radially

(r) uniform film thickness that varies with time (t). Below as in [14] by Dandapat and Ray the usual reduced (in unsteady flow on a rotating disk, see, for instance, [96, 97]) dimensionless equations are:

$$
\begin{aligned}
&2F + \frac{\partial W}{\partial \zeta} = 0,\\
&\mathrm{Re}\left(\frac{\partial F}{\partial \tau} + F^2 + W\frac{\partial F}{\partial \zeta}\right) = \frac{\partial^2 F}{\partial \zeta^2} + G^2 + A,\\
&\mathrm{Re}\left(\frac{\partial G}{\partial \tau} - G\frac{\partial W}{\partial \zeta} + W\frac{\partial G}{\partial \zeta}\right) = \frac{\partial^2 G}{\partial \zeta^2},\\
&\mathrm{Pr}\,\mathrm{Re}\left(\frac{\partial M}{\partial \tau} + W\frac{\partial M}{\partial \zeta} - M\frac{\partial W}{\partial \zeta}\right) = \frac{\partial^2 M}{\partial \zeta^2},\\
&\mathrm{Pr}\,\mathrm{Re}\left(\frac{\partial N}{\partial \tau} + W\frac{\partial N}{\partial \zeta}\right) = \frac{\partial^2 N}{\partial \zeta^2} + 2M,\\
&\frac{\partial A}{\partial \zeta} = \beta\lambda M(\tau, \zeta),
\end{aligned}
\tag{10.25a}
$$

and as boundary conditions we have

$$
F(\tau, 0) = W(\tau, 0) = N(\tau, 0) = 0, \quad G(\tau, 0) = 1, \quad M(\tau, 0) = 1,
\tag{10.25b}
$$

and

$$
\begin{aligned}
&\frac{\partial F}{\partial \zeta} = \alpha\lambda M(\tau, H), \quad \frac{\partial G}{\partial \zeta} = 0, \quad \frac{\partial H(\tau)}{\partial \tau} = W(\tau, H) \quad \text{at } \zeta = H,\\
&\frac{\partial M}{\partial \zeta} = \frac{\partial N}{\partial \zeta} = A(\tau, H) = 0 \quad \text{at } \zeta = H,
\end{aligned}
\tag{10.25c}
$$

with corresponding initial conditions

$$
\begin{aligned}
&F(0, \zeta) = G(0, \zeta) = W(0, \zeta) = M(0, \zeta) = N(0, \zeta) = A(0, \zeta) = 0,\\
&H(0) = 1, \quad \frac{\partial H}{\partial \tau} = 0 \quad \text{at } \tau = 0,
\end{aligned}
\tag{10.25d}
$$

where τ and ζ are non-dimensional time and non-dimensional vertical (normal to disk $\zeta = 0$) coordinates.

In the last equation of (10.25a) and the first one of (10.25c), the parameter λ is a scalar (setting $\lambda = 1$ or -1 we shall obtain the case of a cooling or heating disk) in an *à la* von Karman similarity form for the temperature ($z = h_0\zeta$):

$$T = T_0 - \lambda \left(\frac{r^2}{2}\right) M(t, z) - \lambda N(t, z), \tag{10.25e}$$

where T_0 is the room temperature constant. The parameter β, which is related to the thermal expansion coefficient, acts as a heat sucking parameter, and the parameter α is the measure of thermocapillary force which is induced owing to the variation of surface tension with temperature. We observe, also, that at the free surface, with Newton's law of cooling, the thermal condition is

$$\frac{\partial T}{\partial z} + L(T - T_g) = 0, \tag{10.25f}$$

but below we have only the case when the heat transfer coefficient L at the free surface is assumed to be 0, T_g being the temperature in the passive gas phase.

With similarity form (10.25e) for the temperature T, we write also for the velocity components and the pressure in the starting problem:

$$u = r\left(\frac{U_0}{h_0}\right) F, \quad v = r\Omega_0 G, \quad w = WU_0, \quad p = -\left(\frac{r^2}{2}\right) A + B. \tag{10.25g}$$

In [14] the coupled, nonlinear system of equations (10.25a) with conditions (10.25b–d) is solved by expanding the dependent functions in terms of the powers of low Re, in the form

$$U(\tau, \zeta) = \Sigma\, \mathrm{Re}^n\, U_n(\tau, \zeta), \quad n = 0 \text{ to } \infty,$$
$$H(\tau) = H_0(\tau) + \mathrm{Re}\, H_1(\tau) + \cdots. \tag{10.26}$$

However, noting relations (29), (30) and (31) in [14], we see that the solution we have obtained via (10.26) does not satisfy the initial conditions (10.25d) due to the large-time-scale assumption. Namely, this large-time-scale assumption, linked with the choice of the reference time, $t_l = \frac{\nu}{h_0 \Omega_0^2}$, such that $\tau = \frac{t}{t_l}$, when at $t = 0$, the free surface is at the level $z = h_0$ and Ω_0 is the constant angular velocity. In the framework of non-dimensionalization of the starting physical problem leading to the dimensionless formulation (10.25a–d), a short-time scale analysis is necessary and in [14] such an analysis is consistently performed (according to the 'matching asymptotic expansions – outer and inner – method').

In short-time scale analysis a new time scale is defined, in such a way that the local inertial terms (unsteady terms with $\partial/\partial t$) are of the same order of magnitude as the viscous and centrifugal terms in the governing local-in-time dimensionless equations, which are assumed valid close to initial time. The dimensionless short time is

$$\tau^* = \frac{\tau}{\mathrm{Re}} \tag{10.27}$$

and the corresponding local functions (with an asterisk $*$, and depending on τ^* and $\eta = \zeta$) close to initial time, instead of (10.25a), (10.25b–d), are the solution of the following problem:

$$\begin{aligned}
&2F^* + \frac{\partial W^*}{\partial \eta} = 0,\\
&\frac{\partial F^*}{\partial \tau^*} + \mathrm{Re}\left(F^{*2} + W^*\frac{\partial F^*}{\partial \eta}\right) = \frac{\partial^2 F^*}{\partial \eta^2} + G^{*2} + A^*,\\
&\frac{\partial G^*}{\partial \tau^*} + \mathrm{Re}\left(-G^*\frac{\partial W^*}{\partial \eta} + W^*\frac{\partial G^*}{\partial \eta}\right) = \frac{\partial^2 G^*}{\partial \eta^2},\\
&\mathrm{Pr}\frac{\partial M^*}{\partial \tau^*} + \mathrm{Pr}\,\mathrm{Re}\left(W^*\frac{\partial M^*}{\partial \eta} - M\frac{\partial W^*}{\partial \eta}\right) = \frac{\partial^2 M}{\partial \eta^2},\\
&\mathrm{Pr}\frac{\partial N^*}{\partial \tau^*} + \mathrm{Pr}\,\mathrm{Re}\,W^*\frac{\partial N^*}{\partial \eta} = \frac{\partial^2 N^*}{\partial \eta^2} + 2M^*,\\
&\frac{\partial A^*}{\partial \eta} = \beta\lambda M^*(\tau^*, \eta),
\end{aligned} \tag{10.28a}$$

with as boundary conditions

$$\begin{aligned}
&F^*(\tau^*, 0) = W^*(\tau^*, 0) = N^*(\tau^*, 0) = 0,\\
&G^*(\tau^*, 0) = 1, \quad M^*(\tau^*, 0) = 1,
\end{aligned} \tag{10.28b}$$

and

$$\begin{aligned}
&\frac{\partial F^*}{\partial \eta} = \alpha\lambda M^*(\tau^*, H^*), \quad \frac{\partial G^*}{\partial \eta} = \frac{\partial M}{\partial \eta} = 0, \quad \frac{\partial N}{\partial \eta} = A(\tau^*, H^*) = 0,\\
&\frac{\partial H^*(\tau^*)}{\partial \tau^*} = \mathrm{Re}\,W^*(\tau^*, H^*) \quad \text{at } \eta = H^*,
\end{aligned} \tag{10.28c}$$

and corresponding initial conditions (at $\tau^* = 0$)

$$\begin{aligned}
&F^*(0, \eta) = G^*(0, \eta) = W^*(0, \eta) = M^*(0, \eta) = N^*(0, \eta) = A^*(0, \eta) = 0,\\
&H^*(0) = 1, \quad \frac{\partial H^*}{\partial \tau^*} = 0.
\end{aligned} \tag{10.28d}$$

Expanding the local functions $U^*(\tau^*, \eta)$ and $H^*(\tau^*)$, in terms of the power Re, according to the perturbation scheme (10.26), and solving the local (inner) problem (10.28a–d), Dandapat and Ray in [14] wrote the solution for the zero-order set, when Re $\downarrow 0$ in (10.28a–d) (see relation (36) in [14]) for $H_0^*(\tau^*) = 1$, $M_0^*(\tau^*, \eta)$ and $A_0^*(\tau^*, \eta)$; but the solutions for $F_0^*(\tau^*, \eta)$ and $W_0^*(\tau^*, \eta)$ are dropped! A first-order correction to the film thickness for a short-time scale can be obtained from the kinematic condition

$$\frac{\partial H_1^*(\tau^*)}{\partial \tau^*} = W_0^*(\tau^*, 1). \tag{10.29}$$

Finally, using the technique of composite matched asymptotic expansion (see, for instance the book by Van Dyke [98, sections 5.10, 10.3]), Dandapat and Ray in [14, p. 495] obtained an expression for the film thickness which is uniformly valid for all times.

We observe that large τ^* ($\tau^* \uparrow \infty$) $H_1^*(\tau^*)$, in expansion $H_0^*(\tau^*) + \mathrm{Re}\, H_1^*(\tau^*) + \cdots$, has been calculated and then used as the initial condition for the equations (obtained from the kinematic condition, $\partial H(\tau)/\partial \tau = W(\tau, H)$):

$$\frac{\partial H_0(\tau)}{\partial \tau} = W_0(\tau, H) \quad \text{and} \quad \frac{\partial H_1(\tau)}{\partial \tau} = W_1(\tau, H),$$

along with $H_0^*(\tau^*) = 1$.

According to results obtained in [14], depending on α, a novel feature of flow reversal on the free surface of the film is obtained when the disk is heated from below axisymmetrically; for fixed α, the film thickness increases with β (which plays the role of a heat sucking parameter), and the heating parameter β enhances the film thinning with its increment, whereas for fixed β, α introduces an adverse thinning effect. Thus, the thermocapillary flow which is induced in this case has opposite flow direction, i.e., towards the center. The disk being cooled axisymmetrically, the surface tension is low at the centre, and hence a thermocapillary flow is induced at the free surface in the favourable flow direction. Thus, α enhances the film thinning when the disk is cooled from below.

As a final, personal, comment, I think that the present approach of Dandapat and Ray in [14], very pertinently shows that, for each film problem, with the physically realistic initial conditions, it is, in fact, necessary to investigate (during modeling) an inner-in-time region close to initial time ($t = 0$) and, by matching with the 'usual' outer-film solution, to obtain a solution which is uniformly valid for all time!

In a more recent paper by Kitamura [99], an unsteady liquid film flow of non-uniform thickness on a rotating disk is analyzed. Assuming a small aspect ratio of the initial film thickness to the disk radius and applying Higgins' [96] asymptotic approach, Kitamura obtained the long-time and short-time scale solutions for transient film thickness. Then by matching a composite solution he derived, for an axisymmetrically non-uniform initial film profile, $h^0(r) = h(r, t = 0)$. The small parameter is $\varepsilon = h^0(0)/R_0$, where R_0 is the large disk radius (the peripheral effects being neglected). The author uses, in fact, as starting equation, a 'lubrication' long-time evolution equation, for the shape of the film interface (see, for instance, [100])

$$\frac{\partial h}{\partial t} = -(1/3)\frac{\partial(r^2h^3)}{\partial r} + \varepsilon(1/3r)\frac{\partial}{\partial r}\left\{\mathrm{Re}\left[-(2/5)r^3h^6\frac{\partial h}{\partial r} + (34/105)r^2h^7\right]\right. \\ \left. + \mathrm{Fr}^*\, rh^3\frac{\partial h}{\partial r} - \mathrm{We}^*\, rh^3\left[\left(\frac{1}{r}\right)\frac{\partial}{\partial r}\left(r\frac{\partial h}{\partial r}\right)\right]\right\}, \tag{10.30a}$$

where

$$\mathrm{Fr}^* = \frac{\mathrm{Re}}{\mathrm{Fr}^2} \quad \text{and} \quad \mathrm{We}^* = \varepsilon^3\, \mathrm{We},$$

with

$$\mathrm{Fr}^2 = \frac{R_0^2 h_0^3 \Omega_0^4}{g\nu^2} \quad \text{and} \quad \mathrm{We} = \frac{\sigma}{\rho\Omega_0^2 R_0 h_0^2}. \tag{10.30b}$$

Then he writes (the subscript 'L' is introduced to indicate the long-time scale)

$$h(r, t) = h_{L0}(t) + r^2 h_{L1}(t) + r^4 h_{L2}(t) + \cdots \tag{10.30c}$$

and this expansion is substituted into equation (10.30a). The resulting two equations for $h_{L0}(t)$ and $h_{L1}(t)$ are solved by the variation of constants and the constants ($C_{00}, C_{01}, C_{10}, C_{20}, C_{11}, C_{30}, C_{30}$) appearing in the solutions for $h_{L0}(t)$ and $h_{L1}(t)$ are obtained by matching with the short-time-scale solution up to second order in r,

$$h = h_{S0} + r^2 h_{S1}, \tag{10.30d}$$

to give two composite uniform expansions for the transient film thickness. In results derived in [99] the terms proportional to ε represent the corrections to the Emslie et al. analysis [101]. According to results obtained by Kitamura, the effect of gravitational force tends to promote film thinning, while the other two effects (inertial and surface tension forces) tend to retard it!

Obviously, another approach is possible (different from Dandapat and Ray or Kitamura) via the IBL model equations and the reader who is interested in a such approach can look at the work [102] of Christel Bailly (performed in 1995 at the LML of the University ST of Lille in the framework of a doctoral thesis).

The paper by Needham and Merkin [103] is also interesting, mainly because the authors consider an unsteady liquid film flow on a rotating disk, and consider in the starting formulation very realistic initial conditions at time $t = 0$, the singular time region, near $t = 0$, being taking into account and, again, matching with the outer region, corresponding to evolution of a liquid film after the transition regime has been performed. A simple reasonable conclusion leads to confirmation that

for a practical film problem (inspired from modern technologies) and in comparison with various experimental results, it is necessary during the formulation of the starting problem to also assume pertinent/relevant initial conditions at time = 0 and investigate the inner time region near t = 0, which appears usually because the singular behavior of the solution of the (outer) model equations (in particular, non-isothermal IBL) is used.

10.9 Solitary Waves Phenomena in Bénard–Marangoni Convection

Solitary waves, generally composed of a large maximum and several subsidiary maxima, occur commonly in the nonlinear behavior of liquid films flowing down an inclined (or vertical) plane. These solitary waves, as this is pertinently observed in a paper by Liu and Gollub [104], should not be confused with solitons (concerning 'nonlinear long waves on water and solitons', see, for example, our review paper [105]), because the former are interacting and dissipative. First, it is necessary to mention the pioneering work [106] by 'two Kapitza's', related to an 'experimental study of undulatory flow conditions' for the wave flow of thin layers of a viscous fluid. Later, in [107], Pumir et al. noted the existence of two types of solitary waves running down an inclined plane, positive (because of their solitary humps) and negative (due to their solitary dips). These two (γ_1 and γ_2) families were investigated by Chang et al. [108] using the KS equation (see our Section 7.5)

$$\varphi_t + 2\varphi_x + 4\varphi\varphi_x + (8/15)[\mathrm{Re}_d - (5/4)\cos\beta]\varphi_{xx} + (2/3)\varphi_{xxxx} = 0, \quad (10.31)$$

for Reynolds number (based on the unperturbed film thickness d) $\mathrm{Re}_d \to \mathrm{Re}_{\mathrm{crit}}$ ($= 0$) and an IBL, *à la* Shkadov [83], model for $\mathrm{Re} > 0$, because the KS model equation (10.31) has no singularities, but its applicability is limited to Re very close to $\mathrm{Re}_{\mathrm{crit}}$, and small wave amplitude $\varphi = h - 1$. The KS equation and certain of its extensions (see, for instance, the recent paper by Travelyan and Kalliadasis [85]) can provide a useful starting point for theoretical studies of waves on falling film flows.

It seems that a rational derivation of an evolution equation that can capture most of the nonlinear phenomena accurately is not available; however, a systematic long-wave expansion (as in Sections 7.3–7.5) yields a well-known equation due to Benney [109] that is valid again for Re close to $\mathrm{Re}_{\mathrm{crit}}$, but is successful in describing the initial evolution of nonlinear waves. The dimensionless form of this equation is

$$h_t + 2h^2 h_x + (2/3)\left\{(4/5)\mathrm{Re}\, h^6 h_x - h^3[h_x \cos\beta + \mathrm{We}\, h_{xxx}]\right\}_x = 0, \tag{10.32}$$

but this Benney equation (10.32) can produce singularities in finite time and is of limited applicability.

Solitary wave interactions in film flows are very different from the behavior of solitons in conservative (non-viscous) systems because of their dissipation. Various authors have studied pulse interactions using a KS–KdV (an 'extended KS' equation that includes a dispersive term φ_{xxx})

$$\varphi_t + \varphi\varphi_x + \mu\varphi_{xx} + \delta\varphi_{xxx} + \alpha\varphi_{xxxx} = 0, \tag{10.33}$$

where $\mu > 0$, $\delta > 0$ and $\alpha > 0$ give the relative strengths of instability, dispersion and dissipation, respectively. This equation (10.33) is also called (see, for instance, Velarde and Rednikov [110] and Rednikov et al. [111]) a 'dissipation-modified KdV' equation or a 'generalized (dissipation-modified) KdV–Burgers' equation, since equation (10.33) contains the Burgers, KdV and KS equations as special cases. In Rednikov et al. [111], this 'dissipation-modified KdV' equation is written in the following form, for the elevation, $h(t, y)$, of the free surface in the study of one-directional waves in Bénard-Marangoni layers:

$$h_t + (h^2)_y + h_{yyy} + \delta[h_{yy} + h_{yyyy} + D(h^2)_{yy} + \alpha h] = 0, \tag{10.34}$$

where the coefficient D can be either positive or negative, while α is always non-negative (see [112]). The complicated evolution exhibited by any solution of equation (10.33) can be qualitatively described by the weak interactions of pulses, each of which is a steady solution to equation (10.33).

In numerical simulation of films flowing down a vertical cylinder [113], via the equation (with a modified Weber number S)

$$h_t + 2h^2 h_x + S\{h^3[h_x + h_{xxx}]\}_x = 0, \tag{10.35}$$

for the vertical case, such that the highest-order nonlinearity is proportional to h^3 rather than h^6 – but noting that this equation (10.35) does not include dispersion effects – the collision of two pulses, depending on various conditions, can be either an 'elastic' rebound or an 'inelastic' coalescence.

In a paper by Liu and Gollub [104], the reader can find a description of their experimental method; a detailed description can be found in a previous paper [114].

On the other hand, the various experimental results for development of periodic waves, interactions of solitary waves and generation of solitary waves by interaction are also considered in [104].

An example of a solitary wave profile (a) is given in Figure 10.6 with two (b) and (c) corresponding phase orbits. By plotting $h(x)$ versus $h(x+2d)$ and $h(x+4d)$, where $d = 0.16$ cm, the orbit resembles a nearly homoclinic orbit as in the numerical result (see [107, 115]), and the arrows on the trajectory show the direction in which x increases. In Figures 10.6b and 10.6c, the main wave is visible as a large loop extending far from the fixed point, while the precursor waves are shown as decaying orbits around this point. However, it is unclear whether this representation provides more than a qualitative method of visualization.

In paper [111], the 'dissipation-modified KdV' equation (10.34) is analysed and the authors consider, first, the stationary cnoidal waves propagating with phase velocity c via $x = y - c_t$, and assume $\delta \ll 1$ such that

$$h = h_0 + \delta h_1 + \cdots, \qquad c = c_0 + \delta c_1 + \cdots. \tag{10.36a}$$

For h_0 we obtain a KdV equation and the solution is

$$h_0 = A\left[-\left(\frac{E(s)}{K(s)}\right) + \mathrm{dn}^2\left(\left(\frac{A}{6}\right)^{1/2} x, s\right)\right], \tag{10.36b}$$

where $K(s)$ and $E(s)$ are complete elliptic integrals of the first and second kind, respectively; dn is a Jacobian elliptic function; s is the module ($0 \leq s \leq 1$) and A is some non-negative parameter. The solution for h_1 is also considered in [111]. Admitting that the parameters A and s of (10.36b) depend on the slow-time scale $t_I = \delta t$, an evolution equation for A is derived corresponding to a cnoidal wave of given wavelength. In equation (10.34) the coefficient α is assumed different from zero and in the case of solitary waves we have the relation $c_0 = 2A/3$, and at $x \cong \ln \delta$, large negative x, we obtain the asymptotic form

$$h_0 + \delta h_1 \cong 4A \exp\left[2\left(\frac{A}{6}\right)^{1/2} x\right] - 3\alpha\left(\frac{6}{A}\right)^{1/2} \delta. \tag{10.36c}$$

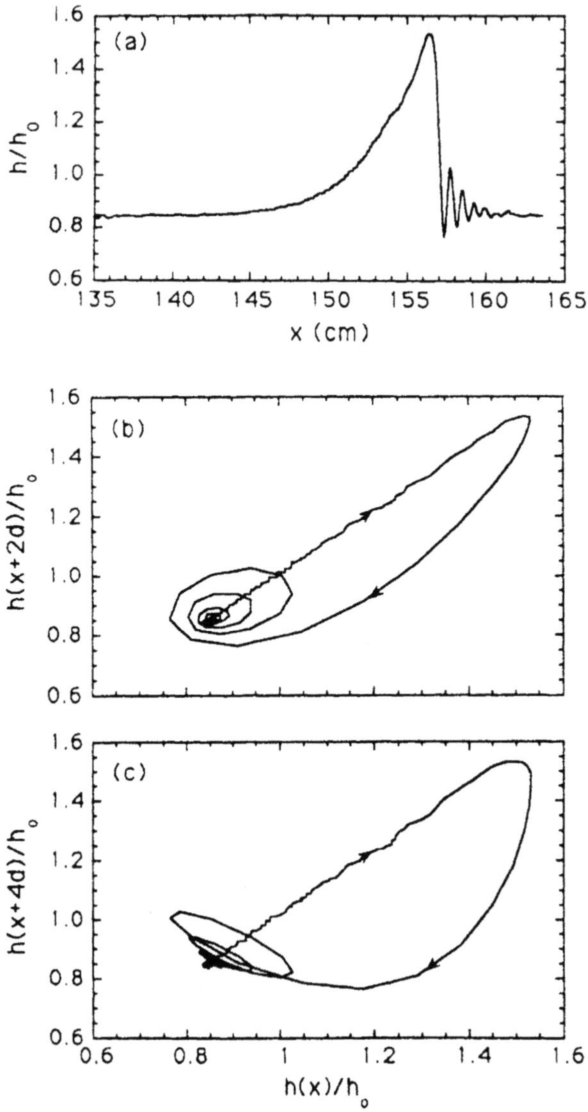

Fig. 10.6 Phase space representation of a solitary wave. Reprinted with kind permission from [104].

In fact, a uniformly valid solution is irrelevant since it cannot satisfy the condition (for a steady solitary wave)

$$\int_{-\infty}^{+\infty} h \, \mathrm{d}x = 0. \tag{10.36d}$$

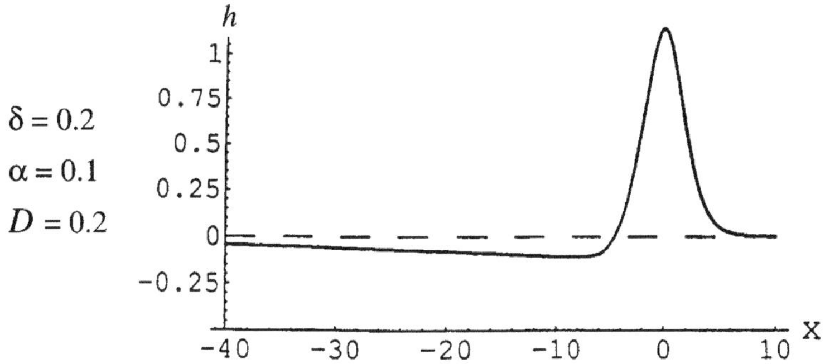

Fig. 10.7 Typical profile of a dissipative solitary wave with asymmetry. Reprinted with kind permission from [16].

The solution $h_0+\delta h_1$, correct for $|x| = O(1)$, must be matched to the solution (with (10.36d)), obtained for large $|x|$. In the region $x \le \ln\delta$ we have as solution

$$h = 4A \exp[(c_0)^{1/2}x] - 3\alpha \left(\frac{6}{A}\right)^{1/2} \delta \exp\left[\left(\frac{\delta\alpha}{c_0}\right)x\right], \qquad (10.36e)$$

the solution in the region $x \ge 1/\delta$, being

$$h = 4A \exp\left\{-\left[(c_0)^{1/2} + \left(\frac{\delta}{2}\right)\left[1 + c_0 + \left(\frac{\alpha}{c_0}\right) + \left(\frac{c_1}{(c_0)^{1/2}}\right)\right]\right]x\right\}, \qquad (10.36f)$$

and the resulting solitary wave is asymmetric. Its profile is schematically outlined in Figure 10.7 above; the head is monotonous, while the tail is not, because $\alpha \neq 0$.

In [16, 110, 112], the reader can find various references to the work of Velarde and his collaborators, where analysis of all possible wave motions have been performed and the book by Velarde [111] is devoted to physicochemical hydrodynamics where as a consequence of Marangoni surface stresses the possibility exists of exciting waves. In [116] the case of a liquid layer heated from the air side is considered and the model equation for $u(t, x)$, the deformation of the free surface in the Bénard–Marangoni instability, is

$$u_t + b_1 u u_x + b_2 u_{xxx} + b_3 u_{xx} + b_4 u_{xxxx} + b_5 (u u_x)_x = 0. \qquad (10.37)$$

Propagating dissipative (localized) structures like solitary waves, pulses or 'solitons', 'bound solitons', and 'chaotic' wave trains are shown to be solutions of the dissipation-modified KdV equation (10.37). In a moving frame,

$\xi = x + c_0 t$, and upon integration from $-\infty$ to the traveling ξ-coordinate we obtain, instead of (10.37),

$$\frac{\mathrm{d}u}{\mathrm{d}\xi} = y, \quad \frac{\mathrm{d}y}{\mathrm{d}\xi} = z, \quad \gamma\frac{\mathrm{d}z}{\mathrm{d}\xi} = -\beta z - \alpha u y - \nu y - F(u), \tag{10.38}$$

where

$$F(u) = \left(\frac{c_0}{b_1}\right) u + (1/2)u^2, \quad \nu = \left(\frac{b_3}{b_1}\right),$$

$$\beta = \left(\frac{b_2}{b_1}\right), \quad \gamma = \left(\frac{b_4}{b_1}\right), \quad \alpha = \left(\frac{b_5}{b_1}\right).$$

For the dynamical system (10.38) Nekorkin and Velarde [116] performed a phase space analysis and considered the homoclinic trajectories and localized structures, and multi-loop homoclinic trajectories and bound solitons states. In Figure 10.8 we have homoclinic trajectories and dissipative localized structures, solitary waves, pulses or 'solitons'.

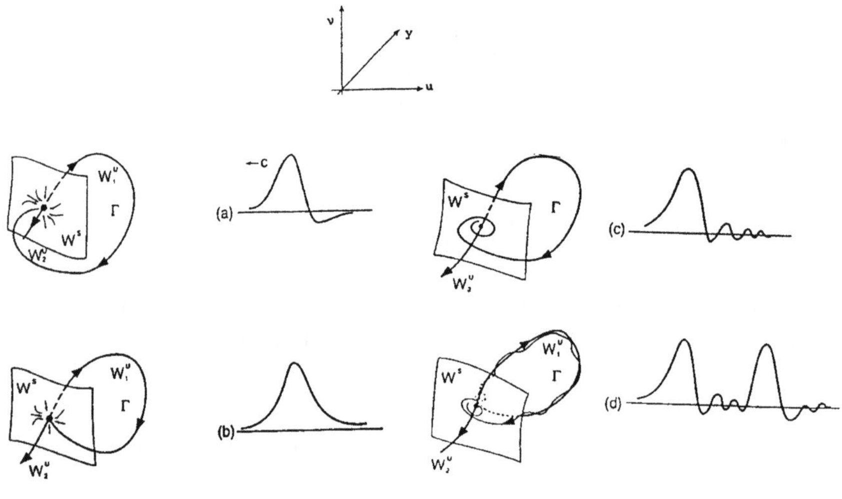

Fig. 10.8 Homoclinic trajectories and dissipative localized structures, solitary waves, pulses or 'solitons' for (a) and (b) 'small' values of γ, (c) 'moderate' values of γ, and (d) 'bound solitons'. Reprinted with kind permission from [116].

In [117], time evolution of a two-hump solitary wave is given for the case of ν positive in (10.38); see Figure 10.9.

In [118] the above Christov and Velarde equation (10.37) is again analyzed. A special finite-difference scheme is devised which faithfully represents the balance law for energy. In this case the numerical simulations

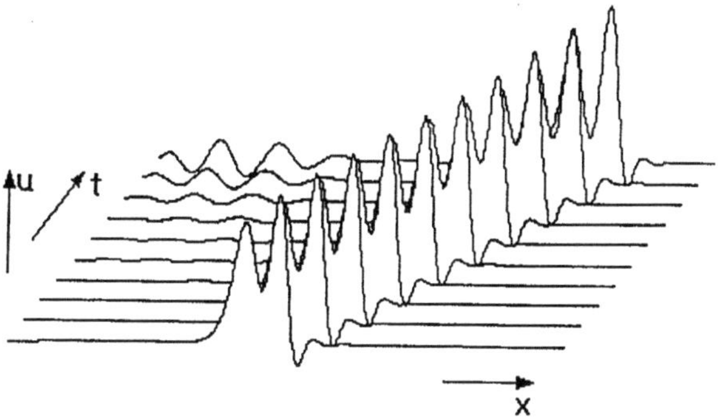

Fig. 10.9 Two-hump solitary wave for ν positive. Reprinted with kind permission from [117].

show that if the production-dissipation rate is of order of a small parameter, the coherent structures upon collisions preserve their localized character and, within a time interval proportional to the inverse of this small parameter, they behave like (imperfect) solitary waves. In Figure 10.10, a head-on collision of the dissipative solitons of *sech* initial shape is presented for two different values of the convective speed of the moving frame.

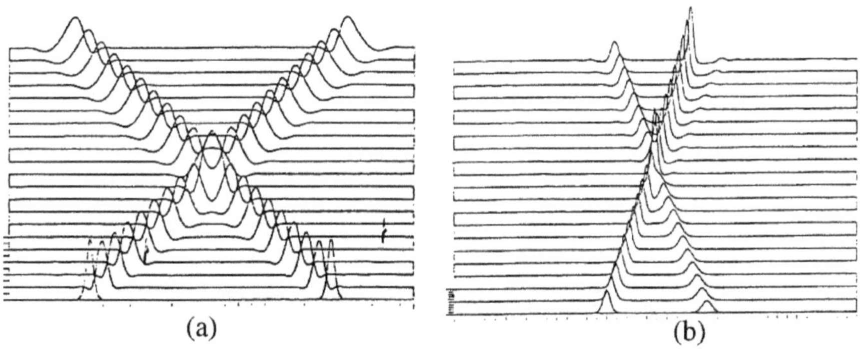

Fig. 10.10 Head-on collision of the dissipative solitons of *sech* initial shape for two different values of the convective speed of the moving frame: (a) for $\gamma = 10$ (dissipation dominated case), and (b) for $\gamma = 1$ (production dominated case). Reprinted with kind permission from [118].

Finally, concerning the 'solitary waves', we note that recently in a paper by Scheid et al. [6, section 3], from the non-isothermal regularized reduced IBL model system (10.15k–m), derived in [5], attention has been concentrated on the travelling wave evolution of system (10.15k–m) and more especially on simple-hump solitary waves. In Figure 10.11 we give some results:

1. for Re $= 2$, Ma $= 50$, Bi $= 0.1$, $\cot\beta = 0$ and $\Gamma = 250$;
2. for Re $= 3$, Pr $= 1$, Ma $= 50$, Bi $= 0.1$, $\cot\beta = 0$ and $\Gamma = 250$;
3. for Re $= 3$, $\cot\beta = 0$ and $\Gamma = 250$, Ma $= 0$.

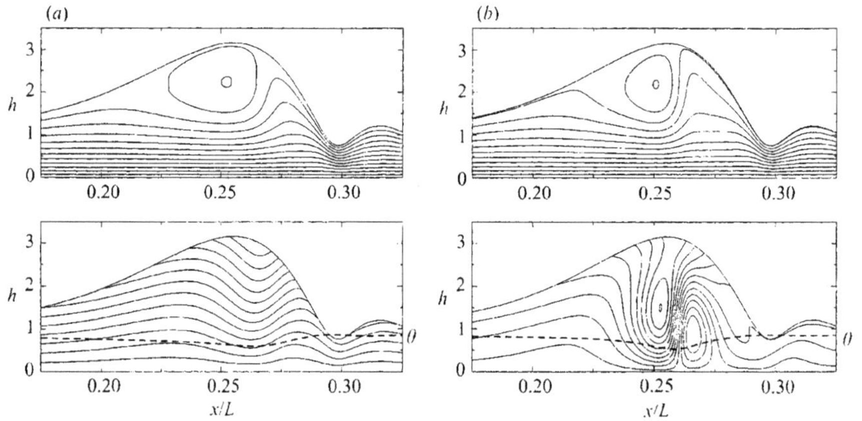

Fig. 10.11 (1) Streamlines (top) and isotherms (bottom); (a) Pr $= 1$, (b) Pr $= 7$; Re $= 2$. Reprinted with kind permission from [6].

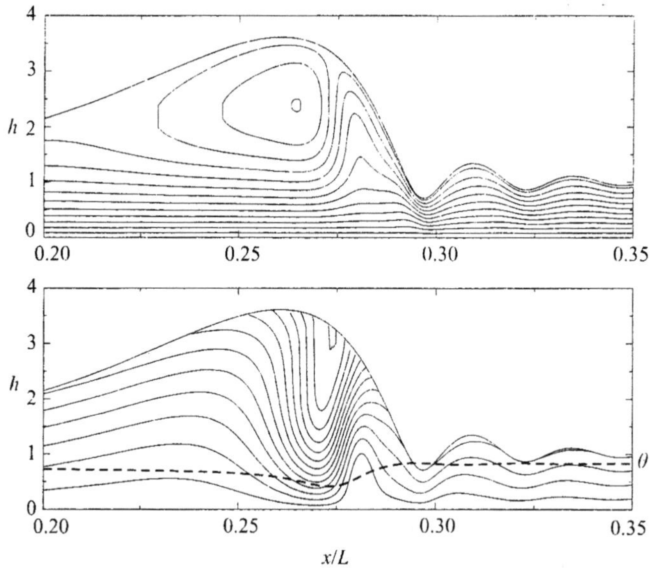

Fig. 10.11 (2) Streamlines (top) and isotherms (bottom), Re $= 3$. Reprinted with kind permission from [6].

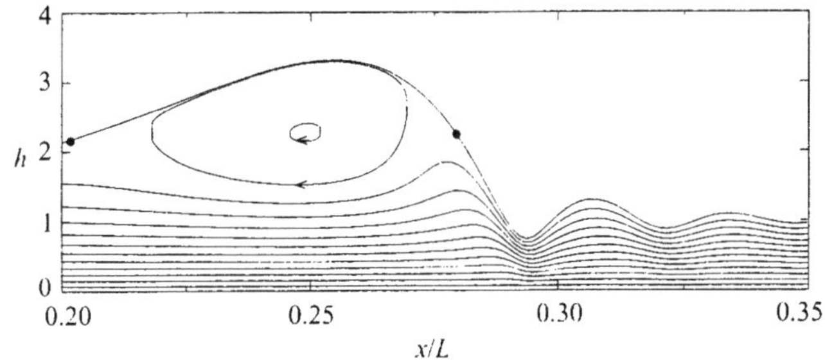

Fig. 10.11 (3) Streamlines; Re = 3 and $M = 0$. Reprinted with kind permission from [6].

The Marangoni effect enhances recirculation in the crest and promotes a strong downward flow there. As a consequence, the transport of heat by the flow contributes to cooling the crest and amplifying the Marangoni effect. Thus, because thermocapillary stresses push the fluid from the rear to the top of the crest, they reinforce clockwise circulation in the crest.

10.10 Some Comments and Complementary References

Although significant understanding of convective flows has been achieved, surface tension gradient-driven (BM) convection flows, in particular, still deserve further study. Indeed, as a paradigmatic form of a spontaneous self-organizing system, the doctrine of the original Bénard problem has not reached the degree of sophistication, in theory and experimentation, attained in buoyancy-driven (R-B) convection. There are still challenging problems like relative stability of patterns (hexagons, rolls, squares, ... labyrinthine convection flows), higher transitions and interfacial turbulence (at low Marangoni number), a case of spacetime chaos with high dissipation. It should be pointed out, also, that the ever increasing number of industrial applications of thin film flows and the richness of behavior of the governing equations make this area a particularly rewarding one for mathematicians, engineers, and industrialists alike. Although Bénard was aware of the role of surface tension and surface tension gradients in his experiments, it took five decades to unambiguously conclude, experimentally and theoretically (see, for instance, the papers by Block [119] and Pearson [120]) that indeed the surface tension gradients rather than buoyancy was the cause of Bénard cells in thin liquid films. Only in 1997 was the almost evident physical fact rigorously proved, through an asymptotic approach, by Zeytounian [121]:

> ... Either buoyancy is taken into account and in this case the free-surface deformation effect is negligible and we have the possibility to take only partially into account the Marangoni effect, or this free-surface deformation effect is taken into account and in this case buoyancy does not play a significant rôle in the Bénard–Marangoni full thermocapillary problem.

It seems that the first author to explain (in a simple linear case) the effect of the surface tension gradients on Bénard convection was Pearson [120]. The review article by Bragard and Velarde [122] has provided salient findings, old and recent, about Bénard convection flows in a liquid layer, heated from below, and open to the ambient air. The Myers [123] paper is a review of work on thin films when (high) surface tension is a driving mechanism. Its aim is to highlight the substantial amount of literature dealing with relevant physical models and also analytic work on the resultant equations. The paper (in two parts) by Ida and Miksis [124] considers also the dynamics of a general 3D thin film subject to van der Waals forces (which plays an essential role for 'ultra-thin' films), surface tension, and surfactants. Using an asymptotic analysis based upon the thinness of the film with respect to its lateral extent, evolution equations for the leading-order film thicknesses, tangential velocities, and surfactant concentrations have been obtained. The scaling was chosen by the above authors such that the surface tension effects occur at leading order in the dynamical model of the thin film. Note that the analysis applies to the breaking of a thin liquid film off of a stable centersurface. Unfortunately, the model equations, as presented, form a complicated set of evolution equations and cannot be solved until the centersurface is prescribed. In part II of their work, Ida and Miksis [124] consider a series of special centersurfaces and in each case consider linear stability and solve the resulting nonlinear equations numerically. In particular, it is shown that increasing surface tension is stabilizing, while increasing the effects of van der Waals forces is destabilizing; the effects of surfactants, although irrelevant in determination of neutral stability curves, is stabilizing; the results obtained by solving the full evolution equations numerically agreed with the stability results obtained analytically. A weakly nonlinear analysis of coupled surface-tension and gravitational-driven instability in thin fluid layer (but, again, with a flat upper, free surface) is presented in [125]. In a weakly nonlinear analysis, it is sufficient to take into acount the modes that are critical at the linear threshold and as a consequence for the critical modes. In the Parmentier et al. paper, a system of three coupled Ginzburg–Landau type equations for the three amplitudes A_1, A_2, A_3 is derived:

$$\tau \frac{\partial A_i}{\partial t} = \varepsilon A_i + a A_j^* A_k^* - b A_i [|A_j|^2 + |A_k|^2] - c A_i |A_i|^2 \qquad (10.39)$$

with $i = 1$ and $j = 2$, $k = 3$; $i = 2$ and $j = 3$, $k = 1$; $i = 3$ and $j = 1$, $k = 2$, wherein the coefficients τ, a, b and c depend generally on the Prandtl and the Biot number and also on the ratio α (the percentage of buoyancy effect with regard to thermocapillary effect $-\alpha = 0$ corresponds to pure thermocapillarity and $\alpha = 1$ to pure buoyancy). The relative distance from the threshold is

$$\varepsilon = 1 - \frac{\lambda}{\lambda_c} \quad \text{with} \quad \lambda = \frac{\text{Ra}}{\text{Ra}^0} + \frac{\text{Ma}}{\text{Ma}^0}, \qquad (10.40)$$

where k is the wave number, and λ_c is related to the critical wave number k_c; Ra^0 is the critical Rayleigh number for pure buoyancy and Ma^0 is the critical Marangoni number for pure thermocapillarity. According to Parmentier et al., when buoyancy is the single responsibility of the convection, only rolls will be observed. As soon as capillary effects are, however, observed it appears that a hexagonal structure is preferred at the linear threshold. The more the thermocapillary forces are dominant with respect to the buoyancy forces, the larger the size of the region where hexagons are stable. It is shown that the direction of the motion inside the hexagons is directly linked to the value of the Prandtl number and for $\text{Pr} > 0.23$, the fluid moves upward at the center of the hexagons, in accordance with experiments. A subcritical region where hexagons are stable has also been displayed by these authors. The region is the largest when buoyancy does not act and in this case, the value found for the subcritical parameter is in excellent agrement with direct numerical simulation performed by Thess and Orszag [126]. But, all these above results correspond (unfortunately) to the case when the upper, free surface is flat.

A detailed analysis of system (10.39) can be found in [127–129]. Instability of a liquid hanging below a solid ceiling (the so-called Rayleigh–Taylor (R–T) instability) has been considered by Limat in a short note [130] according to a lubrication equation derived in Kopbosynov and Pukhnachev (in 1986). He discussed the influence of the initial thickness on the R–T instability and the results are summarized by a diagram giving the different possible regimes. This diagram allows one to predict two different thickness dependencies that are selected by the physical properties of the liquid. For the vertical film, the review paper by Chang [131] gives an excellent survey concerning mostly the various transition regimes on an, isothermal, free-falling vertical film. For an extension of this review, see Chang and Demekhin's survey, [132], but note that in both these review papers, discussion of the Marangoni effect is absent. Nonlinear dynamics and breakup of free-surface flows is reviewed in the paper by Eggers [133]. This film rupture is also considered by Ida and Miksis [134], who considered the dynamics of a lamella in a capillary tube as well [135]. As papers concerning thin films on a rotating disk, we mention those by Sisoev and Shkadov, but again for the isothermal case.

Nonlinear evolution of waves on a vertically falling film (but unfortunately, without the Marangoni effect) is considered by Chang et al. [137].

In the above cited paper by Thess and Orszag [126], devoted to the limit of the infinite Prandtl number, the case of a high Marangoni number is also considered. These authors note that 'the kinematically possible velocity fields can have remarkable complexity when Ma $\gg$ 1', and it is a challenging problem for future studies to understand Bénard–Marangoni convection in the limit Ma $\to \infty$.

Viscous thermocapillary convection at high Marangoni numbers is also considered by Cowley and Davis [138], where a boundary-layer analysis is performed that is valid for large Ma and Pr. In the paper by Nepomnyaschy and Velarde [139], a dissipation-modified Boussinesq-like system of equations governing 3D long wavelength Marangoni–Bénard oscillatory convection in a shallow layer heated from the air side is presented. In this paper, solitary waves and their oblique and head-on interaction are also considered. Marangoni convection and instabilities in liquid mixtures with Soret effects (the inclusion of the so-called 'Soret effect' means that the mass flux is the sum of temperature and concentration gradients, but usually the so-called Dufour effect, by which the concentration gradient would contribute to the heat flux, is ignored) is considered by Joo [140] and more recently by Bergeron et al. [141].

In Oron and Rosenau, [142], the authors show that the quadratic Marangoni instability enables an existence of new stable steady states which in variance with the conventional Marangoni induced patterns, are continuous and do not rupture. It is interesting to note that the inherently unstable viscous liquid film flow down a vertical plate can be stabilized by oscillating the plate at appropriate amplitudes and frequencies, although the stabilization is obtainable only over a relatively small range of Re. The onset of steady Marangoni convection, in a spherical shell of fluid with an outer free surface surrounding a rigid sphere, is analyzed by Wilson [143] using a combination of analytical and numerical techniques. In a doctoral thesis by Vince [144], the propagative waves in convection systems subject to surface tension effects are studied very accurately and a dynamical systems approach is also used, in particularly via amplitude equations *à la* Ginsburg–Landau (see above equations (10.39)).

An interesting and well-documented overview concerning drops, liquid layers and the Marangoni effect is the paper by Velarde [145]. Spatio-temporal instability in free ultra-thin films is considered by Shugai and Yakubenko [146]. The analysis shows that even in the linear approximation, the long-range intermolecular force strongly affects the evolution of initially localized disturbances, but linear theory always overestimates the film life time, due to the explosive nonlinear growth of disturbances at later stages of evolution. The IUTAM Symposium Proceedings (held in Haifa, Israel, 17–21 March 1997) [147] concerned 'Non-Linear Singularities in Deformation and Flow'. Various papers are related to interfacial effects in fluids and also with capillary breakup and instabilities. In [148], RB convection and turbulence is considered in liquid helium. The survey paper by Schechter et al. [149] is devoted to the 'two-component Bénard problem'. In [150] an infinite Prandtl number numerical similation in thermal convection is treated. The paper by Christov and Velarde [151] is an extension of the preceding paper [118]. In the special double issue of the *Journal of Engineering Mathematics* [152], the reader can find various papers devoted to the 'dynamics of thin liquid film'. In [153] the Marangoni convection in a differentially heated binary mixture is studied numerically by continuation, the fluid being subject to the Soret effect. In a recent paper [154], the effects of surfactants on the formation and evolution of capillary waves is considered numerically. The paper by Thual [155] is devoted to 'zero-Prandtl-number convection'; indeed, the Prandtl = 0 convection is a singular problem considered, in particular, in theoretical fluid dynamics.

In another recent paper [156], a set of lubrication models for the thin film flow of incompressible fluids on solid substrates is derived and studied. The models are obtained as asymptotic limits of the NS equations with the Navier-slip boundary condition for different orders of magnitude for the slip-length parameter.

The recent very interesting, but rather 'unusual' book by de Gennes et al. [157] will enable the reader to understand in simple terms some mundane questions affecting our daily lives, questions that have often come to the fore during our many interactions with industry (capillarity and wetting phenomena, drops, bubles, pearles and waves). The strategy in this book is to 'sacrifice scientific rigor' by an 'impressionistic' approach based on more qualitative arguments, which make it possible to grasp things more clearly and to dream up novel situations!

The recent book by Nepomnyaschy et al. [158], is devoted to 'Interfacial Convection in Multilayer Systems'. It is unavoidable (see, the review by T.P. Witelski, Duke University, in *SIAM Review*) that this book has some overlap with some of the authors' previous books [159, 160] relative to 'Nonlinear Dynamics of Surface-Tension-Driven Instabilities' and 'Interfacial Phenomena and Convection'. However, it is clear that this new book represents a significant advance with more depth of analysis and a greatly extended set of models given in a systematic presentation.

The stability of an evaporating thin liquid film is considered (and reconsidered) in a recent paper [161]. The paper [162] is devoted to 'vorticity, free surface and surfactants'. A few-mode Galerkin truncation is used in [163] to set up Lorenz models for convection in rotating binary mixtures. The work [164] develops a boundary-layer model for the thermocapillary feedback mechanism that can occur in material processes, in the limit where convection dominates heat transport but the Prandtl number is small. In [165] the RB convection with a temperature-dependent viscosity is considered from an asymptotic point of view and in [166] the temperature-dependent viscosity is also considered in the stability of a vertical natural convection boundary layer. A systematic description of Marangoni convection in a rotating system is considered in [167]; a long-wave equation is derived, and numerical results are discussed. In the recent paper [168], linear stability of a rotating fluid-saturated porous layer heated from below and cooled from above is studied, when the fluid and solid phases are not in local thermal equilibrium. A route to chaos in porous-medium thermal convection is considered in [169]. Surface-tension-driven Bénard convection in low-Prandtl-number fluids is studied in [170] by means of direct numerical simulations. In another recent paper [171], Marangoni flow around chemical fronts traveling in thin solution layers is considered, and influence of the liquid depth is numerically analyzed. A quantitative description of the combined action of anticonvective and thermocapillary mechanisms of instability is given in [172]. The onset of multicellular convection in a shallow laterally heated cavity is studied in [173]. An asymptotic model of the mobile interface between two liquids in a thin porous stratum is investigated in [174]. In [175], Grossmann and Lohse propose a unified theory for 'scaling' in thermal convection. The film rupture in the diffusive interface model coupled to hydrodynamics is considered in [176] by Thiele et al.

As a 'complement' to the books [159, 160] published in 2001 and 2002, we mention the book [177], published in 2003, by Birikh et al., relative to 'Liquid Interfacial Phenomena – Oscillations and Instability'. Falling films and the Marangoni effect are considerd in [178] by Shkadov et al. The case of large Péclet numbers for a reactive falling film is considered in [179] by Trevelyan and Kalliadasis.

In 2005, Mutabazi et al. [180] published an 'Henri Bénard Centenary Review' concerning 'dynamics of spatiotemporal structures'.

Scheid et al. [181] revisited the validity domain of the Benney equation in a case involving the Marangoni effect. The square patterns in rotating RB convection are considered in [182] by Sánchez-Álvarez et al., and in [183] Sultan et al. considered the diffusion of the vapor and Marangoni instabilities in a problem relative to evaporation of a thin film. Traveling circular waves in axisymmetric rotating convection have been investigated in [184] by Lopez et al.

Now, we mention some more recent papers (2007) by: Lopez et al. ('Onset of convection in a moderate aspect-ratio rotating cylinder: Eckhaus–Benjamin–Feir instability') [185], Nepomnyashchy and Simanovskii ('Marangoni instability in ultrathin two layer film') [186], Merkt et al. ('Short- and long evolution, in the long-wave theory of bounded two-layer films with a free liquid-liquid interface') [187], Pereira et al. ('Dynamics of a horizontal thin liquid film in the presence of reactive surfactants') [188] and Scheel ('The amplitude equation for rotating RB convection') [189].

Finally, as papers published in 2008, I mention, among others, the papers by: Sadiq and Usha ('Thin Newtonian film flow down a porous inclined plane: Stability analysis') [190], Marques and Lopez ('Influence of wall modes on the onset of bulk convection in a rotating cylinder') [191], Busse ('Asymptotic theory of wall-attached convection in a horizontal fluid layer with a vertical magnetic field') [192], Dietze et al. ('Investigation of the backflow phenomenon in falling films') [193].

In an 'Epilogue' to conclude the present book, I should like to stress again that, after working on it for two years, and after an attentive and careful final re-reading of my typescript, I am yet more convinced that:

> There is no better way for the derivation of significant model equations than rational analysis and asymptotic modelling.

A complete, consistent, rational modelling of the various convection phenomena in fluids is a long way in the future, but Chapter 8 in this book shows that, concerning the Bénard, heated from below, convection problem, when we take into account three main effects (buoyancy, free-surface deformation and viscous dissipation) in starting fluid-dynamics formulation via a rational approach, we have the possibility to derive three consistent, leading-order, approximate models; and more, if this is necessary we have at hands a method for derivation of an associated, consistent, second-order model!

It is clear that, at the present time, an essential gap still exists between fluid dynamics modelling and numerical simulation; often the results of computations (which are often 'fascinating') do not correspond, satisfactorily, to experimental results! Rarely in publications do we see the theoretical treatment of a 'full' fluid film problem directly inspired from the technology. For me, one reason is that, in starting physical problems, initial conditions are posed that obviously influence the formation and evolution (in time) of the film flow; thus the approximate models used for the numerical computation are not valid close to initial time. Further, with an *ad hoc* approach, often the model used by the numericians is non-consistent, i.e., not well balanced!

In this book I have been highly selective in my choice of topics and in many cases this choice of topics for rational analysis and asymptotic modelling has been based on my own interest and judgement. In fact, the main purpose of this book is only to give a fluid mechanics description of a certain class of convection phenomena. To that extent the text is a personal expression of my view of the subject.

References

1. B. Straughan, *The Energy Method, Stability, and Nonlinear Convection*, second edition. Applied Mathematical Sciences, Vol. 91, Springer-Verlag, New York, 2004.
2. K. Jöhnk and B. Svendsen, A thermodynamic formulation of the equations of motion and buoyancy frequency for Earth's fluid outer core. *Continuum Mech. Thermodyn.* **8**, 75–101, 1996.
3. R.Kh. Zeytounian, The Bénard–Marangoni thermocapillary-instability problem. *Phys. Uspekhi* **41**, 241–267, 1998
4. S. Kalliadasis, E.A. Demekhin, C. Ruyer-Quil and M.G. Velarde, *J. Fluid Mech.* **492**, 303–338, 2003.
5. C. Ruyer-Quil, B. Scheid, S. Kalliadasis, M.G. Velarde and R.Kh. Zeytounian, *J. Fluid Mech.* **538**, 199–222, 2005.
6. B. Scheid, C. Ruyer-Quil, S. Kalliadasis, M.G. Velarde and R.Kh. Zeytounian, *J. Fluid Mech.* **538**, 223–244, 2005.
7. A.A. Golovin, A.A. Nepommyaaschy and L.M. Pismen, *Phys. Fluids* **6**(1), 35–48, 1994.
8. D. Kashdan et al., Nonlinear waves and turbulence in Marangoni convection. *Phys. Fluids* **7**(11), 2679–2685, 1995.
9. J.S. Turner, Multicomponent convection. *Annu. Rev. Fluid Mech.* **17**, 11–44, 1985.
10. P.H. Coullet and E.A. Spiegel, Amplitude equations for systems with competing instabilities, *J. Appl. Math.* **43**(4), 776–821, 1983.
11. E. Knobloch, D.R. Moore, J. Toomre and N.O. Weiss, *J. Fluid Mech.* **166**, 400–448, 1986.

12. M.R.E. Proctor and N.O. Weiss, Normal forms and chaos in thermosolutal convection. *Nonlinearity* **3**, 619–637, 1990.
13. R.Kh. Zeytounian, *Asymptotic Modelling of Atmospheric Flows.* Springer-Verlag, Berlin/Heidelberg 1990.
14. B.S. Dandapat and P.C. Ray, *Int. J. Non-Linear Mech.* **28**(5), 489–501, 1993.
15. C.I. Christov and M.G. Velarde, Dissipative solitons. *Physica D* **86**, 323–347, 1995.
16. A.Ye. Rednikov, M.G. Velarde, Yu.S. Ryazantsev, A.A. Nepomnyashchy and V.N. Kurdyumov, Cnoidal wave and solitary waves in a dissipation-modified Korteweg–de-Vries equation. *Acta Appl. Math.* **39**, 457–475, 1995.
17. D.R. Fearn and D.E. Loper, Compositional convection and stratification in the Earth's fluid core. *Nature* **289**, 393–394, 1981.
18. P. Melchior, *The Physics of the Earth's Core*. Pergamon Press, Oxford, 1986.
19. D.E. Smylie and A.M.K. Szeto, The dynamics of the Earth's inner and outer cores. *Rep. Prog. Phys.* **47**, 855–906, 1984.
20. F.D. Stacey, Applications of thermodynamics to fundamental Earth physics. *Geoph. Surveys* **3**, 175–204, 1977.
21. R.Kh. Zeytounian, Recent avances in asymptotic modelling of tangent atmospheric motions. *Int. J. Engng. Sci.* **23**(11), 1239–1288, 1985.
22. P.H. Roberts and A.M. Soward, Magnetohydrodynamics of the Earth's core. *Annu. Rev. Fluid Mech.* **4**, 117–154, 1972.
23. W.W. Wood, A note on the westward drift of the earth's magnetic field. *J. Fluid Mech.* **82**, 349–400, 1977.
24. M. Grae Worster, *Annu. Rev. Fluid Mech.* **29**, 91–122, 1997.
25. D.W. Hughes and M.R.E. Proctor, *Annu. Rev. Fluid Mech.* **20**, 187–223, 1988.
26. G. Schubert, *Annu. Rev. Fluid Mech.* **24**, 359–394, 1992.
27. J.C. MacWilliams, *Annu. Rev. Fluid Mech.* **28**, 215–248, 1996.
28. R.W. Griffiths, *Annu. Rev. Fluid Mech.* **18**, 59–89, 1986.
29. T.E. Dowling, *Annu. Rev. Fluid Mech.* **27**, 293–334, 1995.
30. A. Oberbeck, *Ann. Phys. Chem., Neue Folge* **7**, 271–292, 1879.
31. E. Knobloch, Rotating convection: recent developments. *Int. J. Engng. Sci.* **36**(11), 1421–1450, 1998.
32. S. Chandrasekhar, *Hydrodynamic and Hydromagnetic Stability*. Oxford University Press, 1961.
33. P. Germain, Introduction à l'étude de l'aéromagnétodynamique. *Cahiers de Physique* **103**, 98–128, 1959.
34. R.Kh. Zeytounian, Sur la magnétodynamique des fluides pesants. *J. Méc. Appl. Phys. Techn. (PMTF)* **2**, 37–42, 1980.
35. H. Grad, Mathematical problem in magneto-fluid dynamics and plasma physics. in *Proceedings of the International Congress of Mathematics*, Stockholm, pp. 560–563, 1962.
36. R.Kh. Zeytounian, Invariants lagrangiens et intégrales premières en magnétodynamique des fluides. *Appl. Sci. Res.* **32**(6), 593–596, 1976.
37. A.S. Monin, *Weather Forecasting as a Problem in Physics*. Translated from Russian by Paul Superak. The MIT Press, Cambridge, MA, 1972.
38. R.Kh. Zeytounian, La météorologie du point de vue du mécaniciens des fluides. *Fluid Dynam. Trans.* **8**, 289–352, 1976.
39. S.K. Wilson, The effect of a uniform magnetic field on the onset of steady Marangoni convection in a layer of conducting fluid with a prescribed heat flux at its lower boundary. *Phys. Fluids* **6**(11), 3591–3600, 1994.
40. K. Zhang and G. Schubert, Magnethydrodynamics in rapidly rotating spherical systems. *Annu. Rev. Fluid Mech.* **32**, 409–443, 2000.

41. R. Rosensweig, *Ferrohydrodynamics*. Cambridge University Press, 1985.
42. P. Roberts, Electrohydrodynamic convection. *Quart. J. Mech. Appl. Math.* **22**, 211–220, 1969.
43. L. Landau, E. Lifshitz and L. Pitaevskii, *Electrodynamics of Continuous Media*. Pergamon, London, 1984.
44. R. Turnbull, *Phys. Fluids* **11**, 2588–2603, 1968.
45. B. Deo and A. Richardson, Generalized energy methods in electrohydrodynamic stability theory. *J. Fluid Mech.* **137**, 131–151, 1983.
46. M. Cowley and R. Rosensweig, *J. Fluid Mech.* **30**, 671–688, 1967.
47. A. Gailitis, *J. Fluid Mech.* **82**, 401–413, 1977.
48. D. Wollkind and J. Bdzil, Comments on chemical instabilities. *Phys. Fluids* **14**, 1813–1814, 1971.
49. A. Morro and B. Straughan, Convective instabilities for reacting viscous flows far from equilibrium. *J. Non-Equilib. Thermodyn.* **15**, 139–150, 1990.
50. L. Rongy and A. De Wit, *J. Engng. Math.* **59**, 221–227, 2007.
51. S. Lebovich and S. Paolucci, The instability of the ocean to Langmuir circulations. *J. Fluid Mech.* **102**, 141–167, 1981.
52. F. Crisciani, F. Cvallini and R. Mosetti, Stability of analytical solutions of Stommel–Veronis ocean circulation models. *J. Phys. Oceanography* **24**, 155–158, 1994.
53. J. Pedlosky, *Ocean Circulation Theory*. Springer, Berlin/Heidelberg, 1996.
54. G. Veronis, *J. Fluid Mech.* **31**, 113–139, 1968.
55. R.Kh. Zeytounian, *Meteorological Fluid Dynamics*. Lecture Notes in Physics, Vol. m5, Springer-Verlag, Berlin/Heidelberg, 1991.
56. R.Kh. Zeytounian, Weather prediction as a problem in fluid dynamics (On the 100th Birthday of I. A. Kibel). *Izv. Russ. Acad. Sci. Atmosph. Oceanic Phys.* **40**(5), 593–605, 2004.
57. A.S. Monin, *Fundamentals of Geophysical Fluid Dynamics*. Original Russian edition, Hydrometeo-Izdat, Leningrad, 1988 [English translation available].
58. G.I. Marchuk and A.S. Sarkisyan, *Mathematical Modelling of Ocean Circulation*. Original Russian edition. Nauka, Moscow, 1988 [English translation available, Springer-Verlag].
59. G. Küppers and D. Lortz, *J. Fluid Mech.* **35**, 609, 1969.
60. S. Wiggins, *Introduction to Applied Dynamical Systems and Chaos*. Springer, New York, 1990.
61. E. Bodenschatz, W. Pesch and G. Ahlers, Recent developments in Rayleigh–Bénard convection. *Annu. Rev. Fluid Mech.* **32**, 709–778, 2000.
62. K.M.S. Bajaj, L. Jun, B. Naberhuis and G. Alhers, *Phys. Rev. Lett.* **81**(4), 806–809, 1998.
63. R.M. Clever and F.H. Busse, *J. Fluid Mech.* **81**, 609–627, 1979.
64. M. Fantz, R. Friedrich, M. Bestehorn and H. Haken, *Physica D* **61**(1–4), 147–154, 1992.
65. J.J. Niemela and R.J. Donnelly, *Phys. Rev. Lett.* **57**(20), 2524–2527, 1986.
66. Y. Ponty, T. Passy and P.L. Sulem, *Phys. Rev. E* **56**(4), 62–78, 1997.
67. Y. Tu and M.C. Cross, *Phys. Rev. Lett.* **69**(17), 2215–2219 and 2515–2518, 1992.
68. G. Malone and S. Rionero, *Continuum Mech. Thermodyn.* **9**, 347–363, 1997.
69. B. Straughan, *Mathematical Aspects of Penetrative Convection*. Longman Group UK, Harlow, 1993.
70. G. Veronis, Penetrative convection. *Astrophys. J.* **137**, 641–663, 1963.
71. R. Krishnamurti, Convection induced by selective absorption of radiation: A laboratory model of conditional instability. *Dynam. Atmosph. Oceans* **27**, 367–382, 1997.

72. K. Ames and B. Straughan, *Acta Mech.* **85**, 137–148, 1990.
73. B. Cushman-Roisin, *Fluid Dyn.* **19**, 61–91, 1982.
74. R.A. Denton and I.R. Wood, *J. Fluid Mech.* **113**, 1–21, 1981.
75. A.J. Faller and R. Kaylor, *J. Geophys. Res.* **75**, 521–530, 1970.
76. J.H. George, R.D. Gunn and B. Straughan, *Geophys. Astrophys Fluid Dyn.* **46**, 135–158, 1989.
77. R.L. Jennings, *Geophys. Astrophys Fluid Dyn.* **56**, 183–204, 1991.
78. P. Machetel and D.A. Yen, *J. Geophys. Res. (B)* **94**, 10609–10626, 1989.
79. P.C. Matthews, *J. Fluid Mech.* **188**, 571–583, 1988.
80. D.R. Moore and N.O. Weiss, *J. Fluid Mech.* **61**, 553–581, 1973.
81. J.M. Strauss, *Astrophys. J.* **209**, 179–189, 1976.
82. J.P. Zhan, J. Toomre and J. Latour, *J. Geophys. Astrophys Fluid Dyn.* **22**, 159–193, 1982.
83. V.Ya. Shkadov, Wave flow regimes of a thin layer of viscous fluid subject to gravity. *Fluid Dyn.* **2**, 29–34, 1967 [English translation].
84. S. Kalliadasis, A. Kiyashko and E.A. Demekhin, *J. Fluid Mech.* **475**, 377–408, 2003.
85. P.M.J. Trevelyan and S. Kalliadasis, *J. Engng. Math.* **50**(2–3), 177–208, 2004.
86. C. Ruyer-Quil and P. Manneville, *Eur. Phys. J.* **B15**, 357–369, 2000.
87. C. Ruyer-Quil and P. Manneville, *Phys. Fluids* **14**, 170–183, 2002.
88. D. Gottlieb and S.A. Orzag, *Numerical Analysis of Spectral Methods: Theory and Application.* SIAM, Philadelphia, PA, 172 pp., 1977.
89. C. Ruyer-Quil and P. Manneville, *Phys. Rev. Lett.* **93**, 199401, 2004.
90. M. Takashima, *J. Phys. Soc. Japan* **50**(8), 2745–2750 and 2751–2756, 1981.
91. P. Kaloni and Z. Qiao, *Continuum Mech. Thermodyn.* **12**, 185–194, 2000.
92. D.D. Joseph, *Stability of Fluid Motion, II.* Springer-Verlag, Heidelberg, 1976.
93. H.E. Hupper and D.E. Moore, Nonlinear double-diffusive convection. *J. Fluid Mech.* **78**, 821–854, 1976.
94. Y. Ogura and N.A. Phillips, Scale analysis of deep and shallow convection in the atmosphere. *J. Atmosph. Sci.* **19**, 173–179, 1962.
95. D.O. Gough, The anelastic approximation for thermal convection. *J. Atmosph. Sci.* **26**, 448–456, May 1969.
96. B.G. Higgins, Film flow on a rotating disk. *Phys. Fluids* **29**, 3522, 1986.
97. B.S. Dandapat and P.C. Ray, Film cooling on a rotating disk. *Int. J. Non-Linear Mech.* **25**, 529, 1990.
98. M. Van Dyke, *Perturbation Methods in Fluid Mechanics.* Academic Press, New York, 1964.
99. A. Kitamura, Asymptotic solution for film flow on a rotating disk. *Phys. Fluids* **12**(9), 2141–2144, 2000.
100. B. Reisfeld, S.G. Bankoff and S.H. Davis, The dynamics and stability of thin liquid films during spin coating. I. Films with constant rates of evaporation or absorption. *J. Appl. Phys.* **70**, 52–58, 1991.
101. A.G. Emslie, F.J. Bonner and L.G. Peck, Flow of a viscous liquid on a rotating disk. *J. Appl. Phys.* **29**, 858, 1958.
102. Ch. Bailly, Modelisation asymptotique et numérique de l'écoulement dû à des disques en rotation. Thesis, Université des Sciences et Technologies de Lille, No. 1512, 160 pp., April 1995. See also *Publication IRMA, Lille* **33**(XV), XV–2, XV–14, 1993.
103. D.J. Needham and J.H. Merkin, The development of nonlinear waves on the surface of a horizontal rotating thin liquid film. *J. Fluid Mech.* **184**, 357–379, 1987.
104. J. Liu and J.P. Gollub, Solitary waves dynamics of film flows. *Phys. Fluids* **6**(5), 1702–1712, May 1994.

105. R.Kh. Zeytounian, Nonlinear long waves on water and solitons. *Phys. Uspekhi* **38**(12), 1333–1381, December 1995 [English translation from Russian].
106. P.L. Kapitza and S.P. Kapitza, *Zh. Exper. Teor. Fiz.* **19**, 105, 1949.
107. A. Pumir, P. Manneville and Y. Pomeau, *J. Fluid Mech.* **135**, 27, 1983.
108. H.-C. Chang, E.A. Demekhin and D.I.Kopelevich, *J. Fluid Mech.* **250**, 433, 1993.
109. D.J. Benney, Long waves on liquid film. *J. Math. Phys.* **45**, 150, 1966.
110. M.G. Velarde and A.Ye. Rednikov, Time-dependent Bénard–Marangoni instability and waves. In *Time-Dependent Nonlinear Convection*, P.A. Tyvand (Ed.), Adv. Fluid Mech. Series. Computational Mechanics Publications, Southampton, Chapter 6, pp. 177–218, 1998.
111. M.G. Velarde (Ed.), *Physicochemical Hydrodynamics. Interfacial Phenomena*. Plenum, New York, 1988.
112. A.N. Garazo and M.G. Velarde, *Phys. Fluids* **A3**, 2295–2300, 1991.
113. A.L. Frenkel, *Europhys. Lett.* **18**, 583, 1992.
114. J. Liu, J.D. Paul and J.P. Gollub, *J. Fluid Mech.* **250**, 69, 1993.
115. S.P. Lin and O. Suryadevara, Solitary, periodic and chaotic waves in thin films. In: *Transactions of the 2nd Army Conference on Applied Mathematics and Computing*, ARO Report 85-1, pp. 483–499, 1985.
116. V.I. Nekorkin and M.G. Velarde, *Int. J. Bifurcation Chaos* **4**(5), 1135–1146, 1994.
117. M.G. Velarde, V.I. Nekorkin and A.G. Maksimov, *Int. J. Bifurcation Chaos* **5**(3), 831–839, 1995.
118. C.I. Christov and M.G. Velarde, Evolution and interactions of solitary waves (solitons) in nonlinear dissipative systems. *Phys. Scripta* **T55**, 101–106, 1994.
119. M.J. Block, Surface tension as the cause of Bénard cells and surface deformation in a liquid film. *Nature* **178**, 650–651, 1956.
120. J.R.A. Pearson, On convection cells induced by surface tension. *J. Fluid Mech.* **4**, 489–500, 1958.
121. R.Kh. Zeytounian, The Bénard–Marangoni thermocapillary instability problem: On the role of the buoyancy. *Int. J. Engng. Sci.* **35**(5), 455–466, 1997.
122. J. Bragard and M.G. Velarde, Bénard convection flows. *J. Non-Equilib. Thermodyn.* **22**, 1–19, 1997.
123. T.G. Myers, Thin films with high surface tension. *SIAM Rev.* **40**(3), 441–462. 1998.
124. M.P. Ida and M.J. Miksis, The dynamics of thin films I: General theory, and II: Applications. *SIAM J. Appl. Math.* **58**(2), 456–473 and 474–500 1998.
125. P.M. Parmentier, V.C. Regnier and G. Lebon, Nonlinear analysis of coupled gravitational and capillary thermoconvection in thin fluid layers. *Phys. Rev. E* **54**(1), 411–423, 1996.
126. A. Thess and S.A. Orszag, Surface-tension-driven Bénard convetion at infinite Prandtl number. *J. Fluid Mech.* **283**, 201–230, 1995.
127. M.C.Cross, *Phys. Fluids* **23**, 1727. 1980.
128. M.C. Cross, *Phys. Rev. A* **25**, 1065, 1982.
129. M.C. Cross and P.C. Hohenberg, Pattern formation outside of equilibrium. *Rev. Mod. Phys.* **65**, 851, 1993.
130. L. Limat, Instabilité d'un liquide suspendu sous un surplomb solide: Influence de l'épaisseur de la couche. *C.R. Acad. Sci. Paris, Sér, II* **217**, 563–568, 1993.
131. H.-Ch. Chang, Wave evolution on a falling film. *Annu. Rev. Fluid Mech.* **26**, 103–136, 1994.
132. H.-Ch. Chang and E.A. Demekhin, Solitary wave formation and dynamics on falling films. *Adv. Appl. Mech.* **32**, 1–58, 1996.

133. J. Eggers, Nonlinear dynamics and breakup of free-surface flows. *Rev. Mod. Phys.* **69**(3), 835–929, 1997.
134. M.P. Ida and M.J. Miksis, Thin film rupture. *Appl. Math. Lett.* **9**(3), 35–40, 1996.
135. M.P. Ida and M.J. Miksis, Dynamics of a lamella in a capillary tube. *SIAM J. Appl. Math.* **55**(1), 23–57, 1995.
136. G.M. Sisoev and V.Ya. Shkadov, Flow stability of a film of viscous liquid on a rotating disk. *J. Eng. Physics* **52**, 671–674, 1987 [English edition, translated from Russian].
137. H.-Ch. Chang, E.A. Demekhin and E.A. Kopelevich, Nonlinear evolution of waves on a vertically falling film. *J. Fluid Mech.* **250**, 433–480, 1993.
138. S.J. Cowley and S.H. Davis, Viscous thermocapillary convection at high Marangoni number. *J. Fluid Mech.* **135**, 175–188, 1983.
139. A.A. Nepomnyaschy and M.G. Velarde, *Phys. Fluids* **6**(1), 187–198, 1994.
140. S.W. Joo, Marangoni instabilities in liquid mixtures with Soret effects. *J. Fluid Mech.* **293**, 127–145, 1995.
141. A. Bergeron et al., Marangoni convection in binary mixtures with Soret effect. *J. Fluid Mech.* **375**, 143–177, 1998.
142. A. Oron and Ph. Rosenau, On a nonlinear thermocapillary effect in thin liquid layers. *J. Fluid Mech.* **273**, 361–374, 1994.
143. S.K. Wilson, The onset of steady Marangoni convection in a spherical geometry. *J. Engng. Math.* **28**, 427–445, 1994.
144. J.-M. Vince, Ondes propagatives dans des systèmes convectifs soumis à des effets de tension superficielle. Doctoral Thesis, University Paris 7, No, 345, 185 pp., 1994.
145. M.G. Velarde, Drops, liquid layers and the Marangoni effect. *Philos Trans. Roy. Soc. London A* **356**, 829–844, 1998.
146. G.A. Shugai, and P.A. Yakubenko, Spatio-temporal instability in free ultra-thin films. *Eur. J. Mech. B/Fluids* **17**(3), 371–384, 1998.
147. D. Durban and J.R.A. Pearson, *Non-linear Singularities in Deformation and Flow.* Kluwer Academic Publishers, Dordrecht, 1999.
148. R.P. Behringer, Rayleigh–Bénard convection and turbulence in liquid helium, Part I. *Rev. Mod. Phys.* **57**(3), 657–687, 1985.
149. R.S. Schechter, M.G. Velarde and J.K. Platten, The two-component Bénard problem. *Adv. Chem. Phys.* **26**, 265–301, 1974.
150. A. Lenardic and L. Moresi, *Phys. Fluids* **15**(2), 455–466, 2003.
151. C.I. Christov and M.G. Velarde, Dissipative solitons. *Physica D* **86**, 323–347, 1995.
152. D.G. Crowdy, C.J. Lawrence and S.K. Wilson (Guest Editors), *The Dynamics of Thin Liquid Films*. Special Issue of *J. Engng. Math.* **50**(2–3), 341 pp., November 2004.
153. A. Bergeron et al., Marangoni convection in binary mixture with Soret effect. *J. Fluid Mech.* **375**, 143–177, 1998.
154. H.T. Ceniceros, *Phys. Fluids* **15**(1), 245–256, 2003.
155. O. Thual, Zero-Prandtl-number convection. *J. Fluid Mech.* **240**, 229–258, 1992.
156. A. Munch, B. Wagner and T.P. Witelski, Lubrication models with small to large slip lengths. *J. Engng. Math.* **53**, 359–383, 2005.
157. P.-G. de Gennes, F. Brochard-Wyart and D. Quéré, *Capillarity and Wetting Phenomena, Drops, Bubbles, Pearls, Waves*. Springer Science + Business Media, 2004.
158. A. Nepomnyaschy, L. Simanovskii and J.C. Legros, *Interfacial Convection in Multilayer Systems*. Springer-Verlag, New York, 2006.
159. P. Colinet, J.C. Legros and M.G. Velarde, *Nonlinear Dynamics of Surface-Tension-Driven Instabilities*. Wiley/VCH Verlag, Berlin 2001.

160. A. Nepomnyaschy, M.G. Velarde and P. Colinet, *Interfacial Phenomena and Convection*. Monogr. Surv. Pure Appl. Math., Vol. 124, Chapman & Hall/CRC, Boca Raton, FL, 2002.
161. O.E. Shklyaev and E. Fried, Stability of an evaporating thin liquid film. *J. Fluid Mech.* **584**, 157–183, 2007.
162. T. Sarpkaya, *Annu. Rev. Fluid Mech.* **28**, 83–128, 1996.
163. J.K. Bhattacharjee, Convection in rotating binary mixtures. III. Galerkin models. *Phys. Fluids* **A1**(12), 1938–1948, 1989.
164. D. Canright and M. Huber, *Phys. Fluids* **14**(9), 3272–3279, 2002.
165. J. Severin and H. Herwig, *Z. angew. Math. Phys.* **50**, 375–386, 1999.
166. J.-Y. Jang and J.C. Mollendorf, *Int. J. Engng Sci.* **26**(1), 1–12, 1988.
167. A.M. Mancho, H. Riecke and F. Sain, *Chaos* **12**(3), 706–718, 2002.
168. M.S. Malashetty, M. Swamy and S. Kulkami, *Phys. Fluids* **19**, 054102, 2007.
169. S. Kimura, G. Schubert and J.M. Straus, Route to chaos in porous-medium thermal convection. *J. Fluid Mech.* **166**, 305–324, 1986.
170. T. Boeck and A. Thess, Bénard–Marangoni convection at low Prandtl number. *J. Fluid Mech.* **399**, 251–275, 1999.
171. L. Rongi, A. De Witt, *J. Engng. Math.* **59**, 221–227, 2007.
172. A. Nepomnyaschy and L. Simanovskii, *Phys. Fluids* **14**(11), 3855–3867, 2002.
173. P.G. Daniels, P.A. Blythe and P.G. Simpkins, Onset of multicellular convection in a shallow laterary heated cavity. *Proc. Roy. Soc. London A* **411**, 327–350, 1987.
174. M. Panfilov and M. Buès, Asymptotic model of the mobile interface between two liquids in a thin porous stratum. *J. Fluid Mech.* **473**, 59–81, 2002.
175. S. Grossmann and D. Lohse, *J. Fluid Mech.* **407**, 27–56, 2000.
176. U. Thiele, M.G. Velarde, K. Neuffer and Y. Pomeau, *Phys. Rev. E* **64**, 031602, 2001.
177. R.V. Birikh, V.A. Briskman, M.G. Velarde and J.C. Legros, *Liquid Interfacial Phenomena Oscillations and Instability*. Marcel Dekker, New York, 2003.
178. V.Y. Shkadov, M.G. Velarde and V.P. Shkadova, *Phys. Rev. E* **67**, 056310, 2004.
179. P.M.J. Trevelyan and S. Kalliadasis, *Phys. Fluids* **16** (11), 3191, 2004.
180. I. Mutabazi, E. Guyon and J.E. Wesfreid (Eds.), *Dynamics of Spatiotemporal Structures – Henri Bérnard Centenary Review*. Springer, 2005.
181. B. Scheid, C. Ruyer-Quil, U. Thiele, O.A. Kabov, J.C. Legros and P. Colinet, *J. Fluid Mech.* **527**, 303, 2005.
182. J.J. Sánchez-Álvarez, E. Serre, E. Crespo del Arco and F.H. Busse, *Phys. Rev. E* **72**, 036307, 2005.
183. E. Sultan, A. Boudaoud and M. Ben Amar, *J. Fluid Mech.* **543**, 183, 2005.
184. J.M. Lopez, A. Rubio and F. Marques, *J. Fluid Mech.* **569**, 331, 2006.
185. J.M. Lopez, F. Marques, I. Mercader and O. Batiste, *J. Fluid Mech.* **590**, 187, 2007.
186. A.A. Nepomnyashvhy and I.B. Simanovskii, *Phys. Fluids* **19**, 122103, 2007.
187. D. Merkt, A. Potosky, M. Bestehorn and U. Thiele, *Phys. Fluids* **17**, 064104, 2007.
188. A. Pereira, P.M.J. Trevelyan and S. Kalliadasis, *Phys. Fluids* **19**, 112102, 2007.
189. J.D. Scheel, *Phys. Fluids* **19**, 104105, 2007.
190. I.M.R. Sadiq and R. Usha, *Phys. Fluids* **20**, 022105, 2008.
191. F. Marques and J.M. Lopez, *Phys. Fluids* **20**, 024109, 2008.
192. F.H. Busse, *Phys. Fluids* **20**, 024102, 2008.
193. G.F. Dietze, A. Leefken and Kneer, *J. Fluid Mech.* **595**, 435–459, 2008.

Subject Index

V

W

Z

MIX
Papier aus verantwortungsvollen Quellen
Paper from responsible sources
FSC® C105338

If you have any concerns about our products,
you can contact us on
ProductSafety@springernature.com

In case Publisher is established outside the EU,
the EU authorized representative is:
Springer Nature Customer Service Center GmbH
Europaplatz 3, 69115 Heidelberg, Germany

Printed by Libri Plureos GmbH
in Hamburg, Germany